PRESSURE SURGE IN PIPE AND DUCT SYSTEMS

This one is for

Toni *and* Jo

and

Richard *and* Jayne

PRESSURE SURGE IN PIPE AND DUCT SYSTEMS

J A Swaffield

BSc, MPhil, PhD, CEng, MRAeS, MCIBSE, FIWEM
Dean of Engineering
and
William Watson Professor of Building Engineering
Heriot-Watt University, Edinburgh, Scotland

and

A P Boldy

BSc, PhD, CEng, MICE
Senior Lecturer
Department of Engineering
University of Warwick
England

Avebury Technical

Aldershot · Brookfield USA · Hong Kong · Singapore · Sydney

Published by

Avebury Technical
Ashgate Publishing Limited
Gower House
Croft Road
Aldershot
Hampshire GU11 3HR
England

Ashgate Publishing Company
Old Post Road
Brookfield
Vermont 05036
USA

British Library Cataloguing in Publication Data

Swaffield, J.A.
Pressure Surge in Pipe and Duct Systems
I. Title II. Boldy, A.P.
620.1

Library of Congress Cataloging-in-Publication Data

Swaffield, J.A., 1943-
Pressure Surge in Pipe and Duct Systems / J.A. Swaffield and A.P. Boldy.
380 p. 23.4 cm.
Includes bibliographical references (p. 10) and index.
ISBN 0-291-39796-4 : $58.95
1. Pipe-Fluid dynamics. 2. Waterhammer. I. Boldy, A.P. (Adrian P.) II. Title
TJ930.S96 1993
621.8'672 -- dc20 93 - 1532
CIP

Electronically typeset using WordPerfect 5.1™ by Professor J A Swaffield and Mrs A Ormston.

Artwork by Dr D P Campbell using an Apple® Macintosh™ p.c. and Cricket™ software.

ISBN 0 291 39796 4

Contents:

Preface:

The occurrence of pressure surge, waterhammer or pressure transient propagation within any fluid/conduit system is a normal and frequent event that accompanies any change in the system operating condition. The severity of such pressure excursions depends upon the rate of change of the flow conditions at the system boundaries, the properties of the fluid and dimensions and elasticity of its enclosing conduit. Thus the aim of any study of unsteady flow must be to understand the mechanisms of transient propagation so that the possible severity of any event may be determined and suitable design decisions taken. Therefore the objective is to allow the control of transients, rather than their eradication.

This text will attempt to place the study of pressure surge within its historical context, detailing the fundamental understanding of transient propagation causes and effects that developed in the first half of this century, and the intense interest that followed the introduction of fast and reliable computer based methods in the 1960s. The continued development of computing power has led to the current position where complex network simulations may be undertaken, with a confidence based on extensive validation, on 'desk top' computers.

The development of computing methods of solution of the transient propagation equations brought to the forefront of research the method of characteristics, a general mathematical method particularly suited to the solution of the surge defining quasi linear hyperbolic partial differential equations. This technique, and in particular its application to complex network boundary conditions, has become the industry standard for surge simulation. This text will outline the general methodology involved, the representation within the method of a series of common boundary conditions and demonstrate its application in the simulation of pressure transient propagation in a range of fluid engineering systems.

Pressure transients occur in all fluid systems and therefore their study is one of the few areas of research in fluid mechanics that truly crosses the boundaries between the various engineering disciplines. Within this text will be found examples ranging from water supply and distribution networks to in-flight refuelling, from open channel or partially filled pipeflow free surface wave propagation in drains and sewers to air pressure transients in vent systems and tunnels.

The mathematical methods introduced will be shown to be equally applicable to offshore oil platforms, hydroelectric schemes, water distribution networks, long distance oil pipelines, aircraft fuel systems, sewers and vent systems. Such a spread of application is rare within a single research topic and probably explains why over the published literature has demonstrated the continuing interest of an international group of researchers.

The authors have been active in transient research and development over the past 25 years and benefitted from the impetus that the subject received in the 1960s from the introduction of computing methods, notably by Streeter in the United States and Fox in the UK. The authors were therefore fortunate to have been able to work with pressure transient researchers from both the 'pre' and 'post' computing era. The authors are grateful for discussions over many years with their colleagues in this research field, particularly Keith Enever, John Fox, David Thorley and Alan Vardy. Jim Twyman was responsible for introducing one of the authors to this research area, it has remained a source of continuing stimulation.

In preparing this text the authors have sought to place the subject in context and provide a basis for an understanding of the mechanism involved in transient propagation, while at the same time allowing the reader the freedom to embark on independent study.

The simulations presented will, it is hoped, provide the confidence in current modelling techniques that is necessary for the reader to proceed further. The simple computer programs detailed in the appendices, based entirely on material included in the text and utilised to generate many of the demonstrations presented in the text, are included as an aid to further independent study. Therefore it is hoped that the text will be equally useful to the student engineer and the practising system designer who encounters pressure surge problems.

Since 1983 the authors have delivered, initially at the invitation of and in collaboration with Keith Enever, a specialist short course on 'Pressure Surge in Pipe and Duct Systems' at Cranfield Institute of Technology. In many ways this course, and in particular the response of its attendees, almost exclusively professional engineers dealing regularly with pressure surge across the whole spectrum of fluid engineering applications, helped to formulate this text. The authors would wish to acknowledge the support of the School of Mechanical Engineering at Cranfield.

The authors would also like to thank all their colleagues, research associates and students who have contributed to their understanding of pressure surge and therefore to the development of this text. Mrs Anne Ormston is to be thanked for assisting with the preparation of the manuscript and Dr David Campbell for producing the excellent artwork. John Hindley, our publisher, must be thanked for his seemingly inexhaustible patience and support without which this text would forever have remained 'tomorrow's job'. While the authors acknowledge all the assistance and advice of their colleagues in preparing this text, any errors, factual or of understanding, remain theirs alone.

Unsteady flow and transient propagation are fascinating research areas: the authors hope that this text will communicate some of their enthusiasm to the reader.

J.A. Swaffield
A.P. Boldy
Edinburgh and Warwick
February, 1993

Notation:

A	Flow cross sectional area
Acon	% dissolved gas released at any pressure below atmospheric, from Henry's Law
AirVol	Air volume released from a liquid and affecting column separation
A_{in}, A_{out}	Area of valve opening
a_o	Turbine guide vane position
B	Width of rectangular cross section duct
C_1-C_4	Pump characteristic coefficients
C^+	Characteristic curve, transmitting information downstream
C^-	Characteristic curve, transmitting information upstream
C'	Restraint coefficient in wave speed equation
C_d	Valve loss coefficient
C_R	Boundary reflection coefficient
C_T	Boundary transmission coefficient
C_{in}, C_{out}	Air value coefficients
CON	Critical mass rate of inflow of air
COU	Multiplier for critical mass rate outflow of air
c	Wave speed, elastic pipe or open channel flow
c_0	Wave speed in a rigid pipe or an infinite fluid volume
D	Pipe diameter
D	or diameter of turbine runner as appropriate

E	Young's Modulus of Elasticity
E	or open channel specific energy as appropriate
e	Pipe or duct wall thickness
F	Froude Number, ratio of local mean flow velocity to wave speed in open channel flows
F(),f()	Representation of pressure waves in frictionless model
f	Colebrook-White friction factor
FuelVol	Fluid volume assumed to release gas
GasVol	Gas volume released
G	Pipe material Shear Modulus
g	Acceleration due to gravity
H	Local flow head
H	or net head across turbine as appropriate
H_b	Barometric pressure head
H_D	Piezometric pressure head for air in an air vessel
H_p	Net head across pump
H_R	Reference value of net head across pump
H_w	Head in air chamber
h	Local flow depth in open channel applications
h	Non-dimensional net head (H/H_R)
I	Moment of inertia (ωR^2) of rotating masses
I,i	Pipe node
J,j	Pipe number
K	Fluid Bulk Modulus of Compressibility, Bunsen solubility coefficient
K	Constant in perfect gas pressure-density law
K1	Coefficients based on known flow conditions
K2	at time t in order to solve for unknown
K3	conditions at an intermediate
K4	node at time t + Δt
K5	Valve discharge coefficient term
K,k	Separation loss
K_t,K_0	Valve pressure loss coefficient in Schnyder-Bergeron graphical technique
L	Combined equations of continuity and momentum
L_1	Momentum equation
L_2	Continuity equation
ℓ	Length of fluid element in wave speed derivation
m	Hydraulic mean depth, 'reachback' number in time line interpolation procedure

m	Mass rate of air flow through valve
N	Pump speed
N_R	Reference value of pump speed
N	Turbine speed
n	Number of Δx sections in a pipelength
n	Manning loss coefficient in open channel flow
n	Polytropic coefficient of expansion
n_p	Number of pumps in parallel
n_{11}	Turbine unit speed
P	Flow wetted perimeter
P_{atm}	Atmospheric pressure
P_{gas}	Partial pressure of gas released into a vapour pocket or pressure of trapped gas at a pipe dead end
P_{vap}	Fluid vapour pressure
p	Local flow pressure, pressure at a node
p_{irv}	Inwards relief valve opening pressure
p_{orv}	Outwards relief valve opening pressure
p_0	Atmospheric pressure in gas flow applications
Q	Flow rate in a pipe, duct or channel, pump discharge rate
Q_p	Pump discharge
Q_R	Reference value of pump discharge
Q_{RV}	Relief value discharge
Q_t	Turbine discharge
Q_0	Initial flow rate
Q_{11}	Turbine unit discharge
q	Air leakage rate
q_R	Relief valve throughflow
R	Gas constant
R	Pipe radius
Re	Reynolds Number
S	Channel slope
S_0	Open channel steady flow friction slope from Chezy equation
T	Absolute temperature
T	Local flow surface width in open channel flow applications
T,t	Time
T	Torque on pump or turbine
T_f	Tangential force
T_ℓ	Longitudinal force
T_p	Pipe period = 2L/c

T_R	Reference torque on pump
T_{11}	Turbine unit torque
t_c, t'	Valve closure time
u	Local flow mean velocity in gas flow applications and velocity at time $t + \Delta t$ in method of characteristics solution
V	Local flow mean velocity and velocity at time $t + \Delta t$ in method of characteristics solution
Vol_{cav}	Cavity volume at any time t
Vol_{fluid}	Fluid volume releasing dissolved gas at any time t
Vol_{gas}	Air or gas volume released to a cavity volume at time t
V_0 , v_0	Initial flow mean velocity
v	Mean flow velocity, velocity at time t in method of characteristics solution
Z	No of pipes joining at a junction, of which Z0 terminate and Z1 originate at the junction
Z_w	Water surface elevation in an air chamber
X,x	Distance, measured +ve from pipe entry in +ve initial flow direction
y	Free gas volumetric ratio at a given pressure
α	Pipe slope, −ve in initial flow direction assumed in development of basic equations of momentum and continuity
α	Slope of characteristic lines in Schnyder-Bergeron method
α	Non-dimensional pump speed (N/N_R)
β	Non-dimensional torque on pump (T/T_R)
γ	Ratio of specific heat in gas flow applications
ΔD	Change in pipe diameter due to passage of a surge
ΔH	Transient head expressed in m of flowing fluid
$\Delta \ell$	Change in elemental pipe length due to the passage of a surge
Δp	Transient pressure rise expressed in N/m^2
Δt	Time step in method of characteristics solution
Δx	Inter node length in method of characteristics solution
δs	Elemental flow section length in derivation of continuity equation
ε	Interpolation ratio in time line interpolation
ε	Strain
ε_1	Circumferential strain
ε_2	Longitudinal strain
θ	Method of characteristics grid mesh ratio, $\Delta t/\Delta x$

θ	Tan^{-1} (α/υ)
λ	Multiplier in characteristics method
μ	Poisson's Ratio
ρ	Fluid density
ρ_0	Gas density at atmospheric pressure
σ_1	Circumferential stress
σ_2	Longitudinal stress
τ	Stress
τ_0	Shear stress
υ	Non-dimensional pump discharge (Q/Q_R)
ϕ	Function
ω	Stage variable to replace depth/wave speed integral in open channel applications, rotation in pump and turbine treatment

Subscript

A,B,C	Node identification at Δx intervals at time t
atm	Atmospheric conditions
dc	Open channel flow critical depth downstream of a junction
eff	Effective Bulk Modulus for a free gas/liquid/elastic pipe case
f	Liquid density
final	Final value of fluid column velocity on cavity collapse
g	Gas density
gas	Entrapped gas terms, partial pressure terms
j	Junction open channel flow depth
m	Mean values for a liquid/gas mixture
max	Maximum values of wave speed and mean flow velocity in the determination of time step
mc	Open channel normal flow depth upstream of a junction
NTP	Normal temperature and pressure, sea level and 0° C
P	Node to be calculated at time $t + \Delta t$
R	Known condition upstream at base C^+ characteristic at time t
Res	Reservoir
S	Known condition downstream at base C^- characteristic, time t
S′	Known conditions upstream at base C^- characteristic in supercritical flow at time t
t	Total gas and liquid volume
vap	Vapour pressure
0	Conditions at time zero

Superscript

$t, t + \Delta t$	Conditions at the time indicated

Units

SI units are used as standard within the text and apply to all equations and constants unless otherwise stated. As the study of unsteady flow problems applies to a wide range of engineering applications, many within industries, such as oil or aerospace, where Imperial or US units are still standard, these systems of units will be found within the text.

The use of pressure head, measured in m of the flowing fluid, is common in civil engineering applications and is therefore included. However, while this is acceptable for water based systems, care must be taken if this system of units is applied to other fluids. Pressure head relative to atmosphere and above an ordnance datum are utilised as appropriate. Care must be taken to ensure that absolute values of pressure are used in all equations involving forms of the Gas Law.

Similarly, where research reviewed was undertaken prior to the introduction of SI, the original unit system has been retained.

Integer variables

For historical reasons the terms K1 to K4 have been used in the characteristic equations and K5 in the valve discharge coefficient terms. It should be noted that, in FORTRAN, K is an integer variable and the programmed equations, Appendix II, are therefore amended appropriately.

Chapter 1:

Pressure transient mechanisms and historical development

This text will address the problems generated by pressure transient propagation within fluid systems. The first question to be asked is therefore 'what is a pressure transient?'. The simplest definition is that it is the means by which information relating to a change in flow conditions at any point in the system is communicated to all other points within the system. The speed of this information transfer will be dependent upon the properties of both the fluid and the containing pipework; the magnitude of the associated pressure changes, either positive or negative, will depend upon the rate of change of the flow conditions at the source of the transient, the properties of the fluid and the pipe network, the attenuating effect of any relief devices included in the system and the subsequent transmission and reflection of the transient within the system.

Thus the generation of pressure transients, pressure surge or waterhammer, all terms by which this phenomenon is recognised, is entirely normal and will occur in all fluid systems subject to changes in flow conditions. It therefore follows that the study of pressure transient propagation, and the understanding of its controlling variables, is one that truly bridges the engineering disciplines, being as important to the civil engineer designing a hydroelectric plant as to the aerospace systems engineer endeavouring to provide rapid fuel flow rate changes in response to engine thrust demands. These examples chosen at random also illustrate that the consideration of pressure surge may well provide the design condition for the fluid system as a whole.

The objectives of this text will therefore be to provide:

- an understanding of the phenomenon of pressure transient propagation,
- an appreciation of the likely causes of transient propagation and the design alternatives available to limit pressure surge excursions,
- an appreciation of the historical development of the subject and its transformation with the advent of readily available computing power, allowing the application of numerical analysis techniques,
- a familiarity with the application of the method of characteristics finite difference based numerical analysis technique to the prediction of pressure transient propagation for a wide range of commonly met fluid systems,
- an overview of the available surge relief and protection devices,
- and finally a broadening of view to illustrate the fundamental similarities between pressure surge in closed conduit liquid systems and free surface wave and low amplitude air pressure transient propagation.

1.1 Historical development

The literature relating to pressure transient propagation, or waterhammer, as the subject was known, is extensive and covers the last 100 years.

Contributions to the literature range from the purely theoretical, such as early attempts to use Laplace transforms, through discussions of analysis methods to reports of particular transient conditions in individual plant or fluid systems.

It would be inappropriate to include much of this in the present text and the scope of this review will be restricted to the development of the basic theory of transient propagation, the development of modern analysis techniques, concentrating on the use of finite difference based computing models, and the treatment of such specialist areas as column separation.

1.1.1 *Foundation of waterhammer theory*

The first recorded work on waterhammer was due to Weber (1865) and Michaud (1878) who noted the oscillatory nature of the phenomenon and the influence of pipe elasticity; neither realised the connection to the earlier work on sound wave propagation that can be traced back to Euler's solution of the wave equations in 1750. During the 100 years following Euler's work the data on sound waves was extended by D'Alembert, Bernoulli, Young, Savant and Liscovious and E.H. and W. Weber. In 1848 Wertheim noted that the acoustic velocity in water submerged organ pipes was less than predicted by the wave speed expression

$$c_0 = \sqrt{(K/\rho)} \tag{1.1}$$

appropriate for an unconfined mass of fluid; Helmholtz attributed this to the elasticity of the pipe walls in the same year.

Korteweg (1878) established the equation for wave speed in a fluid contained within an elastic conduit that has remained the basis for much later work

$$\frac{c}{c_0} = \sqrt{\frac{1}{1 + \dfrac{DK}{Ee}}} \tag{1.2}$$

by assuming that the longitudinal and bending stresses set up in the pipe wall could be ignored and that the wavelength was long compared to the pipe diameter. Experimental work by Kundt and other workers, confirmed this expression within limits −10% to 4%.

Lamb (1898) extended this work to include longitudinal stresses, obtaining a cubic expression in c, with two finite roots related to a fluid borne wave speed slightly less than that predicted by Korteweg and a pipe wall borne wave speed lower than the acoustic velocity for the wall material, due to the presence of the fluid.

At this time Joukowsky (1900) published the results of a comprehensive experimental programme undertaken in the Moscow Water Works that laid the foundations of all subsequent waterhammer analysis. This work is notable as it derived for the first time the relationship between surge pressure, Δp, or head, ΔH, rise, wave speed and the change in fluid velocity and that for an instantaneous flow stoppage Δp or ΔH may be expressed as

$$\Delta p = \rho c V_0 , \qquad \Delta H = c V_0 / g \tag{1.3}$$

In the development of this expression Joukowsky realised the importance of wave speed and the significance of wave reflections within the pipe system, introducing the concept of pipe period, ie the time taken for a generated wave to travel to a reflecting boundary and return.

The results presented in Simin's 1904 translation show agreement within +/– 15% on pressure measurement and +/– 2% on wave speed, remarkable achievements for the instrumentation employed.

Joukowsky established for the first time a rational explanation for the complex pressure variations experienced within pipe networks by introducing the concept of reflection and transmission at pipe boundaries, by recognising the importance of the pipe period, based upon wave propagation speed, and by developing an expression linking pressure rise to fluid density, wave speed and velocity change.

While Joukowsky is generally credited with the foundation of waterhammer theory, there is evidence, presented by Wood (1970), that an American researcher, J.P. Frizell, developed similar expressions linking pressure variations to wave speed and flow velocity for sudden valve closure in 1897. During this period Allievi was active in Italy and in his Notes 1-IV, (1903-1913), established the Joukowsky relationship by making the same assumptions as Joukowsky and Korteweg, namely a frictionless pipe, uniform pipe dimensions, homogenous wall material and uniform velocity distribution. Allievi developed the wave equations by ignoring the convective terms

$$\frac{\partial^2 p}{\partial x^2} = \frac{1}{c^2} \frac{\partial^2 p}{\partial t^2} \tag{1.4}$$

which may then be solved by a general method proposed by Rieman and D'Alembert yielding

$$p - p_0 = F(t+x/c) + f(t-x/c) \tag{1.5}$$

where x is measured in the initial flow direction and F(),f() represent waves propagated in the –x and +x directions respectively.

Allievi used the above expression to extend Joukowsky's results to 'slow' valve closures, ie valve closures that take longer than one pipe period. This definition of time in terms of pipe periods is central to the understanding of pressure transient propagation. Allievi also extended Joukowsky's work to allow prediction of pressure variations at points along a pipeline, rather than just at the closing valve, introducing the first graphical analysis technique, based upon wave superposition.

Many of the simplifications introduced by Joukowsky and Allievi have now been dispensed with, for example friction is normally included in modern analysis techniques and the influence of longitudinal stresses in the pipe wall is now routinely included via a coefficient, C′, introduced by Parmakian (1955) into Korteweg's wave speed expression

$$\frac{c}{c_0} = \sqrt{\frac{1}{1 + \frac{DK}{Ee}C'}} \tag{1.6}$$

dependent upon pipe restraint and Poisson's Ratio. However Halliwell (1963) has shown that variations in Young's Modulus, E, have a greater effect on wave speed.

Thus the basic concepts of waterhammer theory were established by 1913 and have proved accurate up to the present. The means to analyse transient phenomena in fluid systems however would have to wait a further 50 years until the advent of fast digital computing.

1.1.2 Development of waterhammer analysis from 1910 to 1950

It was 1925 before an English translation of Allievi's work was generally available. In the interim a number of lesser known researchers were active, Gibson, de Sparre and Rateau in Europe, Carpenter and Barraclough (1894) and Peek (1900) in the USA, Kennedy (1964). During this period Johnson (1915) developed a rigid column theory and Gibson (1920) independently developed an equivalent to Allievi's method.

In addition, during the 1910-1925 period, a series of approximate methods were proposed, some developed by researchers working in isolation; others based on partial translations of the earlier European research, Vensano and Warren (1915). The use of the approximations without an understanding of their limitations, or their basic assumptions, led to a wide range of predicted pressure surge levels for any particular case. Contemporary approximations were later compared by Kerr (1968) for a typical valve closure resulted in predictions from 65% to 535% above the observed pressure surge level.

During this period the expansion in the number of large scale water distribution networks and hydroelectric plants, particularly in the USA, made the provision of an accurate design and analysis method vital, and this became available in the early 1930s due to the work of Loewy (1928), Schnyder (1929) and Bergeron (1932), (1961), traceable to the earlier work of Massau (1900). For some reason Loewy's contribution is often neglected and the developed technique is normally referred to as the Schnyder-Bergeron graphical method.

The major advantage of this graphical technique was the inclusion of friction by means of discrete pressure drops, referred to as friction joints, along the length of the pipeline analysed. The disadvantages included the possibility of cumulative graphical errors, the inherent complexity accompanying accurate friction modelling and the inability to use the technique as an iterative design tool prior to the computerisation of the technique in the early 1960s, Thorley and Enever (1979).

This period did however see an enormous increase in interest in waterhammer as an engineering problem, particularly in the USA where ASME/ASCE held symposia in 1933 and 1937. Angus (1935), (1937), (1939), presented a series of papers in this period detailing the application of the graphical technique to a wide range of installations commonly encountered. Marples (1966) and Pearsall (1966) presented papers on the application of the graphical method, while Harding (1966) presented the first computerised version, thereby overcoming many of the technique's shortcomings.

A number of analytical solutions to the complete wave equations were proposed during this period, generally involving the linearisation of the friction term. Wood (1938) introduced Heaviside operational calculus and presented a solution for a simple pipeline and instantaneous valve closure. Rich (1945) proposed the use of the Laplace-Mellin transform for the same simple case. Both solutions linearised the friction term, a poor approximation in the basically turbulent flow regime normally associated with surge events, and both resulted in complex mathematical solutions for even the simplest surge case.

Thus up to the mid 1950s the best available technique was the Schnyder-Bergeron graphical method, indeed texts detailing the application of this methodology were still being published as late as Pickford (1969), well into the next developmental stage, namely the use of digital computing allied to the finite difference based method of characteristics. This technique has now become an industry standard and will be the subject of detailed presentation in this text.

1.1.3 *The introduction of digital computing methods, in particular the method of characteristics, 1950-1970*

The method of characteristics is a general mathematical technique that is particularly suited to the solution of pairs of quasi-linear hyperbolic partial differential equations linking two dependent and two independent variables. The equations defining pressure transient propagation and surge pressure levels within fluid piping systems and wave propagation in open channel or partially filled pipe flow belong to this family of problems.

The method was first proposed by Rieman in 1860 while he was studying finite amplitude sound wave propagation in air. In 1900 Massau employed the method in a study of unsteady flow in open channels. The first application to pressure transient analysis was due to Lamoen (1947), while later contributions by Gray (1953), (1954) received wider attention. Neither author considered the use of a computer. Following Gray's work a considerable number of papers were presented in the USA employing this technique, among the first being Ezekial and Paynter (1957) and most significantly Streeter with a range of co-authors, (1962), (1966), (1967), (1969). Much of the published work rested on the presentation of the method of characteristics by Mary Lister (1960), which remains the basis for the analysis techniques in this text.

In the UK and Europe interest in the use of the method of characteristics developed rapidly with papers by Fox (1968), Evangelisti (1969), Swaffield (1970) and by the early 1970s this technique was established as the standard technique for transient analysis, a fact clearly evident in the wide range of international papers presented at the first BHRA Pressure Surge conference in 1972.

Streeter and Lai (1962) claimed that the method of characteristics was the equivalent of treating the frictional loss as an uniformly distributed loss along the whole pipeline. However this is strictly not the case, rather the method extends and improves upon the graphical technique's friction joint approach by utilising the computer's ability to introduce as many friction joints as the analyst would wish, thus allowing the model to approach the continuously distributed friction loss case. This discussion was active in the 1960s, particularly supported by the last users of the graphical technique. Reference to a description of the graphical method and a comparison to the method of characteristics solution will reveal more similarities than differences between the two approaches, Enever (1970). Both rely on the modelling of the transmission of changes in flow conditions through the pipe network along lines in the time-distance plane whose slopes are determined by fluid density and wave speed, and both represent frictional losses at discrete locations based upon the application of steady uniform flow values of friction factor to the local unsteady flow conditions. The graphical technique and its computerised version are now of historic interest only.

1.1.4 Development of the method of characteristics model, 1970-1990

The introduction and acceptance of the method of characteristics as the most suitable technique for the analysis of pressure transient phenomena opened up the whole pressure surge subject. As will be seen from a study of the literature referred to in this text, this technique has allowed previously difficult transient conditions to be treated and designs for systems modified to prevent potentially damaging transient propagation. Areas that have particularly developed since the introduction of computing methods have included problems involving gas release as line pressure falls in response to negative transient pressure wave propagation and problems involving cavity formation modelling during column separation, both with and without gas release. In this latter case the advantages of the computing methods over their earlier graphical counterparts is particularly clear as the accurate monitoring of cavity growth and collapse is essential in determining the final interface fluid velocity which in turn determines, via Joukowsky's relationship, the resulting pressure surge on cavity collapse. The small time steps inherent in the computing simulation make this possible.

Similarly the introduction of computing methods based around the method of characteristics has allowed progress in the study of the interface between fluid borne transients and the vibration of structures, a subject of obvious importance in the design of pipe system supports and also of importance in the study of structurally transmitted sound waves.

The interaction of machines and pipe flows can also be studied utilising these methods, improving understanding within the design of hydroelectric power stations and the design of surge alleviation devices.

The introduction of rapid computing techniques also has an implication for the use of pressure transient analysis as a design tool rather than as a problem solving technique, only utilised following the demonstration of a surge generated problem within a fluid system. This application of the modelling methods to be described in this text has perhaps the most far reaching consequences and illustrates the correct use of computing power within the engineering system design process.

A review of relevant literature will be included in each of the specialist topic treatments presented in this text so that it would be inappropriate to follow further at this stage the individual progress made in each area mentioned.

1.2 Mechanism of pressure transient propagation

Prior to introducing the analysis techniques and computing methods appropriate to the study of pressure transients, it will be useful to discuss in some detail the mechanisms by which transients are propagated, transmitted and reflected within piping systems and the way in which the resulting pressure variations recorded within systems depend upon such variables as valve closure time, wave speed, initial flow velocity, initial line pressure and overall initial steady flow friction losses.

As already mentioned the propagation of pressure transients is an entirely normal event in any fluid carrying network. The definition of steady flow may be stated as the continuation of constant values of flow variables such as velocity, pressure density etc at any particular point in the system. Unsteady flow may be defined as flow in which the values of these variables change with time at the particular system location being monitored. Similarly uniform and non uniform flow conditions may be defined in terms of the variation, or otherwise, of flow variables with distance along a pipeline. Thus in order for a fluid system's operating point to be moved from one steady condition to another there must be an intermediate flow state that must be both unsteady and non uniform. It is only the severity of these transients that is of concern.

Figure 1.1 illustrates this general case of unsteady flow for the situation where the delivery from a valve is reduced from one steady flow to a second, lower throughflow. In the rapid valve closure case the transients generated may well be damaging, however a much reduced valve closure rate may result in little more than a mild mass oscillation.

In introducing pressure transient propagation mechanisms the problem is to avoid oversimplification while at the same time avoiding the perpetuation of some of the myths generated by past oversimplification. The simplest condition that appears to fit this requirement is illustrated in Figure 1.2, namely a simple uniform pipeline linking two reservoirs and including an internal valve at its midpoint. The closure of this valve will be considered and the propagation of pressure transients both upstream and downstream will be described.

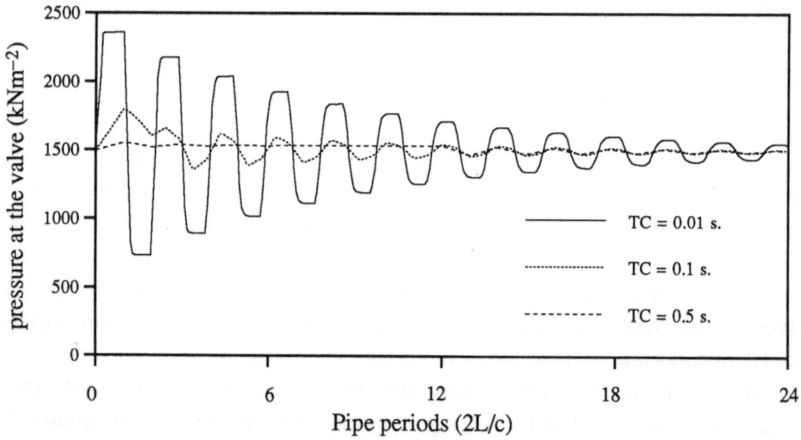

Figure 1.1(a) Pressure history at a valve terminating a simple pipeline as its discharge is reduced by 80%, for a range of valve operation times, TC

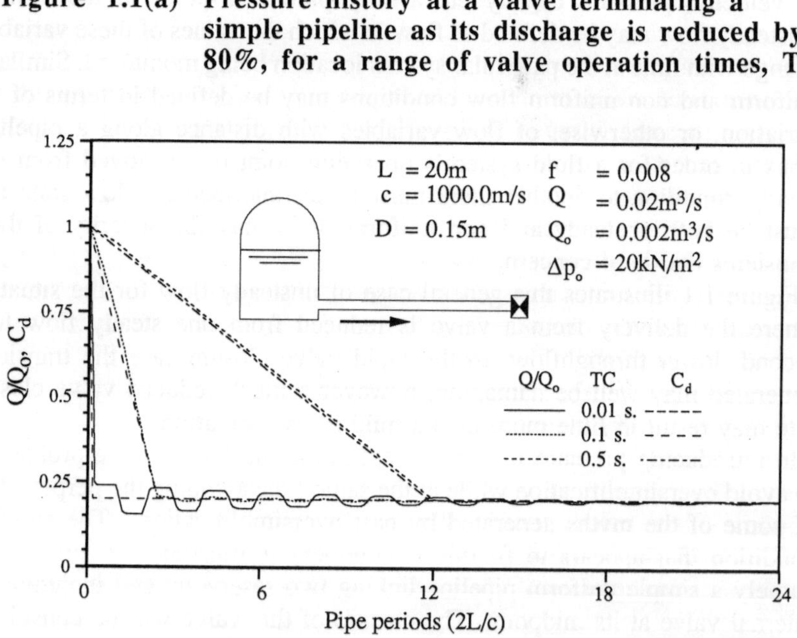

Figure 1.1(b) Flow reduction at the valve as its discharge coefficient is reduced by 80%

As the valve closes the fluid approaching its upstream face is retarded with a consequent compression of the fluid and an expansion of the pipe cross sectional area. The increase in pressure that results on the upstream face of the valve depends upon the relative importance of these two effects, the more flexible the pipe the lower the pressure rise. The pressure change is transmitted upstream as the information concerning the closure of the valve is transmitted. Again the speed at which this information travels depends upon the fluid and pipe wall properties. A rigid pipe would result in information transmission at the fluid only acoustic velocity, while an elastic pipe wall effectively reduces this wave speed.

On the downstream face of the valve a similar mechanism results in the transmission of information relating to the reduced flow through the valve. This is perceived as a drop in pressure on the downstream face of the valve and generates a negative pressure wave that carries this information downstream. The flow velocity in the downstream pipe section also decreases. During this description of the mechanism of transient propagation it will be assumed that the pressure reduction is insufficient to reduce the line pressure to either a gas release or vapour level.

Therefore closure of the valve retards flow in both upstream and downstream pipes, generating pressure waves in both pipes, a positive, or pressure rise, wave in the upstream pipe and a pressure reduction, or negative pressure, wave in the downstream pipe. The pipe itself is affected by the generated wave, a pipe cross section expansion upstream and a contraction downstream. The degree of pipe deformation in turn affects the wave speed, or the rate at which information concerning the changed conditions at the valve is transmitted to the rest of the pipe system.

Once the valve motion is initiated the pressure-time history at the valve depends upon the rate of valve closure and the influence of the system boundaries upon the reflection of the pressure transients generated by the valve motion. This in turn involves the wave speed as this variable determines for any system the time taken for boundary reflections to arrive back at the valve. Thus the relationship between valve closure time and pipe period becomes central to the prediction of the valve pressure-time history.

While this is strictly true it must also be appreciated that the form of the valve characteristic-valve position relationship also has a major role to play in determining the rate of change of conditions at the boundary formed by the closing valve. Many valve types do not retard the flow significantly until the later portions of their closure, thus the measurement of valve closure time with no reference to the throughflow-pressure drop characteristic may well lead to an overestimate of the valve's effective closure time and hence

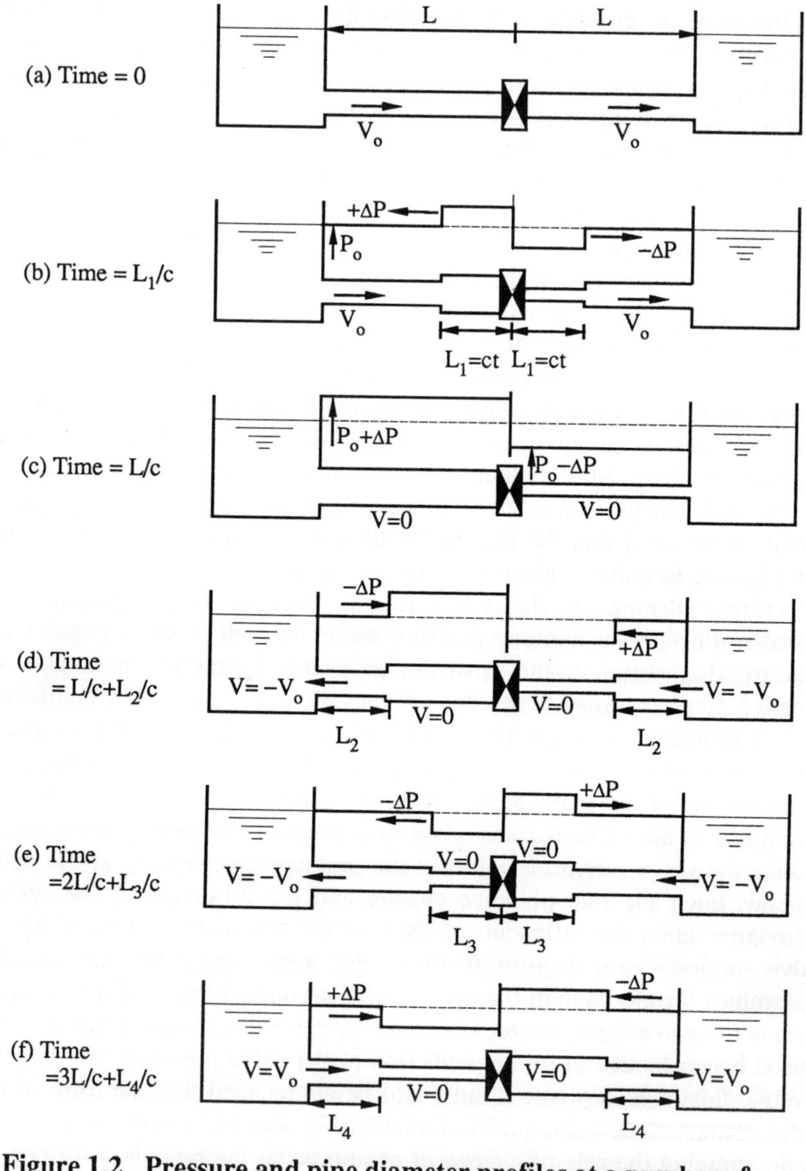

Figure 1.2 Pressure and pipe diameter profiles at a number of instances following an instantaneous valve closure. Frictional effects neglected

a subsequent underestimate of the severity of the resulting transient propagation. (This effect will be returned to as a means of surge limitation when two speed valve closures will be considered, ie rapid over the early stages and slow in the final closing stage.)

Following valve closure the subsequent pressure-time history will depend on the conditions prevailing at the system boundaries. For the case illustrated in Figure 1.2 the sequence of events is most easily described by means of a series of sketches seen as 'snapshots' at particular times during the transient propagation.

Assuming that the valve closure was instantaneous makes the description of events somewhat easier, however this a special case and the differences that arise from a 'slower' valve closure will also be outlined. On valve closure the fluid adjacent to the valve in each pipe would have been brought instantaneously to rest and pressure waves conveying this information would have been propagated along each pipe at the appropriate acoustic velocity. At a later time t the situation is as shown in Figure 1.2(b), the wavefronts having moved a distance L_1=ct along each pipe.

The pressure waves reach the terminating reservoirs at a time t=L/c following valve closure, Figure 1.2(c). (Note that in the simple case considered here both pipes have been taken as of the same length L.) An unbalanced situation then arises where the fluid at the pipe-reservoir interface is at a pressure different to the pressure in the reservoir at that level. This cannot be sustained and a restoring flow is initiated that equalises the pressure at the interface. Pressure waves of a magnitude sufficient to return the pressure in the pipe to the pre-valve closure pressure are thus propagated at time L/c from each reservoir and travel back towards the now closed valve, a negative wave in the upstream reservoir case and a positive pressure wave in the downstream reservoir case.

For the upstream pipe the propagation of this wave also produces a fluid flow from the pipe into the reservoir as the pipe ahead of the transient is at a higher pressure than the reservoir. As the system described has been assumed to be frictionless, the magnitude of this reverse flow is equal to the initial flow velocity, as shown in Figure 1.2(c).

In the downstream pipe the same reasoning explains the establishment of a flow from the downstream reservoir having the same magnitude as the initial flow but opposite sign, Figure 1.2(c).

For the simple upstream pipe considered here the restoring pressure wave reaches the closed valve at a time 2L/c, known as the 'pipe period'. At this time the whole of the upstream pipe length has been returned to its original pressure and a flow has been established from the pipe into the upstream reservoir. At time 2L/c, as the wave reaches the closed valve, there remains no fluid ahead of the wave to sustain this restoring flow and therefore a low pressure region forms at the closed valve upstream face associated with zero flow velocity at the valve. (This assumes that the pressure does not fall to gas release or vapour levels, cases that will be dealt with later.) The imposition of a zero velocity on the fluid at the valve upstream face generates a pressure drop wave that is propagated from the closed valve towards the upstream reservoir, reducing the pipe flow to zero as it passes, as shown in Figure 1.2(e). Again as the system has been assumed frictionless, all the pressure waves propagated will have the same magnitude as that initially propagated by the valve closure, $+/- \Delta p$. Thus the wave propagated upstream from the closed valve at time 2L/c has a magnitude of $-\Delta p$, reducing all points along the pipe to a pressure level Δp below the initial line pressure prior to valve closure.

It is also worth noting that the total instantaneous pressure change at the valve's upstream face at time 2L/c was $-2\Delta p$, ie due to the simultaneous arrival of the $-\Delta p$ wave from the upstream reservoir and its immediate reflection at the closed valve as a $-\Delta p$ wave travelling back upstream. For non instantaneous valve closures in less than one pipe period this arrival and re-reflection process results in the pressure variation at the closed valve displaying a rate of change double that monitored as a single wave passes an observation station remote from either the valve or the reservoir boundaries.

Understanding the form of the pressure variations that accompany the transmission of transients in even the simplest cases is often assisted by a consideration of the actual length of the wavefront making up any transient. For the instantaneous valve closure case this obviously tends to zero, thus all the pressure changes monitored would appear as vertical rises or falls in pressure on a pressure-time plot. If the valve takes a finite time, t', to close, assumed to be still less than one pipe period, then the length of the wavefront is $c\,t'$ and monitoring points closer to the either boundary than $ct'/2$ will record both the passing transient and its reflection from the boundary as a summated pressure change. Thus the maximum Joukowsky pressure change, whether rise or fall, will not be experienced at any point closer to either boundary reservoir than $ct'/2$.

Returning to the description of the propagation of transients following the instantaneous valve closure illustrated in Figure 1.2, the establishment of a line pressure $-\Delta p$ below the initial steady flow value is complete at time 3L/c when the $-\Delta p$ wave generated by reflection at the valve at time 2L/c reaches the upstream reservoir.

Similarly the restoring wave from the downstream reservoir that reached the closed valve's downstream face at time 2L/c had established a reverse flow in the downstream pipe along the pipe's length and this flow is destroyed by the 'dead end' boundary formed by the closed valve. Bringing this restoring flow to rest generates a +Δp transient at the valve's downstream face at time 2L/c and this propagates downstream towards the reservoir, bringing the reverse flow in this pipeline section to rest and increasing the pressure to +Δp above the initial line pressure prior to valve closure. This wave arrives at the downstream reservoir at time 3L/c, at which time the fluid is at rest along the whole pipe section downstream of the closed valve. The observations made for the upstream pipe concerning the rate of change of pressure at any location as the transients and their reflections pass remain true for the downstream pipe, as does the importance of the wavefront length and its relation to valve closure time.

Thus at time 3L/c an unbalanced situation similar to that at time L/c again exists at both the upstream and downstream pipe/reservoir interfaces with the difference that it is the upstream pipe which is at a pressure below the reservoir pressure and the downstream pipe which is at a pressure above the reservoir pressure. In both cases the flow in the pipe sections is zero. However the mechanism for transient propagation is identical to that described for time L/c and thus two restoring transients are again propagated. In the case of the upstream pipe this takes the form of a +Δp pressure wave that travels towards the closed valve, returning the line pressure to its initial steady flow value and imposing a flow towards the closed valve having the same velocity as the initial pre-valve closure flow.

An identical process results in the propagation of a −Δp pressure wave from the downstream reservoir towards the closed valve, again returning the line pressure and flow velocity to their respective pre-valve closure values. These restoring waves arrive at the closed valve at time 4L/c, so that at this time the line conditions are identical to the initial conditions prior to valve closure with the exception that the valve remains closed so that the imposed flows cannot be sustained and the whole cycle described above, and illustrated by Figure 1.2, is forced to repeat indefinitely with a periodic time of 4L/c.

The pipe system chosen to illustrate this cycle was a special case as the pipes were assumed to be the same length and the wave propagation velocity, c, was the same for each. If the pipes were of different lengths and the individual wave speeds were different, the processes described would still apply, however the pressure variations in each pipe would no longer be synchronous. If the valve closure remained instantaneous the pressure transients in each pipe would be independent of each other, the periodic nature of the oscillations being dependent on the pipe length and local wave speed, determined by the properties of the fluid and the pipe wall elasticity.

The period of the pressure cycles described is 4L/c, however a term often met in transient analysis is 'pipe period'; this is defined as the time taken for a transient to travel from its point of inception to a reflecting boundary and return. In this case the pipe period for both pipes was 2L/c based on the time taken for the transient to reach the boundary reservoirs and return as a reflected wave to the closed valve. It will be shown that the use of pipe periods as a measure of time is central to an understanding of transient propagation and the prediction of transient pressure effects.

From the transient cycle described it is possible to infer the pressure time profiles at points along both the pipelines, as shown in Figure 1.3. These variations are based simply on a calculation of the time at which any one of the +/− Δp waves reaches that point in the system, assuming a constant wave propagation velocity, c, frictionless flows and maintenance of line pressures above gas release and vapour levels.

1.2.1 The importance of valve closure time

The most important point illustrated by both Figures 1.2 and 1.3 is that there is no possibility of the pressure attained on valve closure being less than that generated by an instantaneous valve closure unless the valve closure takes longer than one pipe period, 2L/c. The reduction in peak pressure possible by extending valve closure time derives from the fact that the reflected waves from the upstream reservoir are negative and thus reduce the pressure rise that is continuing at the valve. Historically Allievi was able to demonstrate this effect successfully by his graphical method which was based on the principle of superposition of generated and reflected transients. The same reasoning limits the pressure drop experienced downstream of the valve due to the arrival of positive reflections from the downstream reservoir at times greater than 2L/c.

Generally valve closures in less than one pipe period are referred to as 'rapid' while those taking longer than one pipe period are referred to as 'slow'. Figure 1.4, generated for a simple single pipe system terminated by an upstream reservoir and a downstream valve, by the program presented in Appendix II, illustrates the effect of valve closure time on peak pressure levels on the valve upstream face.

The example discussed has also included the simplification that the pressure

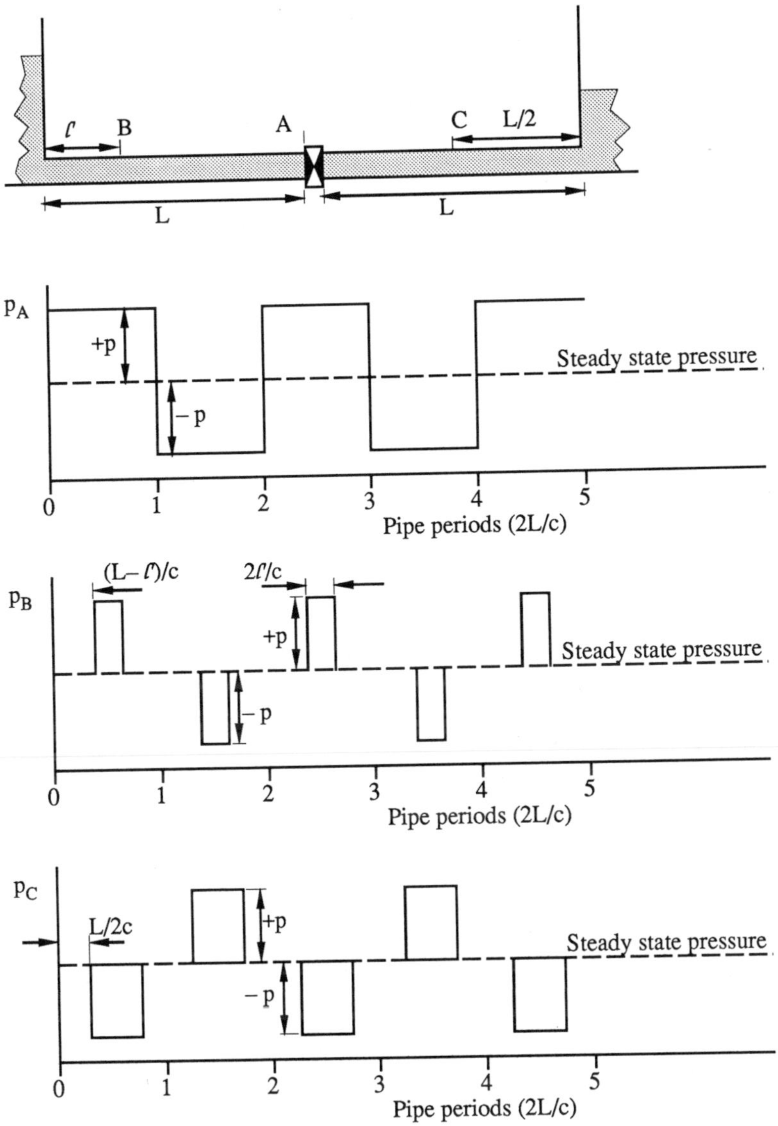

Figure 1.3 Pressure variations following an instantaneous valve closure at points along the two identical pipes linking the reservoirs

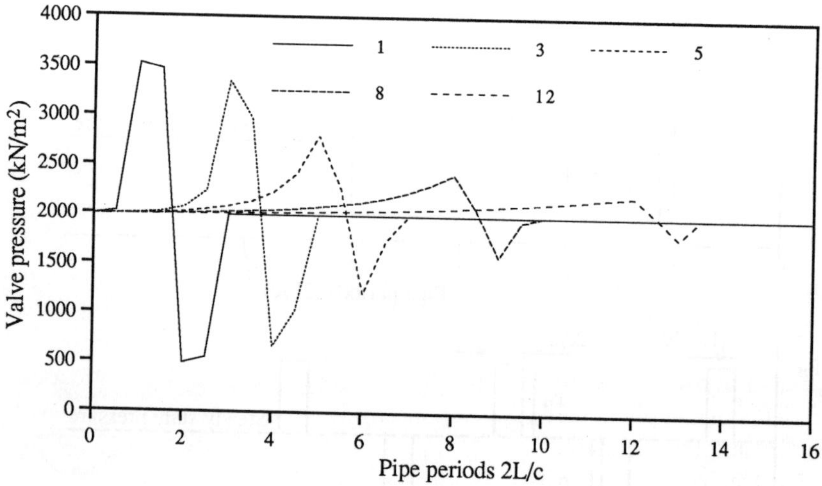

Pipe length	10m	Pipe diameter	0.05m
Wave speed	$1000ms^{-1}$ water	Valve Δp_0	$20kNm^{-2}$
Valve Q_0	$0.004m^3s^{-1}$	Friction factor	0.008
Initial discharge	$0.003m^3s^{-1}$	Upstream reservoir pressure	$2000kNm^{-2}$

Note: for clarity, only the first cycle of the pressure oscillation for each valve closure is shown, the full oscillation in each case mirrors that shown in Figure 1.1

Figure 1.4 Valve pressure rise as a result of valve closure in 1 to 12 *pipe periods*

surge variations recorded on the downstream side of the closing valve do not effect the upstream pipe conditions as the valve was fully shut prior to the arrival of any reflections from the downstream reservoir. In 'slow' valve closure cases these transients would effectively propagate through the still open valve and would affect the upstream pipe condition, and vice versa. This effect will be fully modelled by the analysis techniques to be developed later in the text and are incorporated in the computer program presented in Appendix II. This effect can be important if it is not fully appreciated that the pressure surge generated by a 'slow' valve closure, ie one taking longer than one pipe period based on whichever pipe length is the shorter, will be different from that if the valve were to discharge directly to atmosphere or a constant pressure reservoir. This lack of understanding has in the past been responsible for problems in the acceptance tests for valves to be installed in situations where the valve closure rate was the only effective means of controlling surge generation, for example aircraft fuel systems where weight considerations limit pipe thickness and preclude the introduction of surge alleviation devices.

The description of the transient propagation and reflection mechanism above also allows the identification of the appropriate values of reflection coefficients that apply at the constant pressure boundaries formed by the two reservoirs and the 'dead end' or zero flow velocity boundary formed by the closed valve. While these values will be derived later, by observation it may be seen that the constant pressure boundary possesses a −1 reflection coefficient, while the zero flow boundary may be defined by a +1 reflection coefficient. Use of these boundary reflection coefficients allows the pressure variations illustrated in Figure 1.3 to be explained completely.

1.2.2 *The importance of frictional effects*

In the absence of friction the cycle illustrated by Figure 1.3 would continue indefinitely. In practice friction damps the pressure oscillations within a few cycles. Friction reduces the peak velocity reached during each successive reverse flow imposition described, indeed the periodic nature of the flow reversals has the effect of effectively increasing the value of the appropriate friction factor.

In systems where the initial steady flow frictional losses are high, the neglect of frictional effects can result in a gross underestimate of the peak pressure rise following valve closure. In these cases the initial pressure at the valve is considerably less than the upstream reservoir pressure level, however

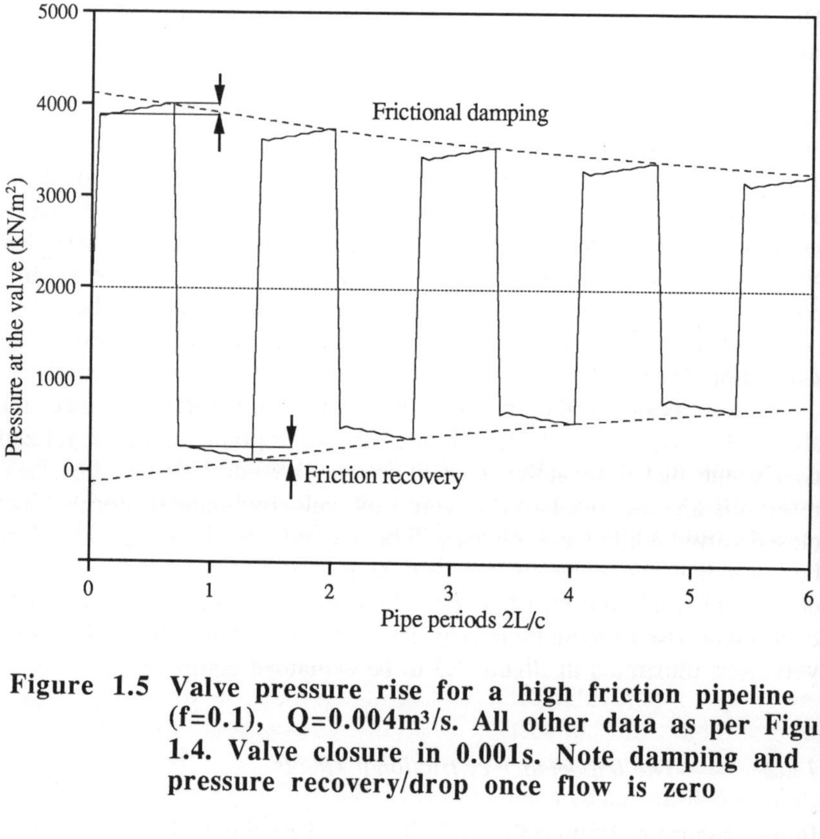

Figure 1.5 Valve pressure rise for a high friction pipeline (f=0.1), Q=0.004m^3/s. All other data as per Figur 1.4. Valve closure in 0.001s. Note damping and pressure recovery/drop once flow is zero

as the flow is retarded by valve closure, frictional pressure loss along the pipe is reduced as flow velocity falls and the pressure at the valve rises towards the reservoir level. Obviously for a zero flow condition the pressure level at the closed valve must become equal to that at the reservoir. As each layer of fluid between the valve and the reservoir is brought to rest by the passage of the $+\Delta p$ pressure transient, so a series of secondary pressure waves, each of a magnitude equal to the frictional pressure loss recovered within that cross sectional layer is transmitted towards the closed valve, resulting in the full friction recovery effect being felt at the closed valve at the end of the first pipe period, 2L/c, Figure 1.5. As the flow reverses in the pipe during the second pipe period, time 2L/c to 4L/c, the opposite effect is recorded at the valve due to the need to re-establish the high friction loss. In certain applications, such as long distance oil or water pipelines, these effects may contribute the larger part of the pressure surge on valve closure.

1.2.3 The importance of initial line pressure, gas release and vapour cavity formation

In addition to the assumptions made with regard to friction, the pressure level at the valve was also assumed to remain above gas release or vapour level. If fluid vapour pressure had been reached at the closed valve, either on the downstream face during closure or on the upstream face at the end of the first pipe period, then the simple repeating cycle described above would have been disrupted by the formation of a fluid vapour cavity at the low pressure location that would have formed a different boundary condition. In particular the presence of a vapour cavity would allow the reverse flows to be sustained as the cavity could expand to allow continuation of flow away from the closed valve. The formation of vapour cavities is followed by a period of time in which the separated fluid columns move under the influence of the pressure gradients set up between the vapour cavity and either the upstream or downstream reservoir pressure, these being the appropriate boundary conditions. Eventually the fluid column separated from the closed valve in either pipe by the vapour cavity is brought to rest by these adverse pressure gradients and is then accelerated back towards the closed valve by the same pressure gradient. The pressure rise following the collapse of the cavity may be severe, and may well exceed a predicted maximum based on the initial flow velocity in the pipe and the initial valve closure time. These effects, known as Column Separation are extremely important and will be returned to later in detail. It will be seen

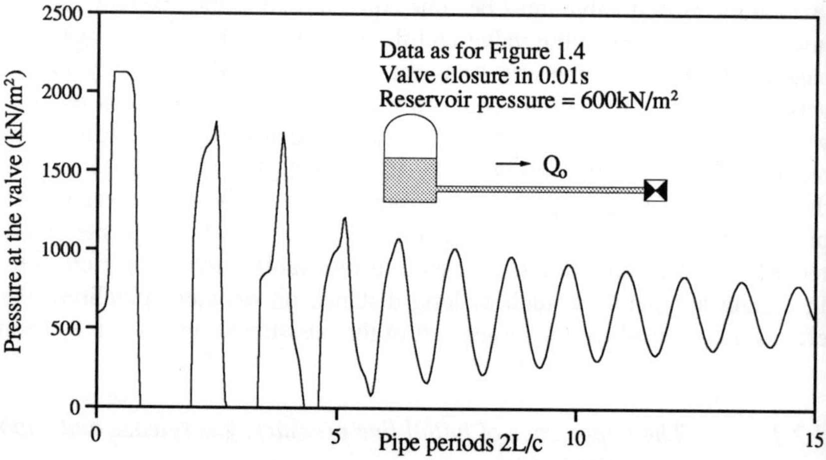

Figure 1.6(a) Pressure at the valve

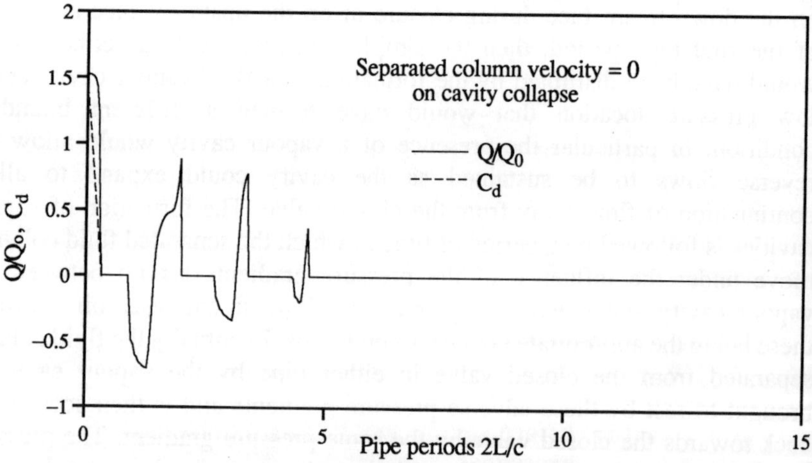

Figure 1.6(b) Nondimensional discharge through the valve

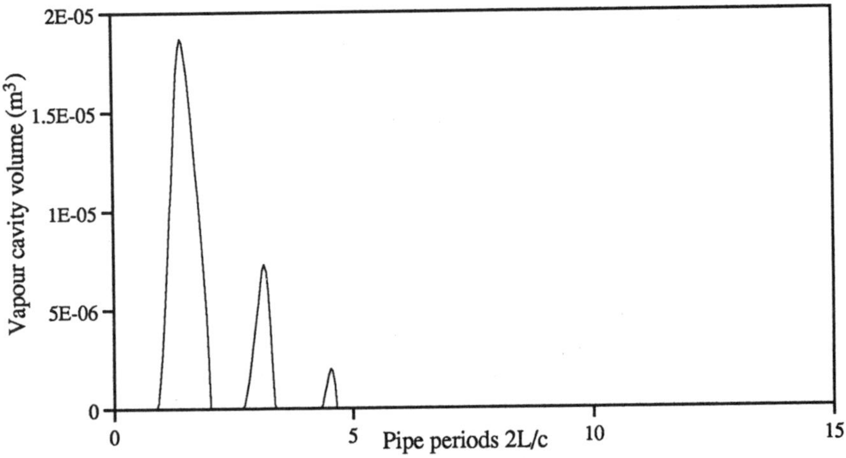

Figure 1.6(c) Vapour cavity volume upstream of closed valve

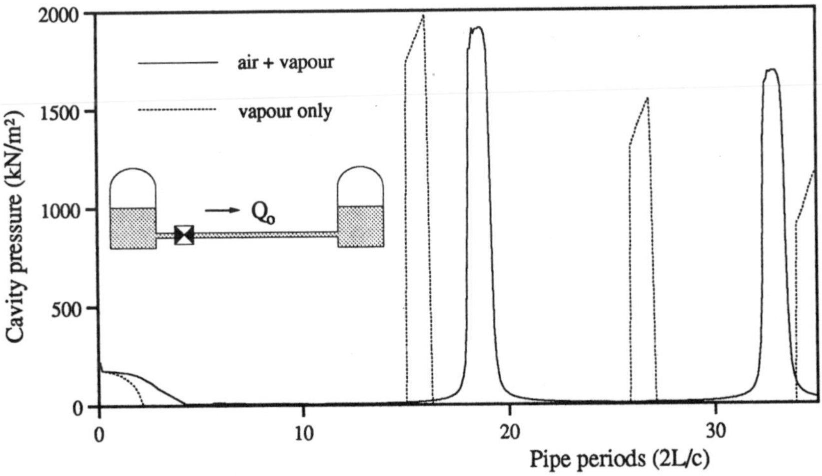

Figure 1.6(d) Effect of air release on cavity collapse pressures downstream of a closing valve

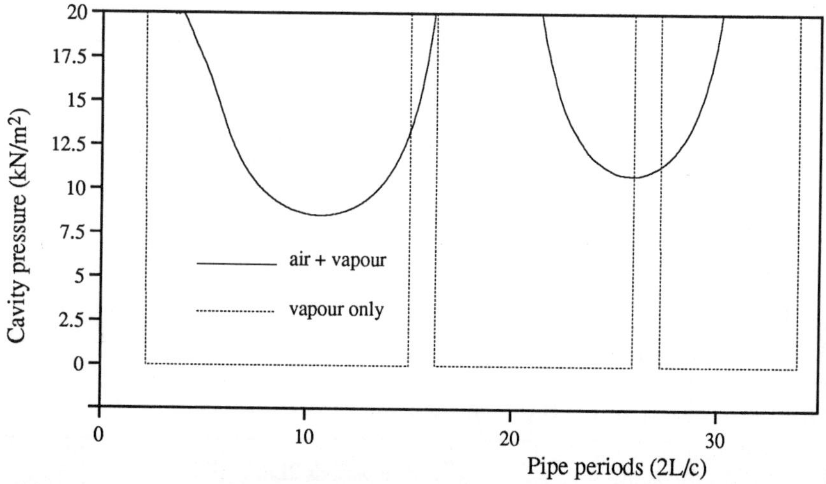

Figure 1.6(e) Effect of air release on the minimum pressure predicted downstream of the closing/closed valve

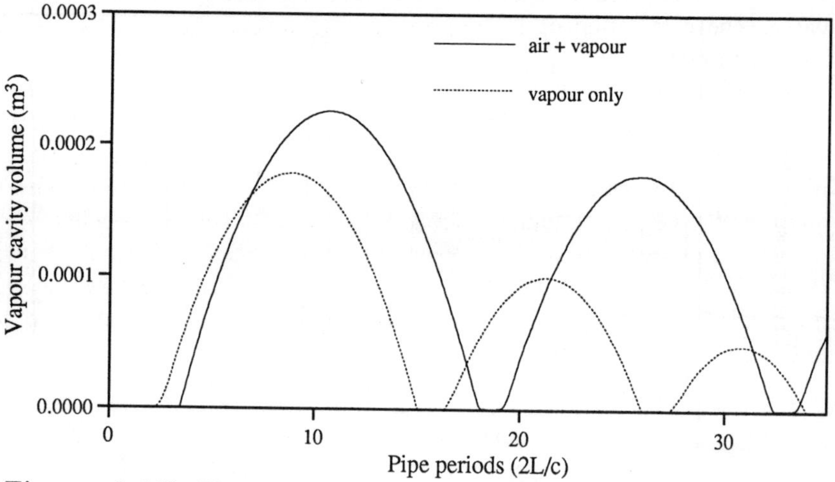

Figure 1.6(f) Vapour cavity volume with and without gas release for Figure 1.6(e)

that the vapour pocket, while in existence, represents a constant pressure boundary at the closed valve, thus replacing the +1 reflection coefficient appropriate for zero flow velocity with a –1 value. The presence of released gas in the cavity region will also be dealt with later, however it will be appreciated that a trapped gas volume will modify the reflection coefficient as its pressure fluctuates.

Figure 1.6 illustrates the effect of column separation both upstream and downstream of a closing valve, generated by the program presented in Appendix II.

1.3 Transient propagation velocity: the effect of pipe elasticity and free gas

It has already been stressed that the wave propagation velocity is a determinant of both the magnitude of the pressure transient generated by a change in the boundary conditions in a fluid flow system and the subsequent form of the pressure variations monitored. The value of wave speed in an unconfined volume of water at atmospheric conditions is of the order of 1500 m/s. The effect of pipe wall elasticity and free gas is to reduce this value. The calculation of the degree of reduction is the objective of this section of text.

The fundamental result that valve closures in less than one pipe period will always generate the full Joukowsky pressure surge, +ve or –ve, has already been developed. The form of the pressure variations at locations along a pipe also depend on the wave speed as this determines the wavefront length and thus the degree of superposition of travelling waves that occurs.

In this section the basic expressions for wave speed will be determined, taking into account the effects of pipe elasticity, wall thickness and pipe diameter, fluid density and compressibility and the presence of any free gas, either entrained or released.

The expression for wave speed in an unconfined fluid is

$$c_0 = \sqrt{(K/\rho)} \tag{1.7}$$

where K is the fluid Bulk Modulus and ρ is the fluid density.

Consider a flow situation where the fluid density is ρ_f and the fluid Bulk Modulus is K_f, the pipe diameter D and the wall thickness e, the material of the wall having a Young's Modulus of Elasticity E and Poisson's Ratio μ, as shown in Figure 1.7.

The mean fluid-gas density, ρ_m may be taken as

$$\rho_m = y\rho_g + (1-y)\rho_f \tag{1.8}$$

where y is the proportion of free gas by volume and subscript g refers to gas properties.

The passage of a positive pressure transient will result in a series of volumetric changes in the fluid-gas-pipe system that may be determined separately and combined by superposition.

For the fluid the volumetric change, Figure 1.7, is given by

$$dVol_f = (Vol_f / K_f)\, dp \tag{1.9}$$

and for the free gas

$$dVol_g = (Vol_g / K_g)\, dp \tag{1.10}$$

The volume of the total fluid and gas element is

$$Vol_t = Vol_f + Vol_g \tag{1.11}$$

The volume of the pipe section containing this initial volume can be distorted both radially and longitudinally by the passage of the transient. If the original volume of the volume is given by

$$Vol_t = \frac{\pi D^2 \ell}{4} \tag{1.12}$$

where ℓ is the length of the element measured along the flow direction, then the distorted volume due to the passage of a transient, dp, may be expressed as

$$dVol_p = \pi D(\frac{\Delta D}{2})\ell \tag{1.13}$$

where subscript p refers to the pipe dimensions and the products of small quantities (eg $\Delta D\ \Delta\ell$) are neglected.

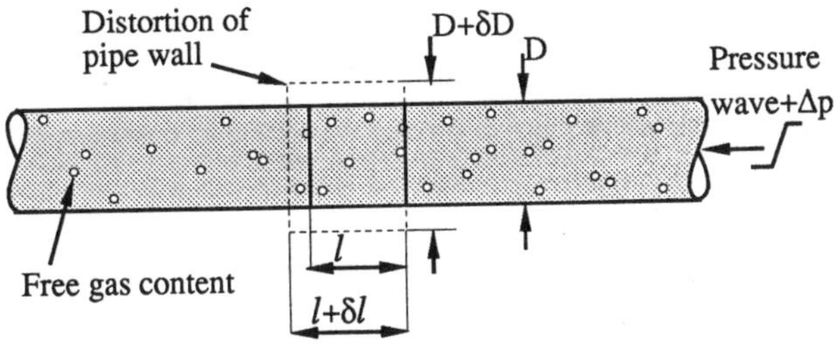

Figure 1.7 Effect of a pressure wave in a gas-fluid pipeline

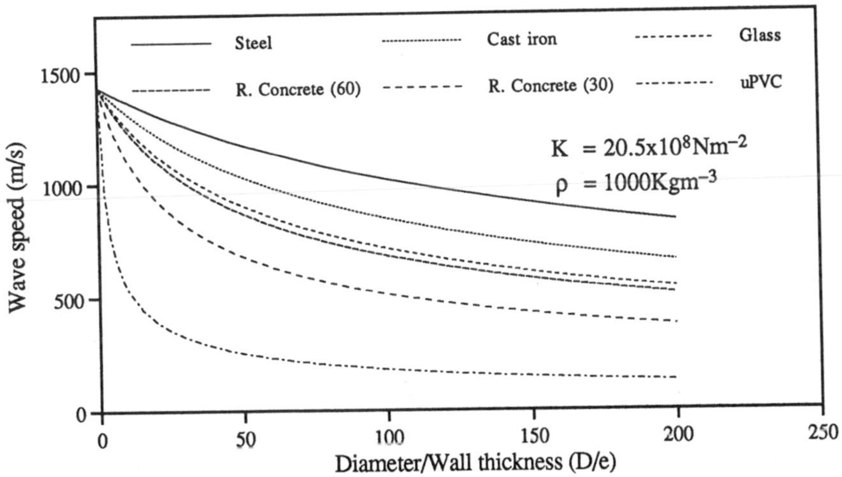

Figure 1.8(a) Effect of the wall thickness to pipe diameter ratio on wave speed for a range of common pipe materials

For thin walled pipes the circumferential stress σ_2 and the longitudinal stress σ_1 are given by

$$\sigma_2 = (\frac{D}{2e})dp \tag{1.14}$$

and

$$\sigma_1 = (\frac{D}{4e})dp \tag{1.15}$$

while the longitudinal strain in the pipe wall is given by

$$\frac{\Delta \ell}{\ell} = \frac{\sigma_1 - \mu\sigma_2}{E} \tag{1.16}$$

and the circumferential strain by

$$\frac{\Delta D}{D} = \frac{\sigma_2 - \mu\sigma_1}{E} \tag{1.17}$$

where μ is the Poisson's Ratio of the pipe material.

Thus

$$dVol_p = \pi D^2 \ell (\frac{\sigma_2 - \mu\sigma_1}{2E}) \tag{1.18}$$

The degree of restraint assumed for the pipe installation will affect the final form of the volumetric change expression

$$\frac{dVol_p}{Vol_t} = 2(\frac{\sigma_2 - \mu\sigma_1}{E}) \tag{1.19}$$

1. for a pipe fully restrained at one end only both circumferential and longitudinal stresses occur and hence from equations 1.14, 1.15 and 1.19 the volumetric change may be expressed as

$$\frac{dVol_p}{Vol_t} = \frac{D}{Ee}(1 - \frac{\mu}{2})dp \tag{1.20}$$

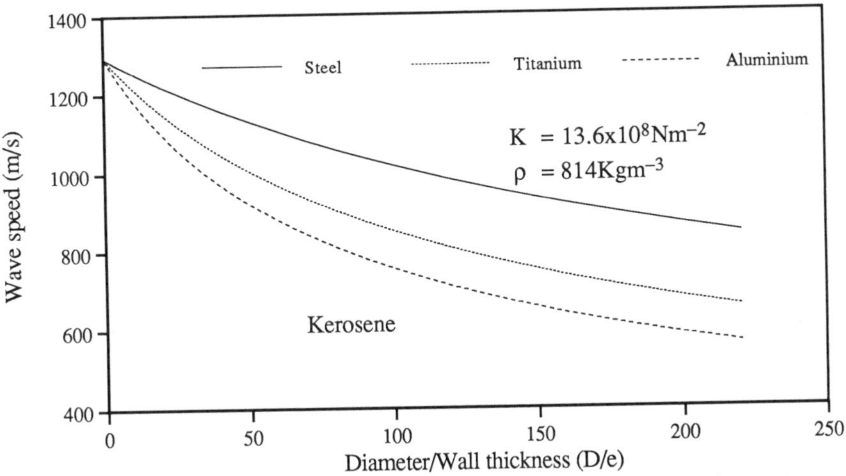

Figure 1.8(b) Dependence of wave speed on pipe D/e ratio for kerosene flowing in steel, titanium or aluminium pipes

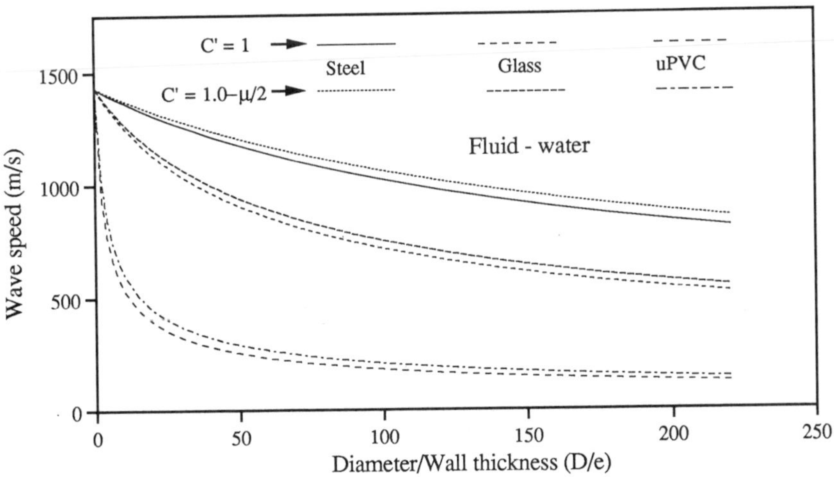

Figure 1.8(c) Effect of pipe restraint upon wave speed for a representative range of pipe materials and D/e ratios

2. for a pipe fully restrained against axial movement along its whole length the longitudinal strain must be zero and hence

$\sigma_1 = \mu\sigma_2$

so that equations 1.14, 1.15 and 1.19 yield

$$\frac{dVol_p}{Vol_t} = \frac{D}{Ee}(1-\mu^2)dp \qquad (1.21)$$

3. for a pipe fitted with expansion joints along its whole length then the longitudinal stress becomes zero. Examples of this type of restraint may be found in aircraft fuel systems and also in systems designed to absorb thermal expansion effects. Equations 1.14, 1.15 and 1.19 thus yield

$$\frac{dVol_p}{Vol_t} = \frac{D}{Ee}(1.0)dp \qquad (1.22)$$

A general form of equations 1.20 through 1.22 may be introduced

$$\frac{dVol_p}{Vol_t} = \frac{D}{Ee}C'dp \qquad (1.23)$$

where the value of the coefficient C′ is self evident.

The total change in element volume as a result of the passage of the dp transient may thus be taken as

$$dVol_t = dVol_p + dVol_f + dVol_g$$

or

$$dVol_t = \left[(\frac{D}{Ee})C'\,Vol_t + \frac{Vol_f}{K_f} + \frac{Vol_g}{K_g}\right]dp \qquad (1.24)$$

where

$Vol_f = (1-y)\,Vol_t$

and

$Vol_g = y\, Vol_t$

Examination of the form of equation 1.24 allows the definition of an effective bulk modulus, cf equations 1.9 and 1.10,

$$K_{eff} = \frac{1}{\dfrac{1-y}{K_f} + \dfrac{y}{K_g} + \dfrac{DC'}{Ee}} \tag{1.25}$$

and a general wave speed expression follows

$$c = \sqrt{\frac{K_{eff}}{\rho_m}} \tag{1.26}$$

incorporating the mean density defined by equation 1.8.

At low temperatures and pressures the gas bulk modulus may be approximated by the local gas absolute pressure, at least as a first approximation as of course the passage of the transient changes the gas pressure.

The presentation and derivation of the wave speed relationships above are independent of the sign of the pressure wave. For convenience it is normal to use a pressure rise as a basis for pipe deformation and gas and liquid 'compression', a reduction in pressure would be defined by the same equations although the pipe would contract and free gas would expand as the line pressure fell.

In the absence of free gas the wave equation is therefore shown to be that derived by Parmakian (1955), equation 1.6, namely

$$c = \sqrt{\frac{K}{\rho(1 + \dfrac{DK}{Ee}C')}} \tag{1.27}$$

Figures 1.8(a,b) illustrate the effect of pipe diameter to wall thickness ratio (D/e) on the wave speed through water flowing in steel, cast iron, glass reinforced concrete and uPVC pipes and aviation kerosene flowing in steel, titanium and aluminium pipes. Increasing the D/e term increases the elasticity of the pipe wall and decreases the wave speed considerably.

Figure 1.8(c) also illustrates the dependence of wave speed upon pipe restraint. It will be seen that the incorporation of longitudinal strain effects has little influence upon the wave speed. For glass and cast iron pipes the value of Poisson's Ratio is of the order 0.25 and thus the value of C′ remains close to unity. In view of the difficulty in determining the true effect of pipe restraint on transient propagation the inclusion of C′ is of dubious value.

If the pipe wall is assumed to be rigid, ie an effectively infinite value of pipe material Young's Modulus, then the wave speed tends to the fluid only value, provided that no free gas is present; in this case the value of wave speed is given by equation 1.1.

The simple program illustrated in Appendix I may be used to replicate Figure 1.8 and to extend it to a range of other pipe materials and pipe wall thickness to pipe diameter ratios.

1.3.1 Non circular ducts

In some situations it may be helpful to introduce non circular sections of pipe or duct into the fluid carrying system subjected to surge propagation. Thorley and Enever (1979) present a generalised equation for a duct whose cross section is a regular polygon that conforms to the general wave speed equation 1.25 with the C′ term unity and the D/Ee replaced by

$$\frac{1}{E}\left[\frac{D}{e}+\frac{\tan^4\alpha}{15}\left(\frac{D}{e}\right)^3\right]+\frac{\tan^2\alpha}{2G}\,\frac{D}{e}$$

where G is the pipe wall material shear modulus and α is half the included angle subtended by a side of the regular polygon, so that tanα = tan(180/ number of sides to the polygon). This expression is limited to small deformations of the non circular ducts. Large deformations tend to alter the cross section shape towards circular and the wave speed will increase as a result of the change.

Figure 1.9(a) illustrates the wave speeds predicted for a range of cross sectional shapes, tending to the circular pipe case as the number of regular polygon sides increases.

An approximation for square and rectangular cross section ducts was proposed by Jenkner (1971), where neglecting the tension term yields an expression

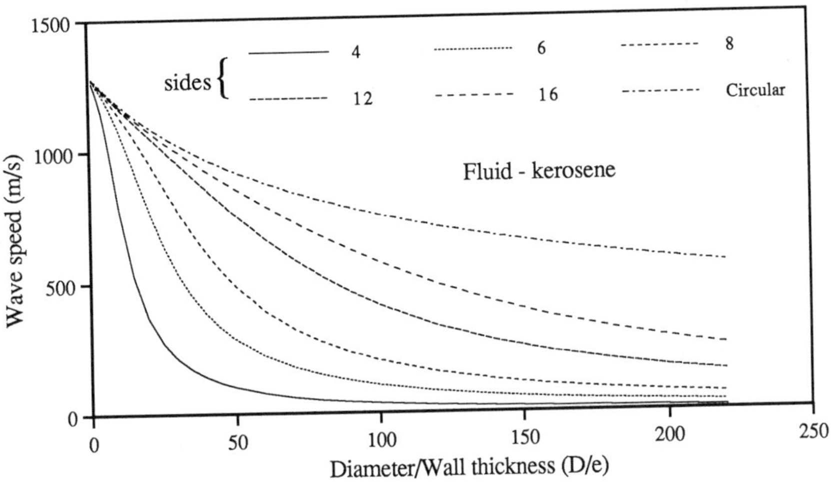

Figure 1.9(a) Wave speed in a range of regular polygon cross section ducts, illustrating the circular cross section pipe as the limiting case

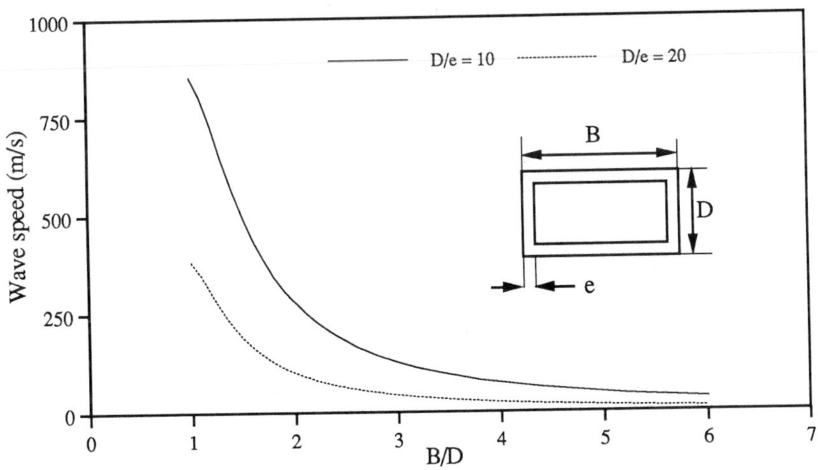

Figure 1.9(b) Effect of width to depth ratio for a rectangular duct carrying kerosene

$$\frac{(B/D)^4 R}{15(e/D)^3 E}$$

in the general wave speed equation, where

$$R = \frac{6-5x}{2} + \left(\frac{D}{B}\right)^5 \left(6-5x\left(\frac{B}{D}\right)^2\right)$$

and

$$x = \frac{1+(B/D)^3}{1+(D/B)}$$

Figure 1.9(b) illustrates the wave speed predictions obtained by this technique for a range of width to depth (B/D) ratios, for two values of pipe diameter to wall thickness ratio (D/e).

1.3.2 Thick walled pipes and tunnels

Halliwell (1963) considered the effect of thick pipe walls or tunnel construction on the wave speed in the confined fluid. In cases where the pipe walls are thick, ie cases where the D/e ratio is less than 25, the assumption that the stress is uniformly distributed through the walls is no longer valid. Halliwell proposed the following values for the C′ coefficient in the general equation 1.27

1. upstream end restraint only,

$$C' = \frac{2e}{D}(1+\mu) + \frac{D}{D+e}\left(1-\frac{\mu}{2}\right) \tag{1.28}$$

2. pipe is fully restrained,

$$C' = \frac{2e}{D}(1+\mu) + \frac{D}{D+e}(1-\mu^2) \tag{1.29}$$

3. expansion joints along whole pipelength,

$$C' = \frac{2e}{D}(1+\mu) + \frac{D}{D+e} \tag{1.30}$$

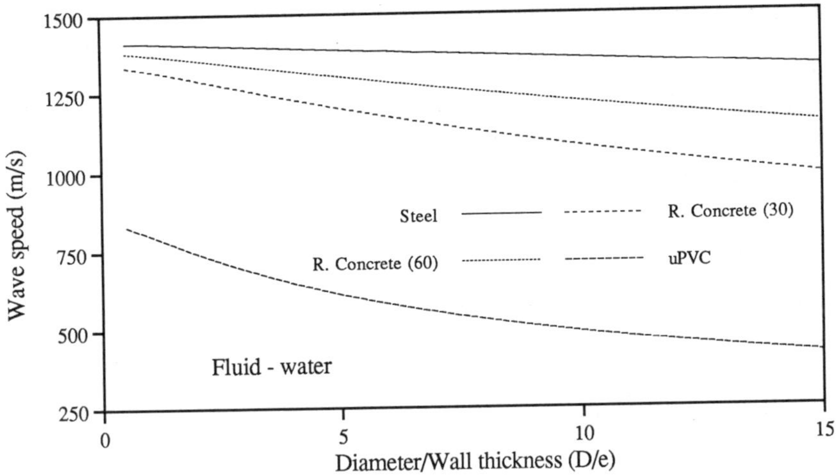

Figure 1.10 Wave speed in thick walled pipelines

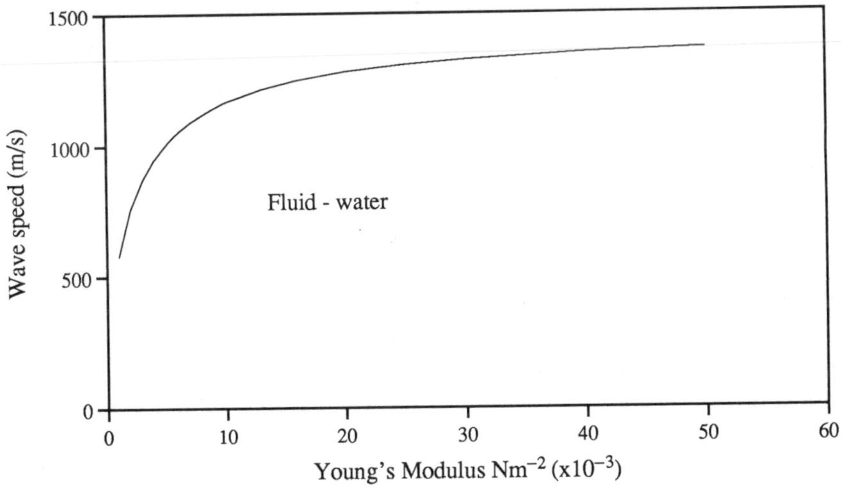

Figure 1.11 Wave speed in rock tunnels based on Youngs Modulus

It will be seen from the above expressions that in thick walled pipes the form of restraint applied has little effect on the wave speed predicted. As the wall thickness, e, becomes small these values tend toward the corresponding thin pipe wall values already developed. In all other respects the equation 1.27 may be used. Figure 1.10 illustrates the resulting wave speed calculations.

Tunnels may be seen as a special case of thick pipe walls. As wall thickness, e, becomes larger the values of C′ all tend towards

$$C' = \frac{2e}{D}(1+\mu) \tag{1.31}$$

and the wave speed equation 1.27 becomes

$$c = \sqrt{\frac{K}{\rho(1 + \frac{2K(1+\mu)}{E})}} \tag{1.32}$$

Typical values are illustrated in Figure 1.11. Halliwell (1963) and Wylie and Streeter (1978) also develop extensions to equation 1.32 to represent lined tunnels, where the wave speed rises compared to an unlined tunnel, and reinforced concrete pipelines where the pipeline is 'replaced' by an equivalent steel pipe whose thickness may be estimated by a comparison of concrete thickness and the size and frequency of the steel reinforcing bars.

1.3.3 Wave speed calculations for a gas-fluid mixture

The presence of even a small quantity of free gas becomes important in reducing the wave speed through the gas-liquid mixture. Figures 1.12(a,b) illustrate the predicted reduction in wave speed through a mixture of kerosene and free nitrogen flowing in an aluminium pipeline. It will be seen that the effect of neglecting pipe elasticity is only important at low gas ratios, obviously the compression of the gas swamps the elastic effects of pipe deformation. A comparison between predicted and measured wave speeds in a free gas liquid mixture is presented in Figure 1.12(c), Swaffield et al (1978). It will be shown later that the effect of free gas is often difficult to quantify and that many transient analysts tend to utilise a combination of wave speeds and a sensitivity analysis approach in order to ensure that these effects do not lead to misleading surge predictions.

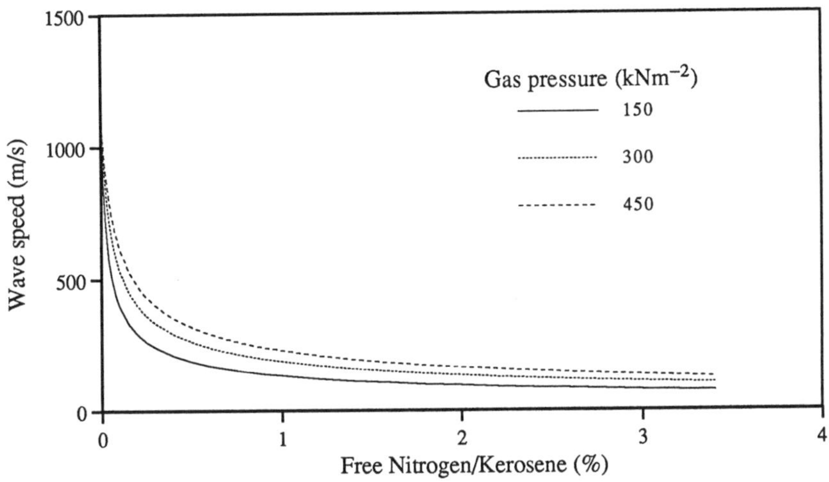

Figure 1.12(a) Effect of free gas volume on wave speed in a nitrogen/kerosene mixture flowing in an aluminium pipe, D/e ratio of 20

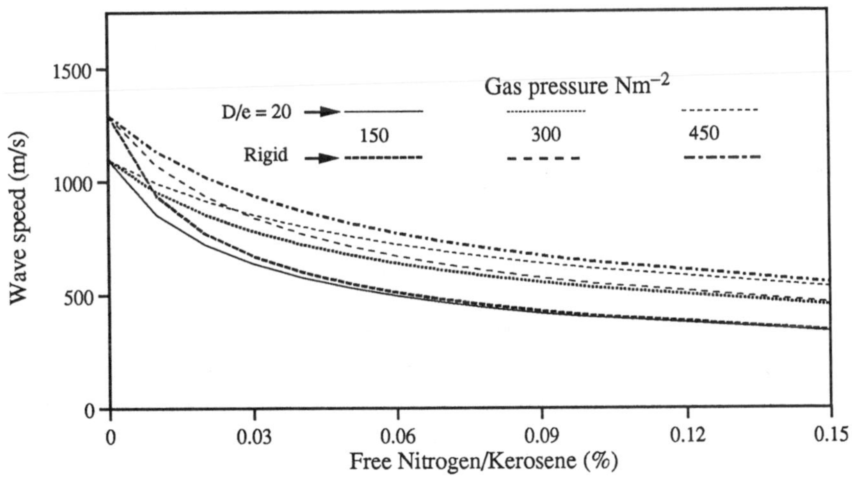

Figure 1.12(b) Expansion of the wave speed variation with free nitrogen volume curve to illustrate the insensitivity of the predictions to pipe elasticity at all but the lowest gas content values

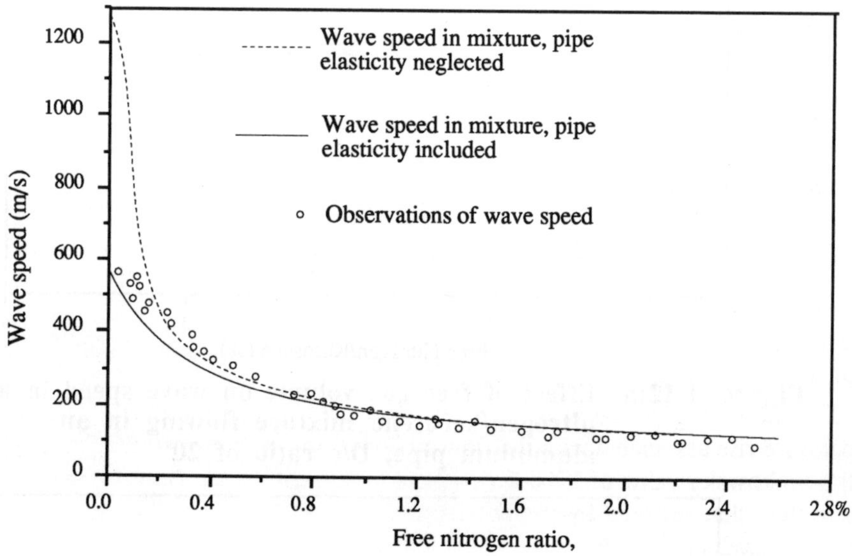

$$y = \frac{Vol_N}{(Vol_K + Vol_N)} \quad \text{at NTP}$$

Figure 1.12(c) Influence of the free gas content on wave speed through a nitrogen/kerosene mixture in a polythene pipeline of 50mm diameter, 6mm wall thickness and 15m length. The free gas pressure has been taken as 175kN/m²

Tables 1.1 and 1.2 present values of fluid density and Bulk Modulus and pipe wall Young's Modulus and Poisson's Ratio for a range of commonly encountered fluids and pipe materials.

In Section 1.2 it was shown that no reduction in surge generation was possible unless the valve closure time exceeded one pipe period, reducing the wave speed effectively lengthens the pipe period and may thus result in a valve closure previously thought 'slow' becoming 'rapid'.

Take as an example two identical 1200 m long pipelines, terminated upstream by a constant pressure reservoir and downstream by a motorised valve. In each case the pipe carries water of density 1000 kg/m^3, at an initial flow velocity of 1 m/s. One pipe is constructed of steel with a value of wave speed assumed to be 1200 m/s, while the other is constructed of polythene with an assumed value of wave speed of 400 m/s.

The pipe period for the polythene pipe is 2L/c, equal to 6 s. For the steel pipe the pipe period is, by the same calculation, 2 s.

If both valves close in 6 s then the pressure rise at the valve in the polythene pipe is the full Joukowsky pressure rise for a flow stoppage, as the valve has closed in one pipe period, in this case 400 kN/m^2 .

For the steel pipe the valve has closed in three pipe periods and thus the pressure rise associated with the flow stoppage will be considerably less than the Joukowsky value of 1200 kN/m^2, and in all probability, from Figure 1.4, less than that suffered by the polythene pipe. (Assuming a constant rate of velocity reduction over the valve closure the peak pressure at the valve will be recorded at the end of the first pipe period and will, in this example only be equal to a third of the full Joukowsky pressure rise, namely 400 kN/m^2.)

The program included in Appendix II may be used by the reader to investigate this effect further. It will also be shown later that the value of wave speed is a determinant of the time step size and the stability of numerical calculations undertaken via the method of characteristics.

1.4 Importance of pipe elasticity, comparison to rigid column theory

The Joukowsky pressure change accompanying a change in flow velocity has been expressed as $\rho c \Delta V$, and this will be derived later in the text. The underlying theory involves the elasticity of the pipe and the interaction between the pipe and fluid properties through the wave speed, as developed in Section 1.3.

Alternatively a rigid column theory may be invoked where no attention is paid to wave propagation, wave reflections or the effect of pipe elasticity on

Table 1.1 Values of Bulk Modulus and density for a range of common fluids

Fluid		Bulk Modulus 10^{-8} N/m²	Density kg/m³
Carbon tetrachloride	at 15°C	11.2	1590.0
Ethyl alcohol	at 15°C	12.0	779.0
Kerosene		13.6	814.0
Oil s.a.e. 10		16.7	880.0-940.0
s.a.e. 30		18.6	880.0-940.0
Sea water	at 10°C	22.0	1026.0
Water	at 0°C	20.5	1000.0
	at 20°C	20.5	998.0
	at 80°C	20.5	972.0

Table 1.2 Values of Young's Modulus of Elasticity, Shear Modulus and Poisson's Ratio for a range of common materials

Material	Young's Modulus 10^{-9} N/m²	Poisson's Ratio	Shear Modulus 10^{-9} N/m²
Aluminium	70.0	0.3	27.6
Cast Iron	80.0-110.0	0.25	40.0-80.0
Concrete	20.0-30.0	0.1-0.3	-
Copper	107.0-130.0	0.34	-
Glass	68.0	0.24	-
GRP	50.0	0.35	3.92
Polythene	3.1	-	1.10
PTFE plastic	0.35	-	-
PVC plastic	2.4-2.8	-	-
Reinforced Concrete	30.0-60.0	0.15	-
Rubber	0.7-7.0	0.46-0.49	-
Steel	200.0-214.0	0.3	82.7
Titanium	103.4	0.34	44.8

the transient propagation velocity.

If the fluid in a pipe, length L, is considered to be brought to rest from an initial velocity V in t seconds by valve action, a case similar to that illustrated in Figure 1.1, then from Newton's Second Law, the resulting pressure rise, Δp, may be expressed as

$$\Delta p = -\rho L \frac{dV}{Dt} \tag{1.33}$$

The pressure rise Δp is often referred to as the inertia pressure resulting from valve closure.

The obvious problems with this approach may be illustrated by a simple calculation. Consider a simple pipe, 1000 m long, carrying a fluid of density 1000 kg/m^3, at an initial velocity of 1 m/s. The wave speed for the pipe-fluid combination may be taken as 1000 m/s, giving a pipe period of 2s.

Table 1.3 below illustrates the divergence between the elastic and rigid column theories for a range of valve closures, assuming that the rate of reduction of flow velocity remains a constant throughout the valve motion.

Table 1.3 Comparison of rigid and elastic fluid column theories for a range of valve closure times

Valve closure time t′ s	t′/(2L/c)	Rigid Column theory $\Delta p = \rho LV/t'$ N/m^2 .10^6	Elastic theory $\Delta p = \rho cV$ N/m^2 .10^6
0.01	0.005	100.0	1.0
0.1	0.05	10.0	1.0
0.5	0.25	2.0	1.0
1.0	0.5	1.0	1.0
2.0	1.0	0.5	1.0
5.0	2.5	0.2	0.4
10.0	5.0	0.1	0.2
20.0	10.0	0.05	0.1
50.0	25.0	0.02	0.04
100.0	50.0	0.01	0.02

Note that for this assumed frictionless case, with a uniform rate of change of flow velocity, the peak elastic theory pressure rise occurs at the end of the first pipe period and will correspond to $\rho cV'$, where V' is the velocity reduction during the first pipe period, a diminishing value as the valve closure time increases.

For valve closures in less than one pipe period the rigid column theory is grossly in error, continuing to yield increasing pressure rises as the valve closure time decreases, while in reality the peak surge is constant for all valve closure times less than one pipe period.

The use of the rigid column theory is only justified at very slow valve closure rates, as illustrated by Table 1.3.

1.5 Pressure transient propagation in fluid carrying systems

Pressure transients will be propagated whenever there is a change in the flow conditions demanded of a fluid system. These demand changes may arise intentionally, for example the need to increase the output of a hydroelectric power station or the requirement to disconnect almost instantaneously in flight a tanker and refuelling aircraft, or may be the result of a system failure, for example a power failure to a pump or a malfunction in a relief valve. These effects are entirely normal and should be considered at the system design stage.

In addition to surges whose propagation may be predicted from a study of the system hardware, care has also to be taken to recognise the possibility of damaging transient propagation due to events that only occur as a result of an initial and apparently innocuous system condition adjustment. An example of this category of surge problem is the generation of a vapour pocket in a pipe following a 'slow' valve closure that did not initially generate high transient pressures. Due to a combination of fluid vapour pressure and a low line pressure, cavity formation may occur and the surge generated on subsequent cavity collapse may be destructive. Similarly the possibility of below atmospheric pressures may result in pipe inwards collapse, a case where pipe strengths may be at their lowest, particularly for concrete pipes.

The objective of this initial chapter has been to identify these possible surge generating conditions and to outline the factors that contribute to the subsequent form of the pressure surge experienced by the system. In addition this chapter has stressed the importance of both fluid compressibility and pipe elasticity in the determination of transient propagation velocity and the subsequent pressure surge history and the role played by the system boundary conditions in determining transient reflection and transmission. Indeed it may be useful to relate the whole process of transient propagation and pressure surge prediction to an information system where the pipes act solely as the links between boundaries that alone are capable of either generating the transient or of subsequently reflecting or re-transmitting incoming transients. The nature of the fluid and the pipe elastic properties alone determine the speed of information transmission between boundaries. This 'information system' model does seem attractive provided the definition of system boundary is sufficiently well understood to allow the inclusion of non hardware boundaries such as trapped gas or vapour cavities that arise as a result of transient propagation.

The historical study identified the necessity to solve the continuity and motion equations defining unsteady flow by numerical techniques designed specifically for use within computer based models of transient propagation within complex systems. The method of characteristics solution technique has now become established as the standard approach and this method will be fully developed within later chapters, with an emphasis on the method's main advantages in the modelling of system boundary conditions.

Chapter 2:

Basic equations defining transient propagation

The transient flow of a fluid through closed conduits is described by the momentum and continuity equations. Since the flow velocity (discharge) and piezometric head (pressure) developed during transient flows are both functions of position and time, these are a set of partial differential equations. In this chapter the appropriate equations of momentum and continuity are derived by considering both the forces acting on a small element of fluid and the mass flow through the element respectively.

The following simplifying assumptions are made in the derivation of the equations

- The flow in the conduit is one-dimensional.
- The conduit wall and the fluid are linearly elastic (stress is proportional to strain). This is true for most fluids and conduit materials such as metal, concrete and lined or unlined rock tunnels.
- The conduit remains full of fluid at a pressure exceeding the vapour pressure of the fluid or the gas release pressure.

It will be demonstrated later that the approach presented will also be generally applicable to free surface unsteady flows in partially filled conduits of a general cross sectional shape.

2.1 Momentum equation

The momentum equation is derived by considering the forces acting on a small element of fluid within the conduit, as shown in Figure 2.1. Applying the equation of motion to the element of fluid gives

$$pA - \left[pA + \frac{\partial(pA)}{\partial x}\delta x\right] + \left[p + \frac{\partial p \delta x}{\partial x\ 2}\right]\frac{\partial A}{\partial x}\delta x$$

$$+ \rho g \delta x\left[A + \frac{\partial A}{\partial x}\frac{\delta x}{2}\right]\sin\alpha - \tau_o P\delta x = \rho\delta x\left[A + \frac{\partial A}{\partial x}\frac{\delta x}{2}\right]\frac{dV}{dt} \tag{2.1}$$

where ρ, V, and p are the density, velocity and pressure of the fluid respectively; τ_o the shear stress between the fluid and the conduit walls; α the conduit angle defined by $\sin\alpha = -\partial z/\partial x$; and P is the conduit wetted perimeter. Ignoring the second order terms and substituting

$$\frac{\partial(pA)}{\partial x} = A\frac{\partial p}{\partial x} + p\frac{\partial A}{\partial x}$$

equation 2.1 reduces to

$$-\frac{\partial p}{\partial x}A\delta x + \rho g\ A\ \sin\alpha\delta x - \tau_o P\delta x = \rho\delta x A\frac{dV}{dt} \tag{2.2}$$

Dividing by ρAδx and rearranging

$$\frac{1}{\rho}\frac{\partial p}{\partial x} - g\sin\alpha + \frac{\tau_0}{\rho m} + \frac{dV}{dt} = 0 \tag{2.3}$$

where m is the flow hydraulic mean depth, defined as the ratio of the cross-section area to wetted perimeter of the conduit.

The shear stress τ_o at the conduit wall is taken as the steady-state stress at the same velocity, defined by

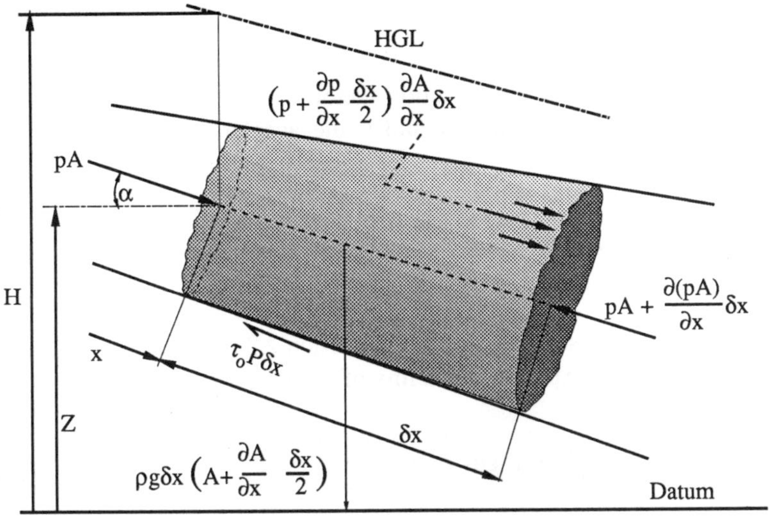

Figure 2.1 Forces acting on a small element of fluid in a conduit

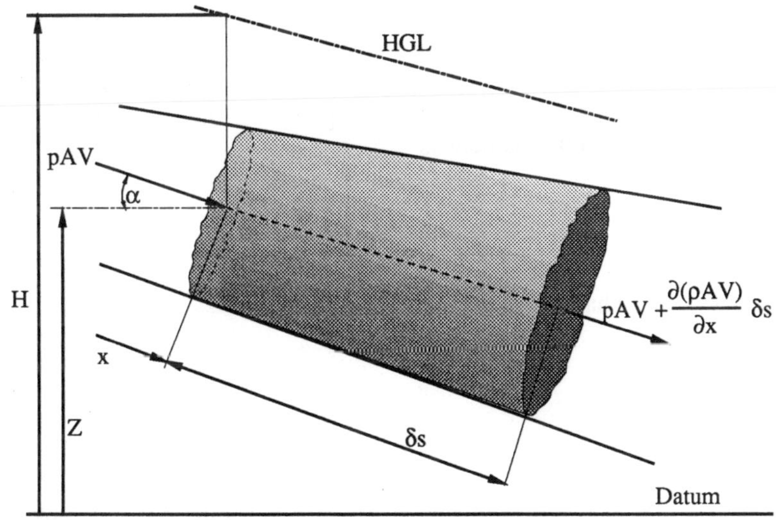

Figure 2.2 Mass flow through an element of fluid in a conduit

$$\tau_0 = \frac{1}{2} \rho f V^2 \tag{2.4}$$

where f is the Darcy-Weisbach friction factor.

Since V is a function of both x and t, the acceleration, dV/dt, may be expressed as

$$\frac{dV}{dt} = V \frac{\partial V}{\partial x} + \frac{\partial V}{\partial t} \tag{2.5}$$

Substituting equations 2.4 and 2.5 into equation 2.3 produces the following momentum equation, expressed in terms of velocity and pressure

$$V \frac{\partial V}{\partial x} + \frac{\partial V}{\partial t} + \frac{1}{\rho} \frac{\partial p}{\partial x} - g \sin \alpha + \frac{f V|V|}{2m} = 0 \tag{2.6}$$

The modulus of the velocity, $|V|$, is introduced so that the friction force opposes the direction of the flow at any given time. (The velocity is defined as positive in the positive x direction).

For some internal fluid flow systems, particularly water distribution and hydroelectric schemes, it is convenient to work in terms of piezometric head, H, defined as the elevation of the hydraulic gradeline above an arbitrary datum, rather than pressure, p. From Figure 2.1 the pressure on the upstream face of the fluid element is

$$p = \rho g(H - z),$$

and, by neglecting changes in density

$$\frac{\partial p}{\partial x} = \rho g \left[\frac{\partial H}{\partial x} - \frac{\partial z}{\partial x}\right]$$

Therefore, since by definition $\sin \alpha = -\partial z/\partial x$, equation 2.6 may be expressed in terms of velocity and piezometric head

$$V \frac{\partial V}{\partial x} + \frac{\partial V}{\partial t} + g \frac{\partial H}{\partial x} + \frac{fV|V|}{2m} = 0 \tag{2.7}$$

2.2 Continuity equation

The partial differential equation of continuity is derived from the law of conservation of mass for an element of fluid flowing in a conduit. The conduit material is assumed to behave in a linearly elastic manner and to be subjected to small deformations and the fluid is assumed to undergo small changes in its density compared with the magnitude of its density. Figure 2.2 shows the mass flow through an element of fluid within a conduit. The two cross sections of fluid are a distance δs apart, fixed in space, not relative to the pipe wall. The pipe and fluid do not move together as a rigid body so the fluid is not directly affected by the change in pipe length. The only effect that the axial movement of the pipe has is to effect the axial stress and hence, through Poisson's ratio, the pipe diameter. The conservation of mass requires that the net inflow per unit time into the elemental space must equal the rate of increase of mass within the space with respect to time, and is expressed by

$$\rho AV - \left[\rho AV + \frac{\partial(\rho AV)}{\partial x}\,\delta s\right] = \frac{\partial(\rho A\delta s)}{\partial t} \tag{2.8}$$

Expanding and dividing by ρAδs

$$\frac{\partial V}{\partial x} + \frac{V}{A}\frac{\partial A}{\partial x} + \frac{V}{\rho}\frac{\partial \rho}{\partial x} + \frac{1}{A}\frac{\partial A}{\partial t} + \frac{1}{\rho}\frac{\partial \rho}{\partial t} = 0 \tag{2.9}$$

Since

$$\frac{d\rho}{dt} = V\frac{\partial \rho}{\partial x} + \frac{\partial \rho}{\partial t}, \qquad \frac{dA}{dt} = V\frac{\partial A}{\partial x} + \frac{\partial A}{\partial t}$$

equation 2.9 becomes

$$\frac{\partial V}{\partial x} + \frac{1}{\rho}\frac{d\rho}{dt} + \frac{1}{A}\frac{dA}{dt} = 0 \tag{2.10}$$

where $\partial V/\partial x$ accounts for the change between inflow and outflow,
$1/\rho\ d\rho/dt$ accounts for the liquid compressibility, and
$1/A\ dA/dt$ accounts for the radial change in conduit size.

Terms 2 and 3 in equation 2.10 may be expressed in terms of the variables velocity and pressure or velocity and piezometric head as follows

Term 2 - the bulk modulus of elasticity K of a fluid may be defined as the ratio between an increase in pressure and the increase in densimetric strain that it causes

$$\frac{dp}{d\rho} = \frac{K}{\rho}$$

therefore

$$\frac{1}{\rho}\frac{d\rho}{dt} = \frac{1}{K}\frac{dp}{dt} \tag{2.11}$$

Term 3 - this term is transformed by considering the stresses and strains acting on an elemental section of the conduit wall, as shown in Figure 2.3.

The total strain in the axial direction is

$$\varepsilon_1 - \mu\varepsilon_2 = \frac{1}{E}(\sigma_1 - \mu\sigma_2)$$

The total strain in the lateral direction is

$$\varepsilon_2 - \mu\varepsilon_1 = \frac{1}{E}(\sigma_2 - \mu\sigma_1)$$

where ε and σ represent the strain and stress respectively.

In the lateral direction, by definition,

$$\frac{1}{E}(\sigma_2 - \mu\sigma_1) = \frac{dR}{R}$$

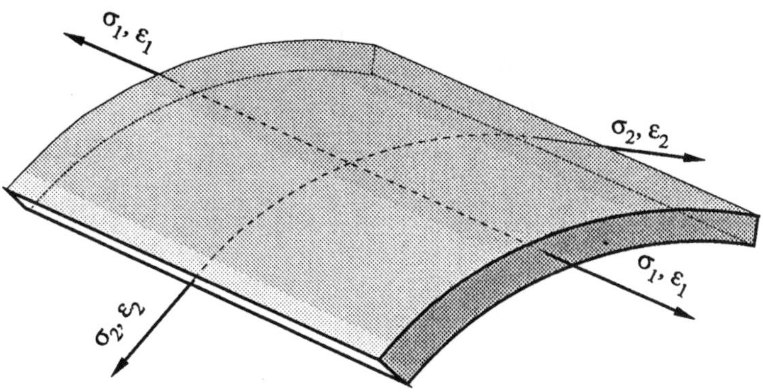

Figure 2.3 Stresses and strains acting on conduit wall

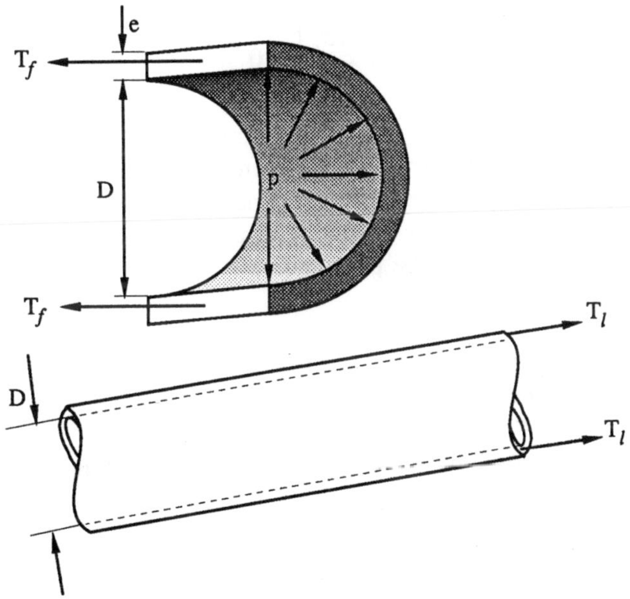

Figure 2.4 Forces and pressures acting on conduit wall

and the time rate of increase of the cross-sectional area is given by

$$\frac{dA}{dt} = 2\pi R \frac{dR}{dt} = \frac{2\pi R^2}{E}\left[\frac{d\sigma_2}{dt} - \mu \frac{d\sigma_1}{dt}\right]$$

therefore

$$\frac{1}{A}\frac{dA}{dt} = \frac{2}{E}\left[\frac{d\sigma_2}{dt} - \mu \frac{d\sigma_1}{dt}\right] \qquad (2.12)$$

Substituting equations 2.11 and 2.12 into equation 2.10 gives the following continuity relationship

$$\frac{\partial V}{\partial x} + \frac{1}{K}\frac{dp}{dt} + \frac{2}{E}\left[\frac{d\sigma_2}{dt} - \mu \frac{d\sigma_1}{dt}\right] = 0 \qquad (2.13)$$

The $d\sigma_2/dt$ and $d\sigma_1/dt$ terms are related to dp/dt in a form which depends on the means of support used for the conduit. Equation 2.13 may therefore be expressed for each of the following three forms of pipeline support that are commonly assumed to be provided.

(a) Conduit supported at the upstream end only and free to move axially as it stretches the conduit wall

Figure 2.4 shows the forces and pressure acting on an unit length of conduit wall. The lateral stress is

$$\sigma_2 = \frac{T_f}{e}, \qquad T_f = pD/2$$

therefore

$$\sigma_2 = \frac{pD}{2e}$$

$$\frac{d\sigma_2}{dt} = \frac{D}{2e}\frac{dp}{dt}$$

The axial stress is

$$\sigma_1 = \frac{T_l}{e}$$

$$T_l \pi D = \frac{p\pi D^2}{4}$$

therefore

$$\sigma_1 = \frac{pD}{4e}$$

and

$$\frac{d\sigma_1}{dt} = \frac{D}{4e}\frac{dp}{dt}$$

Substituting into equation 2.13

$$\frac{\partial V}{\partial x} + \frac{1}{K}\frac{dp}{dt}\left[1 + \frac{DK}{Ee}\left(1 - \frac{\mu}{2}\right)\right] = 0 \qquad (2.14)$$

(b) Conduit anchored throughout its length against axial movement

The lateral stress is the same as in case (a). The axial strain is

$$\frac{1}{E}(\sigma_1 - \mu\sigma_2) = 0$$

therefore

$$\frac{d\sigma_1}{dt} = \mu \frac{d\sigma_2}{dt} = \frac{\mu D}{2e} \frac{dp}{dt}$$

Substituting into equation 2.13

$$\frac{\partial V}{\partial x} + \frac{1}{K} \frac{dp}{dt} \left[1 + \frac{DK}{Ee} \left(1 - \mu^2\right) \right] = 0 \qquad (2.15)$$

(c) Conduit anchored against axial movement but constructed with expansion joints

The lateral stress is the same as in case (a). The axial stress, $\sigma_1 = 0$, hence

$$\frac{d\sigma_1}{dt} = 0$$

Substituting into equation 2.13

$$\frac{\partial V}{\partial x} + \frac{1}{K} \frac{dp}{dt} \left[1 + \frac{KD}{Ee} (1) \right] = 0 \qquad (2.16)$$

Comparing equations 2.14, 2.15 and 2.16 the continuity equation may be expressed as

$$\rho c^2 \frac{\partial V}{\partial x} + \frac{dp}{dt} = 0 \qquad (2.17)$$

where

$$c = \left[\frac{K}{\rho\left(1 + \frac{KD}{Ee}C'\right)}\right]^{0.5} \tag{2.18}$$

and C′ = (1 − μ/2) or (1 − μ^2) or (1) depending on the means of conduit support.

The value of c, defined by equation 2.18, is the speed at which the transient pressure wave propagates through the fluid-conduit combination.

Since p is a function of both x and t then

$$\frac{dp}{dt} = V\frac{\partial p}{\partial x} + \frac{\partial p}{\partial t}$$

and the continuity equation expressed in terms of velocity and pressure becomes

$$\rho c^2\frac{\partial V}{\partial x} + V\frac{\partial p}{\partial x} + \frac{\partial p}{\partial t} = 0 \tag{2.19}$$

Since

$$\frac{\partial p}{\partial x} = \rho g\left[\frac{\partial H}{\partial x} + \sin\alpha\right]$$

and

$$\frac{\partial p}{\partial t} = \rho g\frac{\partial H}{\partial t}$$

the continuity equation may be expressed in terms of velocity and piezometric pressure head as

$$\rho c^2\frac{\partial V}{\partial x} + V\rho g\left[\frac{\partial H}{\partial x} + \sin\alpha\right] + \rho g\frac{\partial H}{\partial t} = 0 \tag{2.20}$$

as a function of the Reynolds number and c can be expressed as a function of the pressure.

Since the nonlinear terms in the above basic partial differential involve only the first power of the derivatives, the equations are called quasi-linear. The equations may be further classified as either elliptic, parabolic or hyperbolic by expressing, for example, equations 2.23 and 2.24 in the following matrix form

$$\begin{bmatrix} \dfrac{\partial V}{\partial t} \\ \dfrac{\partial H}{\partial t} \end{bmatrix} + \begin{bmatrix} V & g \\ \dfrac{c^2}{g} & V \end{bmatrix} \cdot \begin{bmatrix} \dfrac{\partial V}{\partial x} \\ \dfrac{\partial H}{\partial x} \end{bmatrix} + \begin{bmatrix} \dfrac{f\,V|V|}{2m} \\ V\sin\alpha \end{bmatrix} = 0 \qquad (2.25)$$

The eigenvalues of the matrix

$$\begin{bmatrix} V & g \\ \dfrac{c^2}{g} & V \end{bmatrix}$$

determine the type of the equation. The characteristic equation of the matrix is

$$(V - \lambda)^2 = c^2$$

$$\lambda = V \pm c$$

Since thc values of V ± c are real, both eigenvalues are real and distinct, therefore the basic equation of momentum and continuity form a set of hyperbolic partial differential equations.

2.4 Review of methods of solution

Since the basic equations of momentum and continuity are quasi-linear, hyperbolic, partial differential equations, a closed-form solution of the equations is impossible. However by neglecting, or linearising the nonlinear terms, various graphical (Parmakian (1963), Bergeron (1961), Pickford (1969), Harding (1966) and Enever (1970)), and analytical (Allievi (1903), Rich (1963) and Wood (1938)), methods have been developed. These methods are approximate and cannot be used to analyse large systems or systems incorporating complex boundary conditions.

Other techniques are more suitable for computer simulation, such as the implicit finite-difference method (Perkins et al (1964)), and the method of characteristics (Lister (1960), Evangelisti (1969), Abbott (1966), Swaffield (1972), Streeter (1967), Boldy (1975, 1976), Chaudhry (1987), Fox (1977, 1989) and Wylie and Streeter (1978, 1983)).

In the implicit finite-difference method the partial derivatives are replaced by finite differences and the resulting algebraic equations for the whole system are then solved simultaneously. Depending upon the size of the system, this involves the simultaneous solution of a large number of nonlinear equations. The analysis, by this method, becomes even more complicated for systems which incorporate complex boundary conditions, which must be solved by an iterative technique. The method has the advantage of being unconditionally stable, therefore large time steps may be used with a corresponding economy of computer time. However the time step cannot be increased arbitrarily since this will result in a smoothing of the transient pressure peaks.

In the method of characteristics the partial differential equations are converted into ordinary differential equations which are then solved using an explicit finite-difference technique. Because each node point and boundary condition is analysed individually at each time step, the method is particularly suitable for systems containing complex boundary conditions. The disadvantage of the method is that small time steps must be used in order to satisfy the stability criterion (see Chapter 4), but this is now less of a problem due to the rapid advances in the speed of operation of modern computers, in particular PCs.

The method of characteristics is generally accepted as the most appropriate technique for developing computer simulation programs of transient flow in internal fluid flow systems.

Chapter 3:

Transient reflection and transmission within frictionless systems

Friction acts to damp and attenuate transient pressure fluctuations. It may also be responsible, in high friction loss systems, for appreciable pressure rises following valve closure, as described in Chapter 1. Frictional terms may be included in the full equations of motion and continuity that describe transient propagation and may now also be included in the numerical solution of these equations, as will be presented in later chapters.

Initially the inclusion of friction was beyond the analysis techniques developed by the early researchers into pressure transient phenomena and a number of approximations were introduced, for example the 'friction joint' approach in the Schnyder-Bergeron graphical method. Other early methods ignored these effects.

Obviously the inclusion of friction is important and is now unquestioned, however there are interesting and useful results that may be derived from the simplified wave equations that are an indispensable aid in the understanding of transient pressure propagation, reflection and transmission within complex systems. This chapter will present these results as a means of understanding transient pressure fluctuations recorded within fluid systems as a result of boundary condition changes and transient propagation.

In particular the introduction of the concept of transient reflection and transmission coefficients, determined at each system boundary, will be shown to be useful in explaining the mechanisms of transient propagation and the pressure-time histories resulting from a wide range of unsteady flow events.

3.1 Simplification of the wave equations

The full equations of motion and continuity describing the propagation of pressure transients have been developed in Chapter 2. These equations may be reduced if frictional and slope terms are ignored, together with the convective terms, equations 2.22 and 2.23, and are presented below as

Equation of Motion

$$\frac{\partial p}{\partial x} + \rho\frac{\partial v}{\partial t} = 0 \tag{3.1}$$

Equation of Continuity

$$\frac{\partial p}{\partial t} + \rho c^2\frac{\partial v}{\partial x} = 0 \tag{3.2}$$

applicable to a frictionless horizontal pipeline.

The general solution of these equations is due to D'Alembert and may be expressed as

$$p - p_0 = F(t+x/c) + f(t-x/c) \tag{3.3}$$

$$V - V_0 = -\frac{1}{\rho c}\left[\, F(t+x/c) - f(t-x/c) \,\right] \tag{3.4}$$

The F() and f() functions in the solution are entirely arbitrary and may be selected to represent transients propagated through the system as a result of changes in boundary condition. Figure 3.1 illustrates the commonly accepted interpretation of F() and f() as a pair of possible transients respectively propagated in the −x and +x directions as a result of a change in boundary condition. Such a change may be the arrival at the boundary of a transient propagated earlier by some action at a remote system boundary so that this formulation may be seen as capable of representing both transmission and reflection of transients. The form of equations 3.3 and 3.4 conform to the concept of superposition of pressure waves that has already been introduced in Chapter 1 to describe the pressure-time history following valve closure in

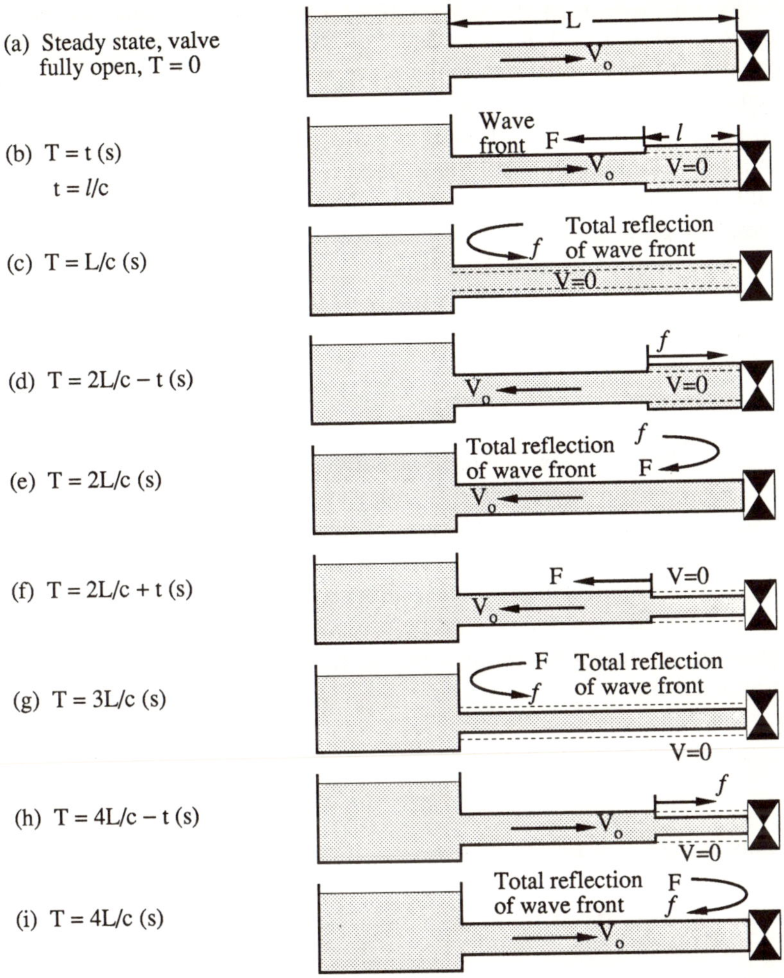

Figure 3.1 Propagation of transient pressure waves caused by an instantaneous valve closure. Note that, in the absence of frictional or vibrational damping or column separation, the reflection coefficient at the reservoir and at the closed valve are –1 and +1 respectively

a simple pipeline. It will be seen from equations 3.3 and 3.4 that the dimensions of F() and f() are those of pressure and that the speed of propagation of both functions is the wave speed c.

From the principle of superposition it follows that the pressure at a point in the system at a time t following the initial transient propagation will be given by the summation of all the F() and f() that have passed that point up to that time. This approach assumes that the waves travel at a constant speed c dependent upon pipe and fluid properties and that they do not attenuate due to the action of frictional damping.

It is therefore possible to derive many of the basic transient relationships from equation 3.3 and 3.4, combined with equations representing conditions at a system boundary.

3.1.1 Pressure rise, or fall, on valve operation

Assuming an instantaneous closure of the valve situated between two pipelines, Figure 3.1, it follows that at time t = 0 a F() wave will be propagated upstream and a f() wave downstream. On the upstream face of the valve no arriving f() wave can be present until a reflection arrives from an upstream boundary at one pipe period following valve closure, hence in equations 3.3 and 3.4

$$f() = 0$$

and thus at time $0 < t < 2L/c$

$$F() = p - p_0 = \rho c\ (V - V_0)$$

and hence the pressure rise at the valve following an instantaneous closure may be expressed as

$$\Delta p = p - p_0 = \rho c V_0 \qquad (3.5)$$

due to the velocity at the valve, V, being zero. This result is generally known as the Joukowsky pressure rise on instantaneous valve closure and remains the pressure rise generated at all valve closure times less than one pipe period, defined as the time taken for the initial generated transient's leading edge to return as a reflection from the first upstream boundary, such a boundary being defined as any constant pressure reservoir, change in pipe condition or machine interface.

This general reasoning also applies on the downstream face of the closed valve, however from the definition of the F() and f() terms it follows that the initial transient generated is a f() wave moving downstream and the first reflection from the downstream pipe boundary takes the form of a F() wave arriving at the closed valve one downstream pipe period later. The magnitude of the initial f() wave generated by valve closure and flow instantaneous stoppage is given by

$$F() = 0$$

and therefore

$$f() = \Delta p = p - p_0 = -\rho c V_0 \tag{3.6}$$

as V = 0 at time t due to the instantaneous valve closure. This result again applies up to the end of the first pipe period calculated for the downstream pipe.

3.1.2 *Transient reflection at a constant pressure boundary*

At a constant pressure boundary it follows that the pressure change caused by the arrival of a transient must be zero so that from equation 3.3

$$f(t-x/c) = -F(t+x/c) \tag{3.7}$$

This implies that an arriving F() wave is reflected as a f() wave with sign reversal, or vice versa at a downstream boundary. This result conforms to the description of transient reflection given for a constant pressure reservoir in Chapter 1. As this is true for all constant pressure zones and therefore applies to the reflections generated at a vapour cavity interface while the local pressure remains at vapour level.

The reflection at a constant pressure zone introduces the concept of a reflection coefficient, defined as

$$C_R = \frac{\textit{Outgoing transient}}{\textit{Incoming transient}} \tag{3.8}$$

In the case of the upstream reservoir in Figure 3.1

$$C_R = \frac{f()}{F()} = -1.0 \tag{3.9}$$

or for the downstream reservoir

$$C_R = \frac{F()}{f()} = -1.0 \tag{3.10}$$

3.1.3 *Transient reflection at a closed valve or dead end*

In the absence of gas release, vapour cavity formation or appreciable pipe vibration the fluid velocity at a closed valve or at a pipe dead end must always be zero. Hence from equation 3.4 it follows that

$$V - V_0 = 0 = -\rho c \left[F(t+x/c) - f(t-x/c) \right] \tag{3.11}$$

and hence

$$f(t-x/c) = F(t+x/c)$$

and the appropriate reflection coefficient is thus

$$C_R = +1.0 \tag{3.12}$$

This result implies that the rate of pressure change experienced at a closed valve, or a pipe dead end, is double that experienced as the waves pass other points in the system, a result indicated by the description of transient propagation in a simple pipeline given in Chapter 1. This result is important in the interpretation of pressure transient measurements following a rapid or near instantaneous flow stoppage, as may occur on the collapse of a vapour cavity formed in a system as a result of some earlier seemingly innocuous change in boundary condition.

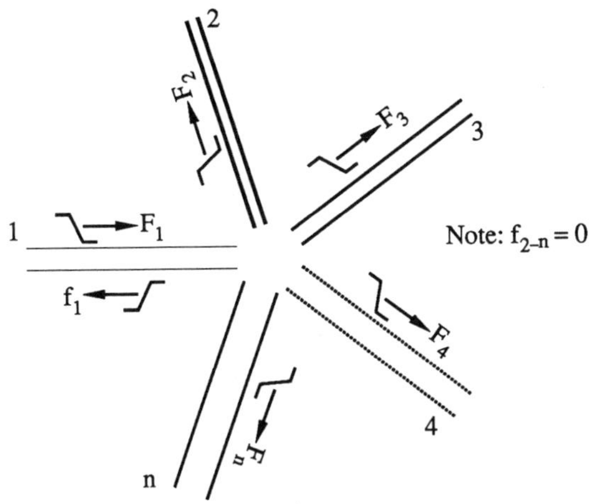

Figure 3.2 General junction case, all pipes have individual wave speed based on cross section dimensions and material properties

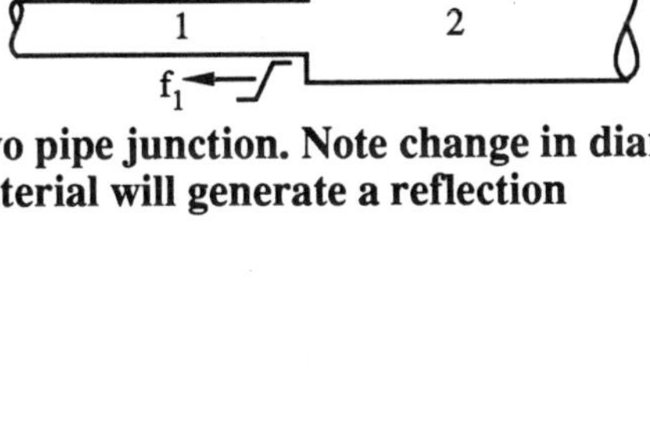

Figure 3.3(a) Two pipe junction. Note change in diameter and/or material will generate a reflection

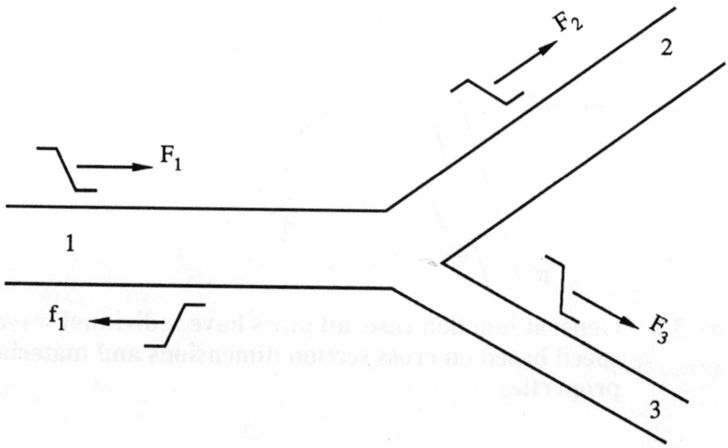

Figure 3.3(b) **Three pipe junction. Note: it may be shown that $F_2 = F_3$ and $f_{2,3} = 0$ as before**

3.2 Transient transmission and reflection at changes in pipe properties and/or pipe junctions

Consider a general junction of n pipes as shown in Figure 3.2, each pipe having its individual diameter, wall thickness and Young's Modulus values, and hence its own characteristic wavespeed. If a transient arrives along any one of the joining pipes, say pipe 1, then ongoing transmissions will be generated in the remaining n−1 pipes and a reflected transient will propagate back along pipe 1 towards the site of the original transient generation.

In the absence of pipe friction and local separation losses at the junction equations 3.3 and 3.4 may be combined with the continuity of flow and pressure at the junction to determine values for reflection and transmission coefficients for the junction. Thus at the junction it follows that

1. Continuity of flow

$$Q_1 = Q_2 + Q_3 + \ldots\ldots + Q_n \tag{3.13}$$

and

$$\Delta Q_1 = \Delta Q_2 + \Delta Q_3 + \ldots\ldots + \Delta Q_n \tag{3.14}$$

2. Commonality of pressure at the junction

$$p_1 = p_2 = p_3 = \ldots\ldots = p_n \tag{3.15}$$

and

$$\Delta p_1 = \Delta p_2 = \Delta p_3 = \ldots\ldots = \Delta p_n \tag{3.16}$$

From equations 3.3 and 3.4

$$\Delta p_1 = F()_1 + f()_1 \quad , \quad \Delta Q_1 = -\frac{A_1}{\rho c_1}(F()_1 - f()_1)$$

$$\Delta p_2 = F()_2 + f()_2 \quad , \quad \Delta Q_2 = -\frac{A_2}{\rho c_2}(F()_2 - f()_2)$$

$$\Delta p_3 = F()_3 + f()_3 \quad , \quad \Delta Q_3 = -\frac{A_3}{\rho c_3}(F()_3 - f()_3)$$

and so on until

$$\Delta p_n = F()_n + f()_n \quad , \quad \Delta Q_n = -\frac{A_n}{\rho c_n}(F()_n - f()_n)$$

As only pipe 1, in this case, carries an incoming transient at time 0 it follows that f() = 0 for all the other pipes, 2 to n, as no reflection can have arrived at the junction from upstream at that time.

Simplification of the above equations thus leads to

$$F()_1 + f()_1 = F()_2 = F()_3 = ... = F()_n \tag{3.17}$$

and

$$F()_1 - f()_1 = \frac{c_1}{A_1}\left(\frac{A_2}{c_2}F()_2 + \frac{A_3}{c_3}F()_3 + ... + \frac{A_n}{c_n}F()_n\right) \tag{3.18}$$

If the junction reflection coefficient is defined as

$$C_R = \frac{f()_1}{F()_1} \tag{3.19}$$

and the junction transmission coefficient, for i = 2 to n, as

$$C_{T_i} = \frac{F()_i}{F()_1} \tag{3.20}$$

it follows that

$$C_{R_i} = \frac{\dfrac{A_1}{c_1} - \dfrac{A_2}{c_2} - \dfrac{A_3}{c_3} - ..... - \dfrac{A_n}{c_n}}{\dfrac{A_1}{c_1} + \dfrac{A_2}{c_2} + \dfrac{A_3}{c_3} + ... + \dfrac{A_n}{c_n}} \tag{3.21}$$

and

$$C_{T_i} = \frac{\dfrac{2A_1}{c_1}}{\dfrac{A_1}{c_1} + \dfrac{A_2}{c_2} + \dfrac{A_3}{c_3} + \ldots\ldots + \dfrac{A_n}{c_n}} \tag{3.22}$$

where $2 \leq i \leq n$.

It must be noted that the transmission coefficient is the same for all the pipes linked at the junction, however the value does depend upon which pipe carries the incoming transient. There is only one reflection coefficient, applicable to the pipe carrying the incoming transient.

This approach will allow a complex series of transients to be studied through the principle of superposition and effectively forms the basis of both the early Allievi graphical method and the later and more sophisticated Schnyder-Bergeron method.

3.2.1 Application of reflection and transmission coefficients to pipe junctions and changes in pipe properties

Consider a simple change of cross section in a single pipeline, Figure 3.3(a). The reflection and transmission coefficients that apply at this interface may be seen to be, from equations 3.21 and 3.22,

$$C_R = \frac{\dfrac{A_1}{c_1} - \dfrac{A_2}{c_2}}{\dfrac{A_1}{c_1} + \dfrac{A_2}{c_2}} \tag{3.23}$$

and

$$C_{T_1} = \frac{\dfrac{2A_1}{c_1}}{\dfrac{A_1}{c_1} + \dfrac{A_2}{c_2}} \tag{3.24}$$

Equations 3.23 and 3.24 continue to apply if the cross section remains constant but the pipe materials change so that the value of wave speed c varies between the two pipe sections thus generating a pressure transient reflection at the junction.

For the case of a three pipe junction these equations become

$$C_{R_1} = \frac{\dfrac{A_1}{c_1} - \dfrac{A_2}{c_2} - \dfrac{A_3}{c_3}}{\dfrac{A_1}{c_1} + \dfrac{A_2}{c_2} + \dfrac{A_3}{c_3}} \tag{3.25}$$

$$C_{T_{2,3}} = \frac{\dfrac{2A_1}{c_1}}{\dfrac{A_1}{c_1} + \dfrac{A_2}{c_2} + \dfrac{A_3}{c_3}} \tag{3.26}$$

provided that the incoming transient arrives along pipe 1. The equivalent expressions for the reflection coefficient and transmission coefficient should the transient arrive along any other pipe are of course identical in form with the only difference being that the incoming transient pipe number replaces 1 in equations 3.25 and 3.26.

Again identical cross section pipes featuring different wave speed due to changes in wall thickness or material properties will generate both reflection and transmission coefficient values. It may also be seen that

$$abs\ (C_R) + abs\ (C_T) = 1.0 \tag{3.27}$$

Figure 3.3(b) illustrates this three pipe junction case.

3.2.2 Application of transmission and reflection coefficients in the interpretation of pressure transducer records

Figure 3.4 illustrates a typical pressure transducer installation featuring a length of small bore tube linking the pressure transducer body to a pipe. The junction between the transducer connection and the pipeline to be monitored may be considered as a three pipe junction, the pipe downstream and upstream of the transducer connection being pipes 1 and 2 respectively and the small bore transducer connection being pipe 3.

Assume that a transient arrives from downstream along pipe 1. This generates transmission into pipes 2 and 3 and a reflection back along pipe 1. However if the cross section of pipes 1 and 2 are equal and much greater than that of 3, and assuming that the wave speeds in each pipe are approximately the same, then it will be seen that the transmission into 2 and 3 will approach 100% while the reflection back along 1 will tend to zero.

Assume further that the wave front length, calculated from the product of wave speed and the time taken to complete the change in the boundary condition that generated the transient to be monitored is less than the length of the transducer connection. When the transmitted transient reaches the closed end of the connection at the transducer a further reflection takes place, with a reflection coefficient of +1.0 as derived earlier.

Thus the pressure recorded at the transducer in the special case of a connecting tube that exceeds the wavefront length of the transient will be double that experienced by the fluid in the monitored pipe. This result, if unexpected, can lead to either transducer damage or misleading interpretation of the transient magnitude, or both.

The transient within the transducer connection is now propagated back towards the junction along pipe 3, Figure 3.4. Reference to equations 3.25 and 3.26 indicates that, as the cross section of the incoming transient carrying pipe, 3, is much less than that of the monitored pipes, 1 and 2, the transmission coefficient from pipe 3 to either 1 or 2 approaches zero while the reflection back along the transducer connection 1 approaches –1.0. Thus the original transient is 'trapped' within the transducer connection and sets up a periodic pressure fluctuation as shown in Figure 3.4.

If the transient wavefront length exceeds the length of the transducer connection then this process still occurs, however the length of the wavefront ensures that the maximum error predicted does not occur, as the 'trapped' transient is analogous to a slow valve closure. Figure 3.4 illustrates both these conditions. The 'length' of the transient may always be calculated from its wave speed and the time taken to generate it, eg valve closure time.

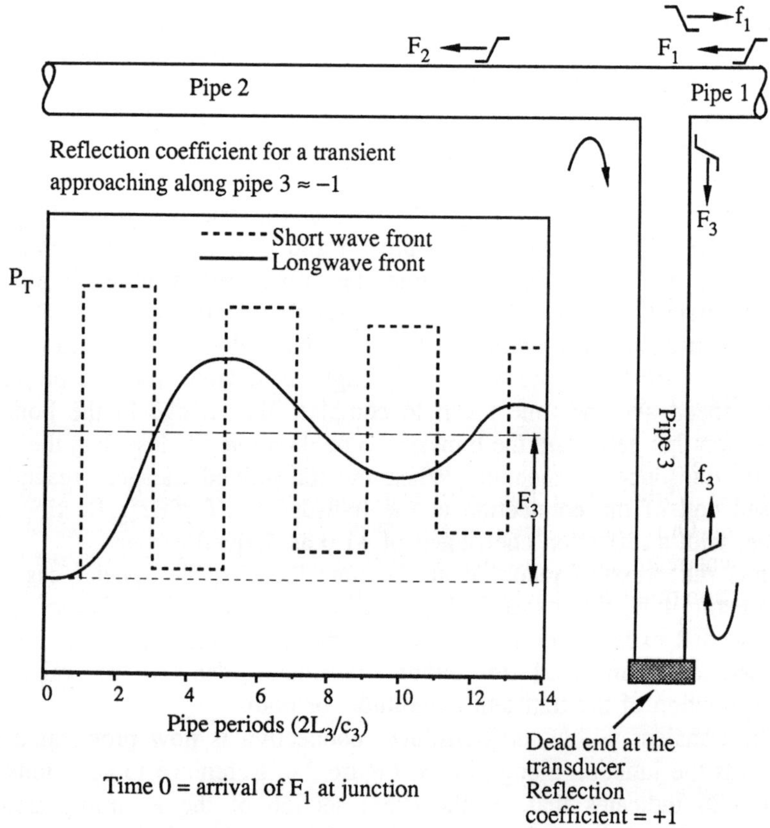

Figure 3.4 Transducer example. Note the effect of lengthening the transient wave front relative to the branch length

However, utilising the closing time of a valve only yields wavefront length and no information on pressure variation along the front.

Further transients arriving at the junction, formed by pipes 1 and 2 and the transducer connection propagate as described above leading to the often complex pressure traces recorded in practice. While it may appear self evident that it will be rare for the wavefront to be shorter than the transducer connection, this cannot be taken as absolute. In cases where column separation occurs at some remote point in the system, subsequent cavity collapse generates a Joukowsky style pressure surge with a near instantaneous pressure rise. Under these special design conditions the potential problem described may well occur.

A further complication with lengthy transducer connections, measured in terms of transient wavefront lengths, is the possibility of vapour cavity formation at the transducer when, in the case of monitoring a pressure reducing transient, the +1.0 reflection coefficient at the transducer is sufficient to drop pressure there to vapour or gas release levels. The output of the transducer would become meaningless for the duration of this column separation and there might be subsequent free gas removal problems.

Figure 3.4 illustrated a case where a transducer connection having a diameter 1/5th that of the monitored pipeline is subjected to a pressure transient. Only one transient arrives at the junction, it may be assumed that the downstream and upstream system boundaries are remote from the sensor connection point and that no reflections return along the monitored pipe. The wave speed in the main pipes and the transducer connection are equal. The relevant values of reflection and transmission coefficients necessary to analyse the subsequent fluctuations monitored by the transducer are therefore as follows

1. Reflection back along main pipe, equation 3.25,

$C_{R_1} = 1 / 51$

2. Transmission from main pipe into the transducer, equation 3.26,

$C_{T_3} = 50 / 51$

3. Reflection at the dead end of the transducer connection, equation 3.12,

$C_{R_{3\,end}} = 1.0$

4. Reflection back into the transducer connection from the junction of a transient approaching along the transducer connection, pipe 3, equation 3.25,

$$C_{R_{3\,junction}} = -\frac{49}{51}$$

5. Transmission from the transducer connection into the main pipe, either upstream or downstream, equation 3.26,

$$C_{T_3} = \frac{2}{51}$$

Application of these coefficients allows the determination of the transducer pressure fluctuations as shown in Figure 3.4, this trace may also generated by the general model program presented in Appendix II.

This approach is extremely useful in providing a basic understanding of transient propagation. It highlights a number of interesting results, for example the fact that a major increase in pipe cross sectional area, with constant wave speed terms, can act as an almost −100% reflector to an incoming transient, equation 3.23 a helpful result if a constant pressure boundary cannot be provided and only a short duration surge investigation is to be carried out under laboratory conditions. However the approach becomes limited if the transients considered have appreciable wavefront lengths, as will be the case for all slow valve closures, or where the effect of friction is seen as important. These shortcomings led to the development of graphical techniques that, while originally tedious, could be made to include these effects.

3.3 Justification for the simplification of the wave equations

As already mentioned the wave equations have been simplified by the exclusion of friction and the deletion of the so called convective terms, $V\partial V/\partial x$ and $V\partial p/\partial x$, in comparison to the $\partial V/\partial t$ and $\partial p/\partial t$ terms.

The justification for this simplification may be demonstrated by reference to a typical transient propagation condition.

Consider the closure of a valve terminating a section of a hydrant fuelling system in 0.05 s, the wave speed in the fuel/pipe material combination being 1200 m/s and the flow velocity destroyed being 2.0 m/s.

The resulting wavefront has a ‘time length’ of 0.05 s and a physical length, ignoring any attenuation effects, of (1200 m/s * 0.05 s), or 60 m.

The accompanying maximum pressure change accomplished in 0.05 s is given by the Joukowsky expression, assuming a fuel density of 800 kg/m^3, of (800 kg/m^3 * 1200 m/s * 2.0 m/s), or 1920 kN/m^2.

Thus the compared terms in the wave equations referred to above become

$\partial p/\partial t$ equal to (1920 / 0.05), or 38400 $kN/m^2/s$

while

$V\partial p/\partial x$ equal to (2 . 1920 / 60.0) or 64 $kN/m^2/s$

and

$\partial V/\partial t$ equal to (2 / 0.05) or 40 m/s^2

while

$V\partial V/\partial x$ equal to (2 . 2 / 60.0) or 0.066 m/s^2

Thus the simplification presented is clearly justified in most applications. If the wave speed becomes very small, possibly due to thin pipe walls or low Young's Modulus values for the pipe material, then the terms reviewed above become comparable and the simplification is no longer justified, Enever (1970).

The further simplification of friction neglect cannot be so conveniently supported. The early graphical methods did include frictional approximations mainly by either concentrating the whole pipeline frictional loss at exit or by distributing it along the length of the pipe by means of a number of discrete 'friction' joints, each of which would concentrate the frictional loss over a given section of pipe.

It may be shown that the modern numerical methods developed specifically for computer application do fundamentally the same thing, however the pipe section lengths become very short and thus an impression is formed that the friction loss is represented as a distributed loss, Harding (1966).

This representation of frictional loss was a subject of some discussion in the 1960s, following the Streeter and Lai (1962) paper introducing the method of characteristics in its now accepted format. However there is in fact little difference in the modes of frictional representation between the earlier non-computing methods and the method of characteristics.

3.4 Graphical solution techniques

The graphical solution of pressure surge problems is now of historical interest only as the use of digital computing techniques, particularly the method of characteristics, have become the industry standard for surge analysis and simulation. However it is perhaps interesting to briefly present the basis for the best known of the graphical methods, the Schnyder-Bergeron method, that remained in use up to the introduction of the computing methods, and even survived into that era as a translation into a computing method, Harding (1966), Pickford (1969).

The solution of the simplified wave equations due to D'Alembert has been given in equations 3.3 and 3.4 above, where the F() and f() functions represent pressure wave propagating within a pipeline system at the appropriate wave speed, the F() waves in a negative direction and the f() waves in a positive direction relative to the direction of initial steady state flow. It has been shown that these waves may be superimposed to give the cumulative pressure variation at any particular location.

Eliminating the f() term from equations 3.3 and 3.4 yields

$$p_{x,t} - p_0 = \rho c(V_{x,t} - V_0) + 2F(t+x/c) \tag{3.28}$$

For a particular location in space and time, coordinates X,T, this expression becomes

$$p_{X,T} - p_0 = \rho c(V_{X,T} - V_0) + 2F(T+X/c) \tag{3.29}$$

This F() wave propagates in the negative x direction, ie opposite to the direction of initial system steady flow. An imaginary observer travelling with this transient would view this wave as a constant, therefore

$$F(T+X/c) = F(t+x/c) \tag{3.30}$$

and so the F() function and the p_0, V_0 terms may be eliminated from equation 3.28,

$$p_{X,T} - p_{x,t} = \rho c(V_{X,T} - V_{x,t}) \tag{3.31}$$

This expression obviously links pressure and flow velocity, or flow rate, as shown in Figure 3.5 by means of a linear relationship in a p-V plane. (It will be useful later to contrast this approach with the method of characteristics use of a x-t plane.) The slope of the straight line relationship is ρc.

An identical process may be utilised in the positive x direction, acting upon the f() wave to generate an expression

$$p_{X,T} - p_{x,t} = -\rho c(V_{X,T} - V_{x,t}) \tag{3.32}$$

Equations 3.31 and 3.32 are represented by lines I and II in Figure 3.5, note that these are lines linking pressure change to velocity change and are therefore compatible, in a p-V plane, with curves, representing experimental or manufacturers data, linking pressure loss across the valve, at any particular open setting, to the flow rate through that valve.

Lines I and II also represent the transmission of information regarding pressure and velocity changes and are therefore analogous to the characteristic lines to be dealt with later in this text. This linkage was clearly defined by Harding (1966).

Equations 3.31 and 3.32 may be solved to yield pressure and velocity solutions at points along a pipeline. Two types of solution may be identified, namely for conditions at points along the pipe, referred to as internal calculations, and for conditions at the boundary to the system. Boundaries, as will be seen later in the treatment of the method of characteristics analysis, provide the greatest challenge and require the solution of either 3.31 or 3.32 with some equation defining the p-V conditions at the boundary considered. Boundaries are formed by any pipe entry, termination, junction, or by any valve or machine interface with the pipeline.

Internal calculations are carried out by identifying the intersection points of equations 3.31 and 3.32 in the p-V plane. The model described here is assumed frictionless, however friction may be included graphically by the introduction of friction joints along the pipeline, represented in the p-V plane by parabolae linking frictional pressure loss along that section of pipe to the flow velocity. The number of friction joints along a pipe is severely limited, by the need to minimise both the number of graphical constructions and the time involved.

Boundary conditions may be represented by lines drawn in the p-V plane that then intersect the appropriate I or II line that represents the wave arriving at that boundary. For example in a frictionless system with no local

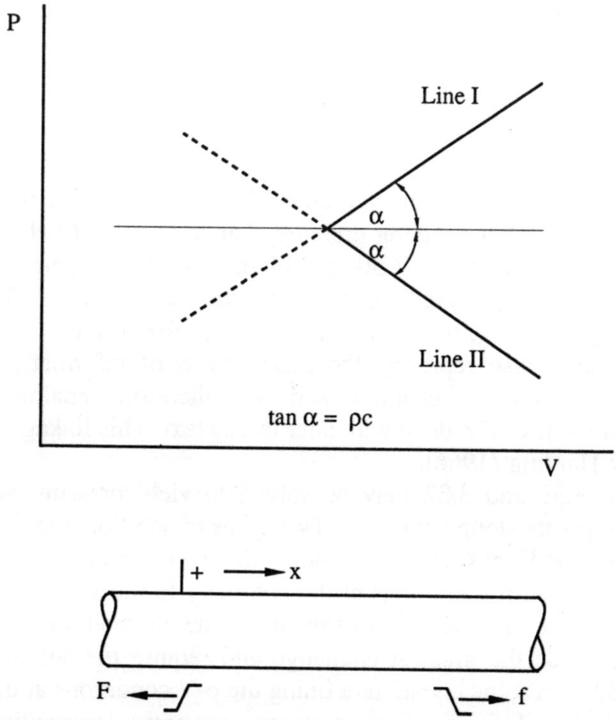

Figure 3.5 Characteristic transient lines in the p-V plane in the Schnyder-Bergeron graphical method

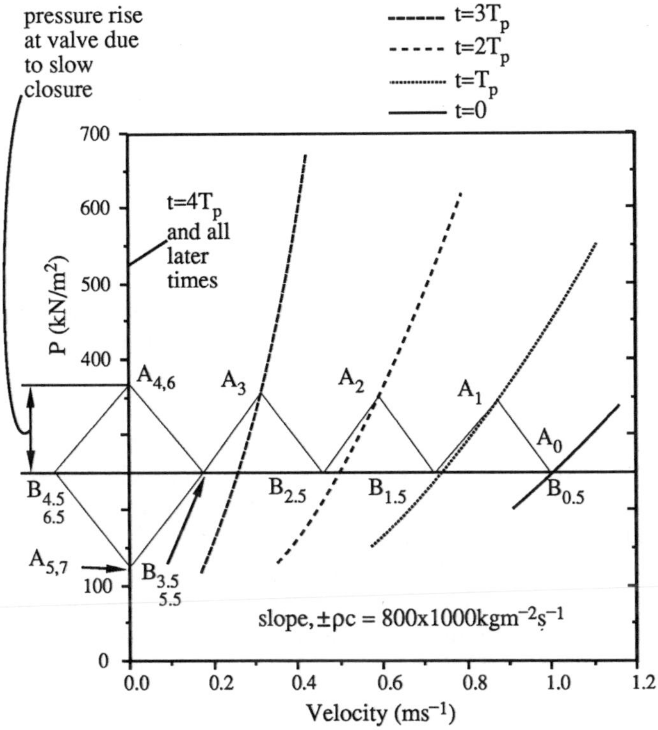

Figure 3.6 Schnyder-Bergeron construction for slow valve closure

separation loss at entry to a pipe from a constant pressure reservoir, the conditions at the boundary are represented by a horizontal line drawn in the p-V plane at the appropriate reservoir pressure coordinate. Similarly the zero flow velocity boundary for a dead end or at a valve following closure is represented by the vertical axis at V = 0 in the p-V plane. A combination of these two conditions could be used to represent the development of a vapour cavity at a pipe termination as discussed in Chapter 1.

The boundary condition represented by a closing valve depends upon the degree of closure at the instant that the transient arrives at the valve. The choice of a p-V plane does possibly make this somewhat more difficult to visualise, however for each valve position during closure an individual parabola may be drawn in the p-V plane linking pressure drop across the valve to flow through it. The intersection of this curve with either the I or II line yields the pressure-velocity solution at the closing valve at that instant. It will be appreciated that the series of parabolas representing valve pressure loss become increasingly steep in the p-V plane as the valve position moves toward fully closed and that this limiting condition is represented by the vertical axis at V = 0.

Figure 3.6 illustrates the simplest application of the graphical method to a slow valve closure in 4 pipe periods in a single pipeline 500 m long carrying a fluid of density 800 kg/m^3 at an initial flow velocity of 1.0 m/s and a constant, frictionless, line pressure of 250 kN/m^2. The appropriate wave speed for the system is 1000 m/s. The valve pressure loss coefficient under steady flow conditions is assumed to apply during transient propagation, a standard assumption in the absence of any better information, the value of the loss coefficient as a ratio of the fully open valve loss coefficient is assumed to be linearly related to the time to valve closure time ratio.

If the valve discharges to either atmosphere or to a constant pressure reservoir, then the pressure on the valve upstream face under steady flow conditions is related to the through flow velocity by

$$p_0 = K_0 V_0^2$$

where the suffix 0 indicates time zero.

As the valve closes, the value of K at any time is assumed to be linearly related to the elapsed valve operation time through a ratio of total closure time, t_c, thus

$$\frac{K_t}{K_0} = \frac{t}{t_c}$$

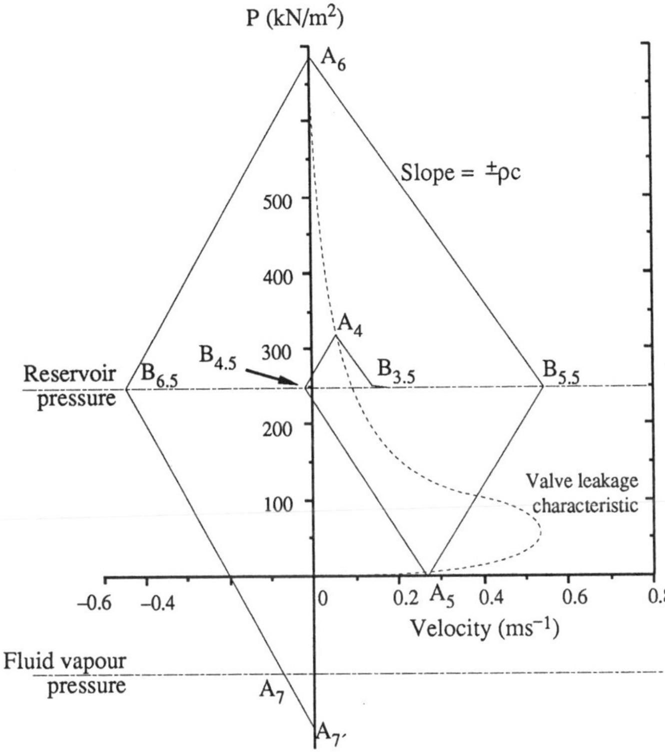

Figure 3.7 Schnyder-Bergeron construction following valve closure and subsequent leakage from the valve

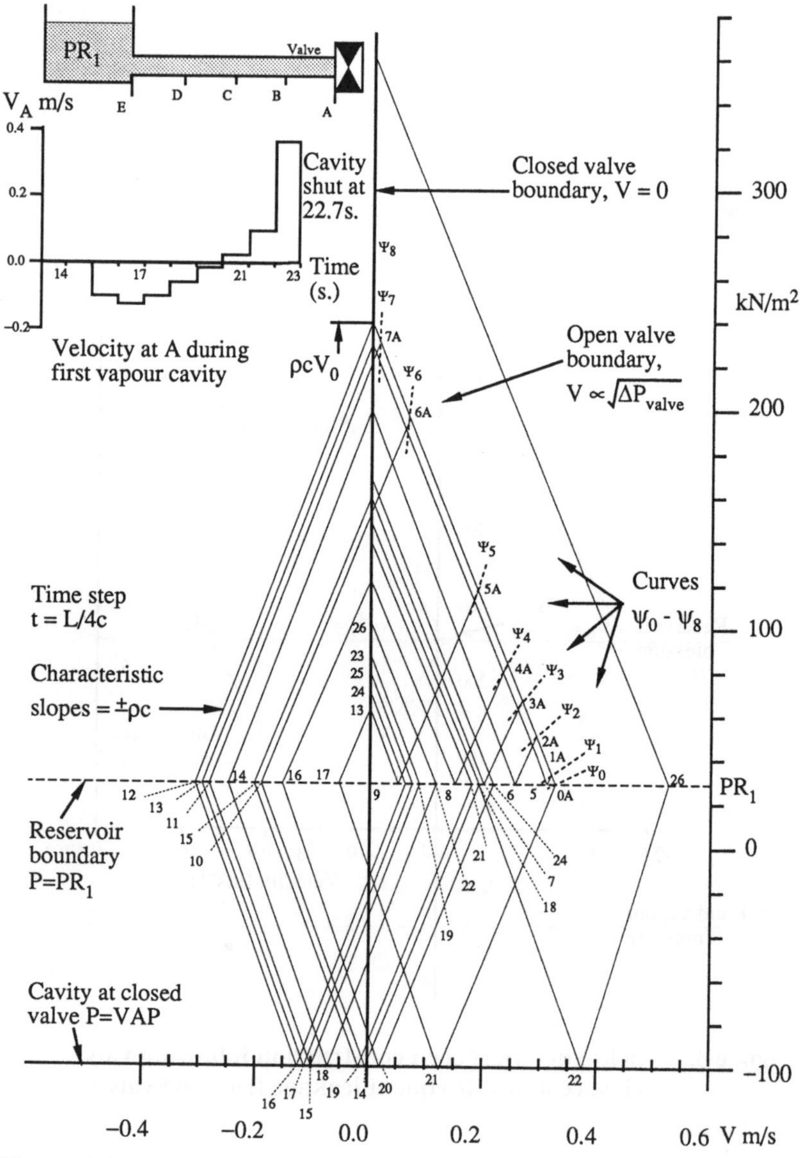

Figure 3.8(a) **Schnyder-Bergeron analysis of pressure variations at the valve following a rapid closure in 0.03s. $V_0 = 0.29\ ms^{-1}$, $PR_1 = 130\ kNm^{-2}$ abs**

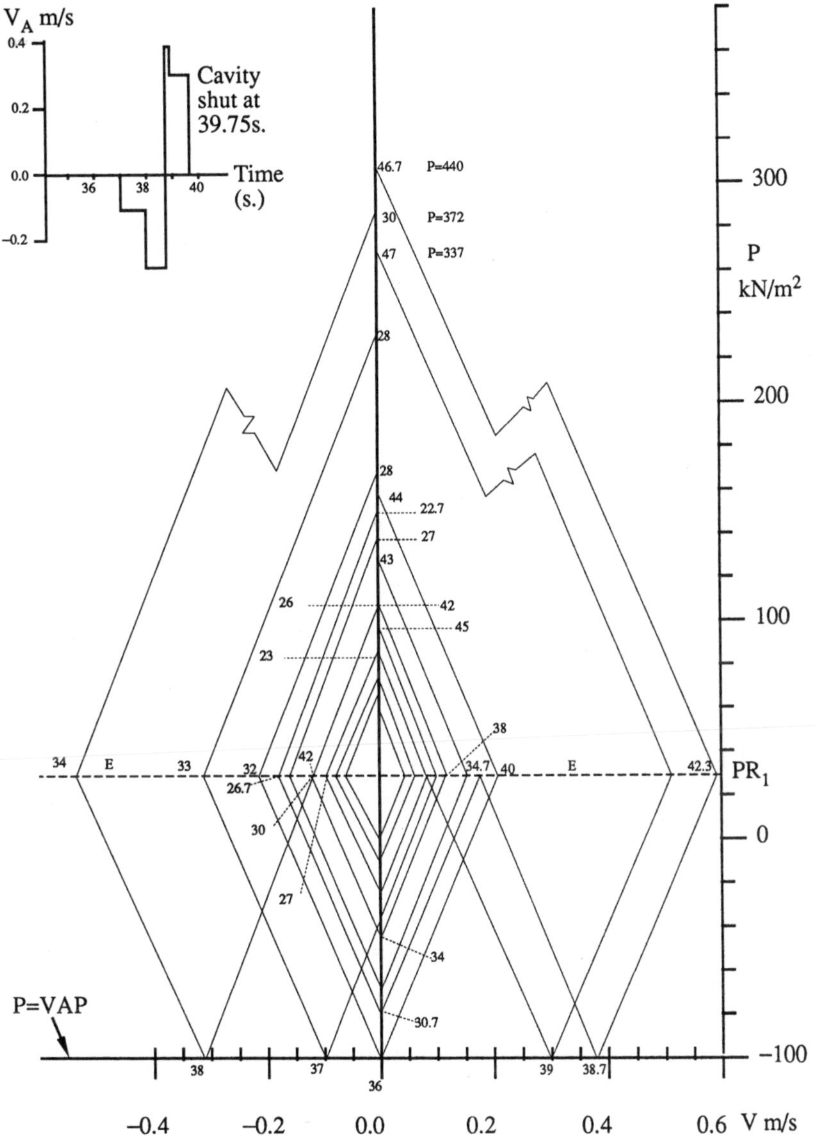

Figure 3.8(b) Continued Schnyder-Bergeron analysis of pressure variations at the valve following a rapid closure in 0.03s. $V_o = 0.29\ ms^{-1}$, $PR_1 = 130\ kNm^{-2}$ abs

For the four pipe period case considered in Figure 3.6 there may thus be drawn four parabolas indicating the conditions at the closed valve at the end of each pipe period. The boundary at time t_C is represented by V = 0.

Referring to Figure 3.6, conditions at the reservoir B remain constant at 250 kN/m^2 at all times. Conditions at valve A are represented by the parabolas described above, indicated in Figure 3.6 by a time identification. For the case presented conditions will only be considered at one pipe period intervals. If a closer time grid was required then the conditions at the valve would require to be represented by further parabolas indicating the pressure - velocity conditions there at these time intervals, an example of the rapidly escalating complexity of even the simplest graphical solution.

Steady flow conditions prevail at the valve, A, up to time = 0 and at the reservoir, B, up to time 0.5 t_p, where t_p is the pipe period 2L/c. Thus the initial conditions at these points may be labelled A_0 and $B_{0.5}$ on Figure 3.6, both points being coincident.

The reflected wave leaving the reservoir at time 0.5 t_p travels towards the closing valve and arrives there at a time 1.0 t_p. Thus the conditions at the valve are represented by the parabola labelled $t = t_p$ and the intersection of the II line and this curve at point $A_{1.0}$ indicates the conditions at the valve at that time. This wave is reflected at the closing valve and arrives at the constant pressure reservoir at a time 1.5 t_p. The wave's propagation may be represented by the I line leaving point $A_{1.0}$ and intersecting the reservoir boundary condition, represented by a horizontal boundary line drawn at the reservoir pressure of 250 kN/m^2 at point $B_{1.5}$.

This process repeats with the remaining intersection points being identified at $A_{2.0,3.0,4.0}$ and $B_{2.5,3.5}$. Once the valve has fully shut the boundary conditions are represented for the valve by V = 0 and for the reservoir by p = 250 kN/m^2, resulting in a repeating 'diamond' construction illustrated in Figure 3.6 by the points $A_{4,6,8}$, $B_{4.5,6.5}$, $A_{5,7,9}$, $B_{6.5,8.5}$.

The difficulty with the p-V plane presentation of the graphical method is that great care and understanding is needed to keep track of the conditions at the boundaries of the system, particularly in complex applications.

The x-t plane format of the method of characteristics is a major improvement on this from an ease of understanding viewpoint.

Figure 3.7 illustrates a simple extension to this example as an illustration of the scope of the graphical method. Assume that the valve design is such that a leakage occurs at t = 4.0 t_p, ie on final closure. Figure 3.7 illustrates the valve pressure loss-throughflow velocity characteristic for this leakage and this curve thus replaces the V = 0 axis as the boundary curve representing the valve from t = 4.0 t_p onwards.

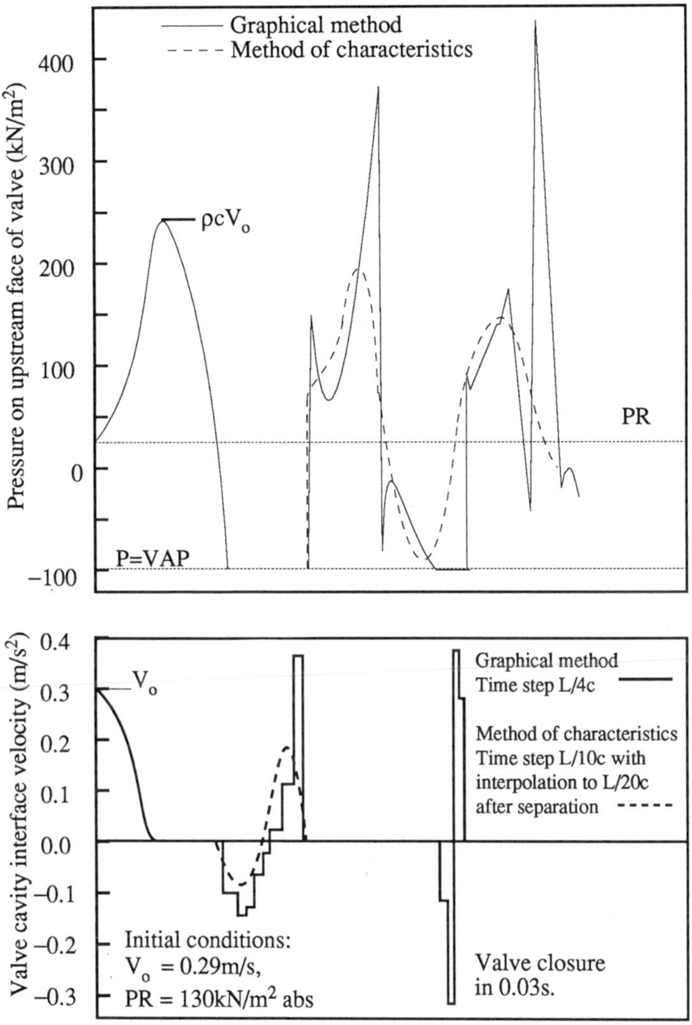

Figure 3.9 Comparison between a graphical and numerical analysis

The incoming transient, arriving at the valve at time t = 4.0 t_p now intersects this curve at point A4 and the resulting subsequent pressure history at the valve becomes A_5 A_6 A_7. It should be noted that this valve leakage characteristic has been chosen to illustrate an escalating pressure variation at the valve that eventually would predict pressure below vapour level at the valve so that a vapour pocket forms on column separation and the valve boundary becomes itself a constant pressure zone until such time as the cavity collapses. It is possible to represent separation graphically, as shown by Figure 3.8, however the monitoring of cavity size is crude due to the time step size and the complexity of the construction increases rapidly following the first cavity collapse as recurrent column separation events must be allowed for. Figure 3.9 illustrates a comparison between the effects of column separation predicted by both the graphical and computing methods to be introduced later.

3.5 Summary of the frictionless model

This chapter has presented a simple frictionless solution to the wave equations for a horizontal pipeline. The techniques discussed, particularly the concepts of reflection and transmission coefficients are most helpful in providing a better understanding of the mechanisms of transient propagation. Taken together with the application of superposition this technique can be used to explain many of the transient phenomena encountered and will be utilised later in this text to explain transient pressure variations recorded following the collapse of a vapour or trapped air pocket.

However there is a definite limit to the scope of these techniques due to the inherent difficulty of 'keeping track' of the transmissions and reflections that exist within a real system, where the time taken to generate pressure transients would normally exceed one pipe period. There was a need to introduce a numerical methodology to solve the wave equations that would also allow a better 'visual' appreciation of the transient propagation mechanisms than that offered in a p-V plane.

Dating from the late 1940s, and becoming the main stream methodology from the advent of the reliable and fast digital computer in the early 1960s, the method of characteristics filled this requirement and is now the industry standard technique. This technique will be fully developed in the remainder of this text, with an emphasis upon the ease of representation of boundary conditions possible with this method and its application as a tool for the better understanding of transient pressure propagation mechanisms.

Chapter 4:

The method of characteristics

The prediction of pressure transient propagation within complex pipe networks, possibly incorporating pumps, valves, turbines and surge relief devices, as well as numerous junctions and possible changes of pipe material or dimension, requires an orderly and flexible mathematical model in order to provide the accuracy needed for system design. While simple numerical methods and later graphical methods have been available since the end of the 19th century, Allievi (1903), Bergeron (1935), the complexity of more modern fluid systems dictates a computer based numerical approach.

The basic equations of pressure transient theory, the St Venant equations of momentum and continuity, developed in the mid 19th century, St Venant (1870), may be shown to be a pair of quasi linear hyperbolic partial differential equations. As such these equations may be transformed into a pair of total derivative equations suitable for solution by a finite difference technique via the method of characteristics. This mathematical technique was used in 1860 by Riemann to study sound wave propagation in air and in 1900 by Massau in a study of open channel flows. The first known application to pressure transient analysis was due to Lamoen (1947), followed by Gray (1953), although the method was well established in the treatment of supersonic flows by this date.

The method of characteristics is now established as the standard solution technique for the St Venant equations applied to pressure transient analysis. However this predominance had to await the arrival, in the mid 1960s, of efficient high speed digital computers. As will be shown, the method has the great advantage of being conceptually descriptive of the transmission of transients through a pipe network. The restriction that the method only applies along 'characteristic' curves drawn in an x-t plane having a slope dependent upon the local wave propagation and flow velocities leads naturally to a linkage between the propagation of the 'real' transient in the pipe network, Chapter 1, and the numerical model.

This restriction also explains why the method was embryonic until the advent of high speed digital computers. The characteristic slopes determine the model time step and in general these are very small. The time step depends on both the wave speed appropriate to the fluid-pipe combination, commonly in the region of 1000 m/s, and the length of the pipe section chosen as the internode distance, possibly as short as 1 m if the pipe network is small scale and complex, or as great as 1 km in long distance pipeline problems. The choice of an internode distance depends on the number of locations at which a knowledge of the transient pressure oscillation would be useful to the designer, however it cannot be greater than the actual length of the pipe being considered, setting an upper limit on time step equal to half the pipe period. In the two cases mentioned the time step would be 0.001 s and 1 s respectively. However in the latter case the total simulation time required would of necessity also be much greater, so that a simulation of transient propagation would require many thousands of computing steps. This calculation alone illustrates why the method was not exploited prior to the 1960s.

The exploitation of the method of characteristics for transient analysis, in conjunction with the use of digital computing techniques, was introduced in the mid 1960s by a number of authors, most notably in the USA by Streeter and Lai (1962), Streeter and Wylie (1967), Streeter (1969), and in the UK by Fox (1968). In each case the basis for the analysis was the method of characteristics as described by Lister (1960). Subsequently the technique has become the standard analysis technique with many authors applying and modifying the basic method to model associated problems, including gas release, Kranenberg (1974), column separation and vapour generation, Doyle and Swaffield (1972), Swaffield (1972), machine interface boundary problems, Boldy (1976), and the application of the method to investigate related areas such as air pressure transients in high speed train tunnels, Henson and Fox (1972), Vardy (1976) and open channel and partially filled pipe flow transients, Price (1974), Song (1976), Galowin and Swaffield (1989). The interest in the application of the method continues and many useful and directly relevant papers may be found in the succession of B.H.R.A. Pressure Surge Conference Proceedings from 1972 onwards.

4.1 Development of the unsteady flow equations of continuity and momentum

The equations of momentum and continuity for an elemental strip of flow under unsteady flow conditions have been developed in Chapter 2, illustrated in Figures 2.1 and 2.2.

It will be appreciated from that derivation that no restrictions were placed upon the fluid type, beyond its homogeneous and Newtonian nature, and the enclosing conduit was also treated as a general conduit so that the developed equations may be seen to apply equally to liquid or gas flows or to both full bore or free surface liquid flow situations.

The momentum equation was shown to be

$$L_1 = V\frac{\partial V}{\partial x} + \frac{\partial V}{\partial t} + g\frac{\partial H}{\partial x} + \frac{fV|V|}{2m} = 0 \qquad (4.1)$$

in velocity-head terms, or

$$L_1 = V\frac{\partial V}{\partial x} + \frac{\partial V}{\partial t} + \frac{1}{\rho}\frac{\partial p}{\partial x} - g\sin\alpha + \frac{fV|V|}{2m} = 0 \qquad (4.2)$$

in velocity-pressure terms.

As pressure transient analysis crosses the more traditional disciplinary boundaries within fluid mechanics both sets of units remain in current use and therefore the derivation of the method of characteristics in both will be presented.

A further modification to present the equations in terms of flow rate-head (or pressure) terms will be left to the reader.

In the above equation it will be noted that the convention introduced in Chapter 2, namely that the pipe has a negative slope in the initial flow direction, has been maintained, as shown by the sign of the gravity force term, thus $\sin\alpha = -\partial z/\partial x$.

Similarly it will be seen that a generalised definition of the frictional resistance term has been maintained, expressed in terms of the flow hydraulic mean depth, m, ie the ratio of flow cross section to wetted perimeter; this will simplify the later application of the equations derived, initially for full bore flow transient situations, to free surface or non circular cross section conduits.

The use of the absolute flow velocity, $|V|$, in the frictional representation is standard practice as it ensures that the frictional forces always act to oppose the local flow direction. It will be shown later that it is acceptable to utilise the steady equivalent flow rate values of friction factor within the method of characteristics solution.

The continuity equation may be expressed as

$$L_2 = \rho c^2 \frac{\partial V}{\partial x} + V\rho g \left[\frac{\partial H}{\partial x} + \sin\alpha) \right] + \rho g \frac{\partial H}{\partial t} = 0 \tag{4.3}$$

in velocity-head terms, or

$$L_2 = \rho c^2 \frac{\partial V}{\partial x} + V \frac{\partial p}{\partial x} + \frac{\partial p}{\partial t} = 0 \tag{4.4}$$

in velocity-pressure terms.

As shown in Section 2.4, equations L_1 and L_2 are a pair of quasi linear, hyperbolic, partial differential equations and as such are not readily solvable. The method of characteristics allows the transformation of these equations into a pair of total differential equations suitable for solution by finite difference based numerical methods, and particularly by the computerisation of such techniques.

4.2 The method of characteristics applied to the solution of the transient propagation equations

The equations of continuity and momentum may be combined as

$$L = L_1 + \lambda L_2 \tag{4.5}$$

where for any value of λ the conditions and assumptions inherent in the original derivation presented in Chapter 2 remain inviolate.

The derivation presented in Chapter 2 was of a generalised nature and so the utilisation of the method of characteristics presented below will also serve as a model for the application of the method in later chapters to solve the transient equations applicable in low amplitude air transient propagation and partially filled pipeflow examples.

4.2.1 Velocity-pressure formulation

From equations 4.2 and 4.4, collecting like terms yields

$$L = \left[\frac{\partial V}{\partial x}(V + \lambda\rho c^2) + \frac{\partial V}{\partial t}\right] + \lambda\left[\frac{\partial p}{\partial x}\left(\frac{1}{\lambda\rho} + V\right) + \frac{\partial p}{\partial t}\right] - g\sin\alpha + \frac{fV|V|}{2m} = 0$$

Now since

$$V = \phi_1\,(x,t)$$

and

$$p = \phi_2\,(x,t)$$

then

$$\frac{dV}{dt} = \frac{\partial V}{\partial x}\frac{dx}{dt} + \frac{\partial V}{\partial t}$$

and

$$\frac{dp}{dt} = \frac{\partial p}{\partial x}\frac{dx}{dt} + \frac{\partial p}{\partial t}$$

Substitutions into L allows the combined expression to be rewritten as

$$L = \frac{dV}{dt} + \lambda\frac{dp}{dt} - g\sin\alpha + \frac{fV|V|}{2m} = 0 \qquad (4.6)$$

provided that

$$\frac{dx}{dt} = V + \lambda\rho c^2 = \frac{1}{\lambda\rho} + V$$

which implies that

$$\lambda = \pm \frac{1}{\rho c}$$

so that the combined equation may be written, in velocity-pressure terms,

$$\frac{dV}{dt} \pm \frac{1}{\rho c}\frac{dp}{dt} - g\sin\alpha + \frac{fV|V|}{2m} = 0 \tag{4.6}$$

provided that

$$\frac{dx}{dt} = V \pm c \tag{4.8}$$

These equations are the characteristic equations and, as a pair of total derivative equations, are solvable via the application of finite difference techniques. The two relationships obtained from equation 4.7 are dependent upon dx/dt = V+c or dx/dt = V−c, from equation 4.8, and are known respectively as the C^+ and C^- characteristics and will be shown to represent the way in which information is transmitted through unsteady flow.

4.2.2 *Velocity-head formulation*

Following a similar process in velocity-head terms yields a combined expression

$$L = \left[\frac{\partial V}{\partial x}(V + \lambda\rho c^2) + \frac{\partial V}{\partial t}\right] + \lambda\rho g\left[\frac{\partial H}{\partial x}\left(\frac{1}{\lambda\rho} + V\right) + \frac{\partial H}{\partial t}\right] + \lambda V\rho g\sin\alpha + \frac{fV|V|}{2m} = 0$$

and if

$$\frac{dx}{dt} = V + \lambda\rho c^2 = \frac{1}{\lambda\rho} + V$$

which implies as before that

$$\lambda = \pm \frac{1}{\rho c}$$

then the combined equation may be written, in velocity-pressure terms,

$$\frac{dV}{dt} \pm \frac{g}{c}\frac{dH}{dt} \pm \frac{g}{c}V\sin\alpha + \frac{fV|V|}{2m} = 0 \tag{4.9}$$

provided that

$$\frac{dx}{dt} = V \pm c \tag{4.10}$$

The form of the combined equations may be seen to be suitable for finite difference treatment, however the dx/dt expressions inherent in the transformation of the original partial derivatives into the more convenient total derivatives place restrictions on the ensuing solution. It is these dx/dt expressions that give the method its name as to comply with them the solution is restricted to a pair of 'characteristic' directions in the x-t plane.

Referring to Figure 4.1 the combined equations 4.7 and 4.8 relate the known conditions at R and S in an x-t plane to unknown conditions at P. In the general case illustrated the spatial and temporal increments between R and P and between S and P may differ and strictly the characteristic lines linking RP and SP are curved. Such a representation will allow the prediction of conditions at P and, by subsequent forward movement of the base times, will allow the solution to proceed further.

However the generalised grid in this form is unwieldy and does not readily lend itself to an orderly solution. Figure 4.2 illustrates the more widely used method of specified time intervals. Here a rectangular grid in time and distance increments is set up and the characteristic equations are assumed, at least in a first order approximation, to be straight lines linking known conditions, at a particular time in the past, at two points to unknown conditions at an intermediate point at some time in the future; the respective slopes of these characteristics lines are based, at least in a first order approximation, on wave speed and flow velocity at R and S. Such a scheme makes possible an orderly solution for unknown conditions at P.

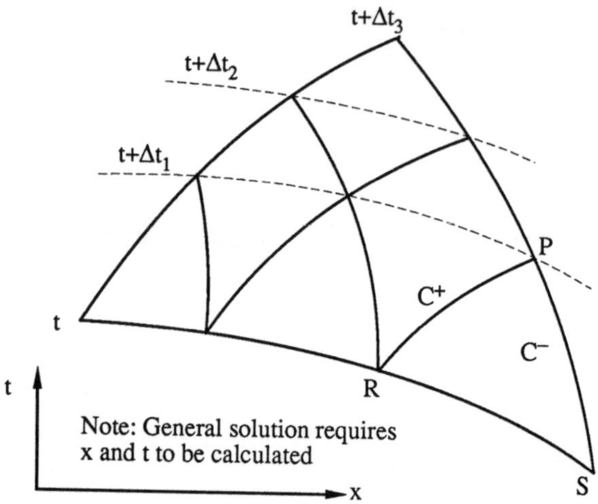

Figure 4.1(a) General solution by the method of characteristics

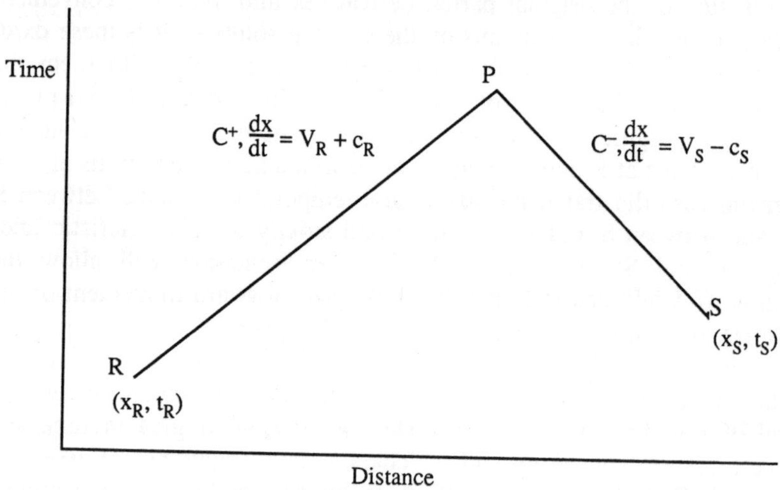

Figure 4.1(b) First order approximation to the C^+ and C^- characteristics in an x-t plane

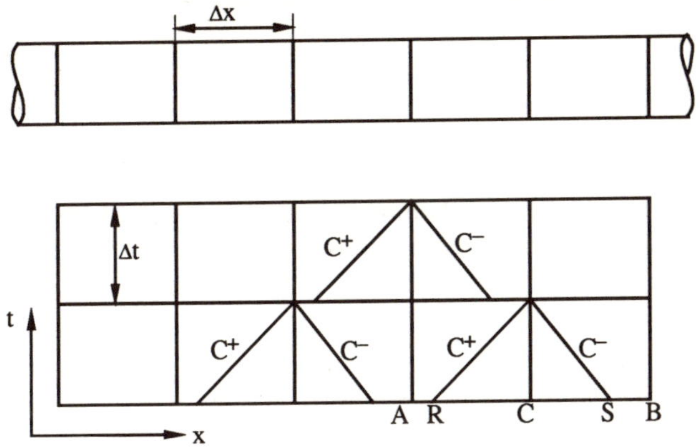

Figure 4.2 Method of specified time intervals. The time and distance increments are linked to the flow velocity and wave speed by the Courant criterion

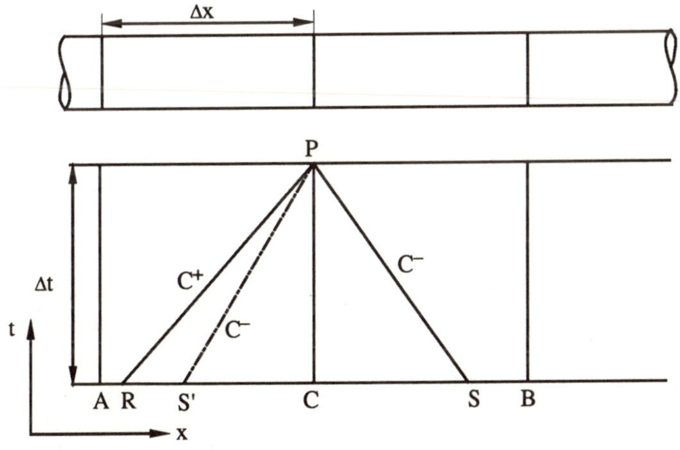

Figure 4.3 C^+ and C^- characteristics in an x-t plane for the method of specified time intervals. Note that characteristic PS' lies upstream of point C if the fluid velocity exceeds the wave speed; normal conditions in supercritical open channel flow

The generalised slope of the characteristic lines is

$$\frac{dx}{dt} = V \pm c$$

and it will be seen that it is therefore a necessary condition that the base points for these two characteristics, known generally as C^+ and C^- lines, fall within $\pm\Delta x$ of the position of the point for which conditions are to be calculated, as shown in Figure 4.2. This clearly implies a relationship between the time and distance increments for the method of specified time intervals, usually defined as the Courant Criterion, that

$$\Delta t \leq \frac{\Delta x}{V+c}$$

In a great many pressure transient applications the Courant Criterion may be reduced to

$$\Delta t \leq \frac{\Delta x}{c}$$

as the wave propagation velocity, c, is much greater than the local mean flow velocity. For example wave speeds are often in the region of 1000 m/s while flow velocities are often less than 10 m/s. However there are specialist applications where this simplification cannot apply, for example in conditions where there is a high free gas content, in very thin walled pipes or in gas or free surface flows. These cases will be returned to later.

A second limitation upon the method of characteristics may also be demonstrated by reference to Figure 4.2. If the condition that $V = c$ is met then the characteristic line becomes vertical and no solution is possible. This condition is equivalent to the introduction of a discontinuity, such as a shock wave or a hydraulic jump, into the flow as a boundary between two types of flow regime. The method of characteristics remains valid on either side of the discontinuity, however additional equations would be necessary to allow the solution to proceed across it.

If the condition that $V \lll c$ is accepted then it will be seen that both the C^+ and C^- characteristics pass through node points $\pm\Delta x$ on either side of the node currently being treated, with reference to Figure 4.2 this implies that points A and R coincide and S and B coincide.

The wave speed may not of course remain constant along the whole length of a pipe or throughout a network. Local changes in pipe wall thickness, pipe material, pipe diameter or free gas content will all affect wave speed and it may be necessary to take these variations into account. It is however imperative that, as will be discussed further, the time step utilised in the calculation remains a constant for all the constituent pipes within the network. Methods of achieving this constancy will be presented.

The equations derived are only valid along the C^+ and C^- characteristics in the x-t plane. The development process has however left the equations of momentum and continuity otherwise unrestricted in their application.

These characteristic lines may conceptually be considered to define the transmission of information through the pipe network or unsteady flow being considered. Changes in flow conditions are only transmitted along these C^+ and C^- lines and it is this transmission process that is perceived as the propagation of pressure transients. Thus it would be correct to see the propagation of transients as a normal occurrence in any flow situation where the demands made upon the system are changed, for example either by a change in pump delivery, turbine load or discharge valve setting.

An interesting aspect of this approach to transient propagation concerns the already mentioned situation where $V > c$, ie supersonic gas flows or supercritical free surface flow. In this case both the C^+ and C^- characteristics lie upstream of the point P, as shown in Figure 4.3, and thus information cannot propagate upstream, a well known phenomena in both the supersonic and supercritical flow regimes.

Figure 4.2 illustrates an x-t grid set up under the method of specified time intervals with the characteristic lines passing through the adjacent nodes as shown. If conditions at R and S are known at time t, then the unknown conditions at P one time step later may be calculated from the following first order finite difference approximations:

1. the C^+ characteristic

$$V_P = V_R - \frac{1}{\rho c_R}(p_P - p_R) + g\sin\alpha_R \Delta t - \frac{f_R V_R |V_R| \Delta t}{2m_R} \tag{4.11}$$

when

$$\frac{dx}{dt} = V + c$$

2. the C^- characteristic

$$V_P = V_S + \frac{1}{\rho c_S}(p_P - p_S) + g\sin\alpha_S \Delta t - \frac{f_S V_S |V_S| \Delta t}{2 m_S} \qquad (4.12)$$

when

$$\frac{dx}{dt} = V - c$$

These two equations are often reduced for simplicity to

$$V_P = K1 - K2\, p_P \qquad (4.13)$$

for the C^+ characteristic, and

$$V_P = K3 + K4\, p_P \qquad (4.14)$$

for the C^- characteristic, where

$$K1 = V_R + \frac{p_R}{\rho c_R} + g\sin\alpha_R \Delta t - \frac{f_R V_R |V_R| \Delta t}{2 m_R} \qquad (4.15)$$

$$K2 = \frac{1}{\rho c_R} \qquad (4.16)$$

$$K3 = V_S - \frac{p_S}{\rho c_S} + g\sin\alpha_S \Delta t - \frac{f_S V_S |V_S| \Delta t}{2 m_S} \qquad (4.17)$$

$$K4 = \frac{1}{\rho c_S} \qquad (4.18)$$

as all the relevant conditions are known at R and S. This form of the characteristic equations will be utilised later in Chapter 5 to simplify the presentation of boundary calculations.

The equivalent velocity-head relationships may be stated as

$$V_P = K1 - K2\ H_P \tag{4.19}$$

for the C^+ characteristic, and

$$V_P = K3 + K4\ H_P \tag{4.20}$$

for the C^- characteristic, where

$$K1 = V_R\left(1 - \frac{g}{c_R} \sin\alpha_R \Delta t - \frac{f_R|V_R|\Delta t}{2m_R} \right) + \frac{gH_R}{c_R} \tag{4.21}$$

$$K2 = \frac{g}{c_R} \tag{4.22}$$

$$K3 = V_S\left(1 + \frac{g}{c_S} \sin\alpha_S \Delta t - \frac{f_S|V_S|\Delta t}{2m_S} \right) - \frac{gH_S}{c_S} \tag{4.23}$$

$$K4 = \frac{g}{c_S} \tag{4.24}$$

In the case where the added simplification that c >>> V is present then the C^+ and C^- characteristics pass through A and B, as shown in Figure 4.2. The above equations remain valid with the obvious notation change that values identified as at R and S become values at A and B. This point will be automatically incorporated into the computing expressions by identifying R and S in terms of the node P, Chapter 5.

The implication of a first order approximation is that the flow velocity and wave speed values vary slowly between R, S and P, an assumption justified by the small time increments normally dictated by the method of characteristics; thus it follows that the characteristic lines are straight. It also follows that the velocity at P does not enter into the frictional terms, allowing the avoidance of an unknown squared velocity term and the retention of a simplistic simultaneous solution of the above equations. Similarly the retention of a first order approximation considerably simplifies the modelling of the system boundary equations.

In general terms the first order approximation may be expressed as

$$\int_{x_0}^{x_1} f(x)dx = f(x_0)\,(x_1 - x_0)$$

while the second order approximation may be expressed as

$$\int_{x_0}^{x_1} f(x)dx = 0.5\,(\,f(x_0)+f(x_1))\,(x_1 - x_0)$$

With a second order finite difference scheme the C^+ characteristic becomes

$$V_P = V_R - \frac{1}{2\rho}\left[\frac{1}{c_P}+\frac{1}{c_R}\right](p_P - p_R) + \frac{1}{2}g(\,\sin\alpha_P + \sin\alpha_R\,)\Delta t$$

$$- \frac{1}{2}\left[\frac{f_P V_P |V_P|}{2m_P}+\frac{f_R V_R |V_R|}{2m_R}\right]\Delta t$$

when

$$\frac{dx}{dt} = 0.5\,(V_P + c_P + V_R + c_R)$$

and the C^- characteristic becomes

$$V_P = V_S + \frac{1}{2\rho}\left[\frac{1}{c_P}+\frac{1}{c_S}\right](p_P - p_S) + \frac{1}{2}g(\,\sin\alpha_P + \sin\alpha_S\,)\Delta t$$

$$- \frac{1}{2}\left[\frac{f_P V_P |V_P|}{2m_P}+\frac{f_S V_S |V_S|}{2m_S}\right]\Delta t$$

when

$$\frac{dx}{dt} = 0.5\,(V_P - c_P + V_S - c_S)$$

resulting in the introduction of a quadratic in unknown velocity at point P.

Generally the use of a second order approximation has not found favour in the application of the method of characteristics to pressure transient modelling. Equally valid predictions may be obtained from the first order approximation without the complexity introduced by the need for an iterative solution as the unknown wave speed and velocity at P appear in the friction term (velocity) and the characteristic slope (wave speed and velocity).

It is of historical interest that, in the early stages of the acceptance of the method of characteristics as the standard technique for the analysis of transients, considerable discussion revolved around the representation of friction, Harding (1966), and the acceptability of the first order approximation, Streeter and Lai (1962) discussion by Paynter and Gray.

4.3 Interpolation as a means of defining base conditions

Variable wave speed along the length of a pipe, duct or channel leads to different slopes for the C^+ and C^- characteristics. However the Courant Criterion must be satisfied so that, as shown in Figure 4.4, the point R and S become closer to P and separate from the nodes on either side of P.

Assuming that initial conditions are known at all nodes at time t, then it is necessary to determine the conditions at R and S at time t in order to apply the characteristics solution already developed. The most common solution to this situation is to interpolate linearly between conditions at A and C to obtain conditions at R and similarly between C and B to obtain conditions at S. (Note here that if $c < V$, ie the supercritical free surface or supersonic flow condition, then both R and S lie between A and C in Figure 4.3.)

The technique is set out below. However interpolation must be recognised as a potential source of serious rounding errors in the numerical modelling of pressure transient propagation.

Although this problem will be returned to later, it is worth mentioning at this stage that interpolation effectively implies that a pressure transient arriving at A or B at time t determines conditions at R or S at that time. This effectively increases the speed of propagation of the transient and also decreases the rate of change of pressure or velocity that it imparts to the flow it passes through. Both effects lead to a rounding in the predicted transient, Vardy (1976).

Figure 4.4 illustrates the conditions that result from a variable wave speed and comparable flow velocity magnitudes along the length of a pipe. The characteristic slopes change along the pipe depending on the local c and V values, this results in the positions of R and S, relative to P, varying along the pipe. The degree of interpolation therefore also varies along the pipe length and care must be taken to ensure that the resulting rounding errors do not become excessive.

Referring to Figure 4.3 as the general case where both c and V vary along the pipe and also where the possibility that V > c is allowed to exist so that both C^+ and C^- characteristics slope downstream, a series of equations may be presented linking conditions at R and S (or S′) to conditions at the nodes A, C and B.

For the C^+ characteristic passing through R and linking conditions at R at time t to conditions at P at time t+Δt, consideration of the velocity variation yields

$$\frac{V_C - V_R}{V_C - V_A} = \frac{x_C - x_R}{\Delta x} = (V_R + c_R)\frac{\Delta t}{\Delta x}$$

since

$$x_C = x_P$$

and

$$\Delta x = x_C - x_A$$

Similarly the wave speed terms yield

$$\frac{c_C - c_R}{c_C - c_A} = \frac{x_C - x_R}{\Delta x} = (V_R + c_R)\frac{\Delta t}{\Delta x}$$

Simultaneous solution of these equations results in a series of interpolation relationships that allow the determination of base conditions at point R

$$V_R = \frac{V_C + \theta(V_A c_C - V_C c_A)}{1 + \theta(V_C - V_A + c_C - c_A)} \tag{4.25}$$

$$c_R = \frac{c_C + \theta V_R(c_A - c_C)}{(1 + \theta(c_C - c_A))} \tag{4.26}$$

where

$$\theta = \frac{\Delta t}{\Delta x}$$

and for the pressure or head terms

$$p_R = p_C - (p_C - p_A)\theta(V_R + c_R) \tag{4.27}$$

$$H_R = H_C - (H_C - H_A)\theta(V_R + c_R) \tag{4.28}$$

A number of points need to be stressed concerning the above equations

1. Quite obviously if the wave speed is a constant, but still comparable in magnitude to the local flow velocity, the above equations will yield the interpolated values of flow velocity V with no modification.

2. In cases where the velocity is negligible with respect to a constant wave speed, so that

$$c_P = c_A = c_B$$

These interpolation equations may be simplified as

$$\frac{V_C - V_R}{V_C - V_A} = \theta c_A$$

thus

$$V_R = V_C - \theta c_A(V_C - V_A)$$

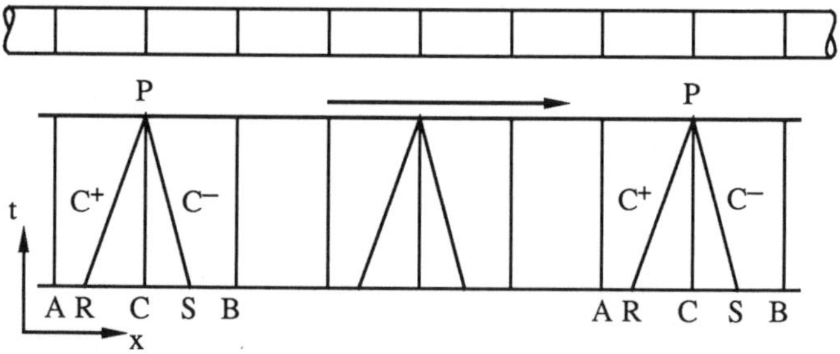

Figure 4.4 Time step based on the maximum value of the wave speed and velocity leads to significant interpolation errors in the flow regions exhibiting lower V and c values. Relative positions of R and S illustrate the problem

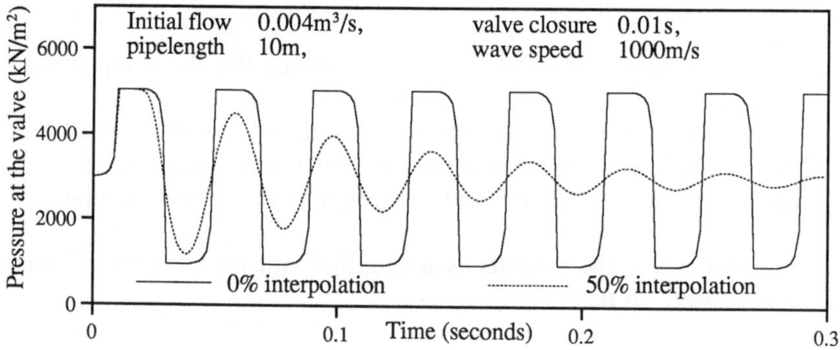

Figure 4.4(a) Effect of a 50% time step reduction by interpolation on the pressure variation predicted at a rapidly closing valve in a frictionless pipeline

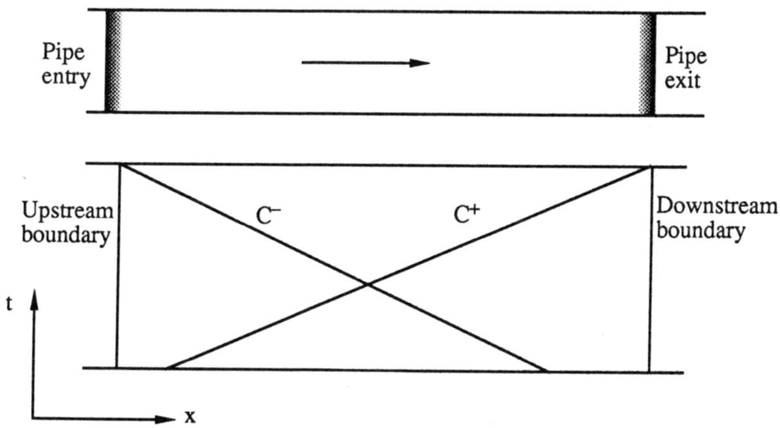

Figure 4.5 Example of a short pipe where the pipelength becomes a single Δx. C^+ and C^- characteristics are available for solution at the boundaries as shown

and pressure, p, or head, H, is given by

$$p_R = p_C - \theta c_A(p_C - p_A)$$

$$H_R = H_C - \theta c_A(H_C - H_A)$$

When the wave speed exceeds the local flow velocity, the C^- characteristic slopes upstream and the base conditions are found at point S in Figure 4.3. By a similar series of substitutions to those above the following expressions may be derived

$$V_S = \frac{V_C - \theta(V_C c_B - V_B c_C)}{1 - \theta(V_C - V_B - c_C + c_B)} \tag{4.29}$$

$$c_S = \frac{c_C + \theta V_S(c_C - c_B)}{(1 + \theta(c_C - c_B))} \tag{4.30}$$

and for the pressure or head terms

$$p_S = p_C + (p_C - p_B)\theta(V_S - c_S) \tag{4.31}$$

$$H_S = H_C + (H_C - H_B)\theta(V_S - c_S) \tag{4.32}$$

If c >> V, and is assumed to be constant, then these equations reduce as before to

$$\frac{V_C - V_S}{V_C - V_B} = \theta c_B$$

thus

$$V_S = V_C + \theta c_B(V_B - V_C)$$

and pressure, p, or head, H, is given by

$$p_S = p_C + \theta c_B(p_B - p_C)$$

$$H_S = H_C + \theta c_B(H_B - H_C)$$

If the local flow velocity exceeds the wave speed then the C⁻ characteristic also slopes downstream and the required base point, S′, is to be found upstream of point P. This condition would not be expected in the analysis of pressure transients, however it is the norm in many free surface flow applications of the method of characteristics as it represents the supercritical flow regime. In this case the interpolation equations are derived in the same manner as set out above, resulting in the following expressions

$$V_{S'} = \frac{V_C - \theta(V_A c_C + V_C c_A)}{1 + \theta(V_C - V_A + c_A - c_C)} \tag{4.33}$$

$$c_{S'} = \frac{c_C + \theta V_{S'}(c_A - c_C)}{(1 + \theta(c_A - c_C))} \tag{4.34}$$

and for the pressure or head terms

$$p_{S'} = p_C - (p_C - p_A)\theta(V_{S'} - c_{S'}) \tag{4.35}$$

$$H_{S'} = H_C - (H_C - H_A)\theta(V_{S'} - c_{S'}) \tag{4.36}$$

The effect of interpolation, in the case where the wave speed, c, is constant along a pipelength, and where the value of c clearly exceeds the flow mean velocity, may be clearly demonstrated for a frictionless pipeline by application of the surge program included in Appendix II. Figure 4.4(a) illustrates the predicted pressure variations at the exit valve of a 10 m long pipeline, appropriate wave speed 1000 m/s, following a valve closure in one pipe period. In the absence of friction and without interpolation, ie with a time step equal to Δx/c, where Δx is 2 m, the pressure at the valve follows the periodic cycle discussed in Chapter 1, ie no attenuation with time due to the absence of frictional damping. When a reduced time step is introduced with the same Δx, by interpolation, the effect of numerical damping becomes obvious; in the case illustrated a 50% interpolation is introduced. While the effects of interpolation become less dramatic as the rate of change of the boundary conditions decreases, rapid changes in conditions can still occur within systems, particularly as a result of cavity collapse following column separation and, as the onset of such conditions is not always predictable from a study of a system design, it is advisable to avoid interpolation.

4.4 Implications of interpolation within a transient analysis

Satisfaction of the Courant Criterion requires that

$$\Delta t \leq \frac{\Delta x}{V+c} \tag{4.37}$$

at all points along a particular pipe or channel. An orderly simulation of a network consisting of a large number of pipes joined in reality by pipe junctions, and in the model by the solution of a series of suitable boundary equations, requires that the same time step is utilised for all the constituent pipes.

If wave speed and flow mean velocity varies along the length of a pipe or channel then the Courant Criterion clearly must be rewritten as

$$\Delta t \leq \frac{\Delta x}{V_{max} + c_{max}} \tag{4.38}$$

where V_{max} and c_{max} are the maximum values of wave speed and velocity along that pipe at that particular time step. It is worth noting that these maximum values may well change at the next time step. Similarly at any time step the maximum values of wave speed and flow velocity may not be found at the same node. This point is illustrated by Figure 4.4.

Thus the degree of interpolation necessary may well vary along the length of the pipe or channel, from the 'acceptable' close to the location of the maximum values of V and c, to the 'unacceptable' where these values are low. In this context 'unacceptable' may be taken to imply that the interpolated base condition, R or S in Figure 4.4, lies close to the node P being calculated, with subsequent unacceptable rounding errors being introduced by the mechanism already described.

In most pressure transient applications, ignoring the influence of a large free gas content, the wave speed along a particular pipe length may be taken as constant, calculated on the basis of pipe wall dimensional and elastic data and fluid density and compressibility values. (This convenient simplification may not be assumed in either gas flow transients or in the study of free surface surges.) In this simplified case therefore it is sufficient to ensure that the Courant Criterion is satisfied by writing

$$\Delta t = \frac{\Delta x_1}{(V+c)_1} = \frac{\Delta x_2}{(V+c)_2} = \ldots\ldots = \frac{\Delta x_n}{(V+c)_n} \tag{4.39}$$

ie the distance increment varies between individual pipes as a means of ensuring a constant time increment for the network simulation as a whole.

This solution is both convenient and simplistic. However, as shown in Figure 4.5, it is always necessary for any one pipe to have at least one Δx increment, ie to have two nodes, one at entry and one at exit. If a pipe network consists of a number of long pipes with some very short side branches, possibly leading to storage tank entry valves or pump discharge connections, then this solution can become unduly costly and slow if the short pipe lengths are allowed to determine the time step for the network as a whole.

For example if Δx_S is the length of a short stub pipe serving as a tank entry connection from a long manifold distribution gallery, then the number of nodes in the gallery can easily become excessive if the system time step is based upon Δx_S.

Doubling the number of nodes in a particular pipe length quadruples the number of calculation steps necessary to achieve a given total simulation time as the distance increment affects the time step magnitude. In this particular case the 'art' of modelling dictates that the stub pipe be incorporated into a boundary condition that describes the flow condition in the distribution gallery at the point where flow is both diverted into the stub pipe, and subsequently the tank, and also continues to flow along the distribution main. Figure 4.6 illustrates this condition.

An alternative solution often proposed where variations in the calculated wave speeds are small between the network's constituent pipes is to recognise that the accuracy of the pipe and fluid data utilised to calculate the wave speed may not justify adherence to unique wave speed values having a small numerical spread. In such cases it is often sufficient to assume a constant wave speed for the system as a whole and then carry out a number

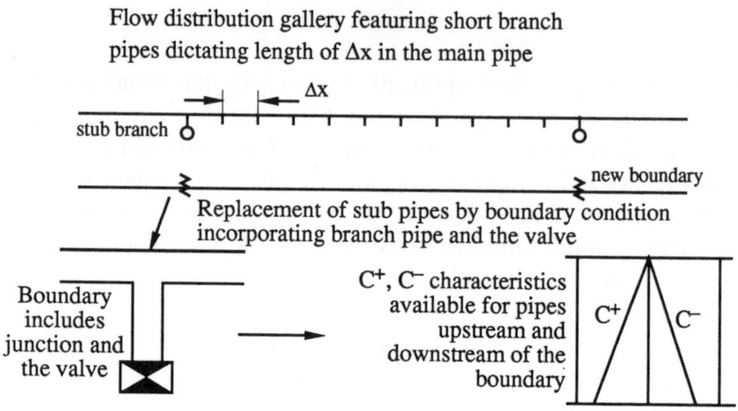

Figure 4.6 Illustration of technique to avoid the use of a short Δx for a complex network

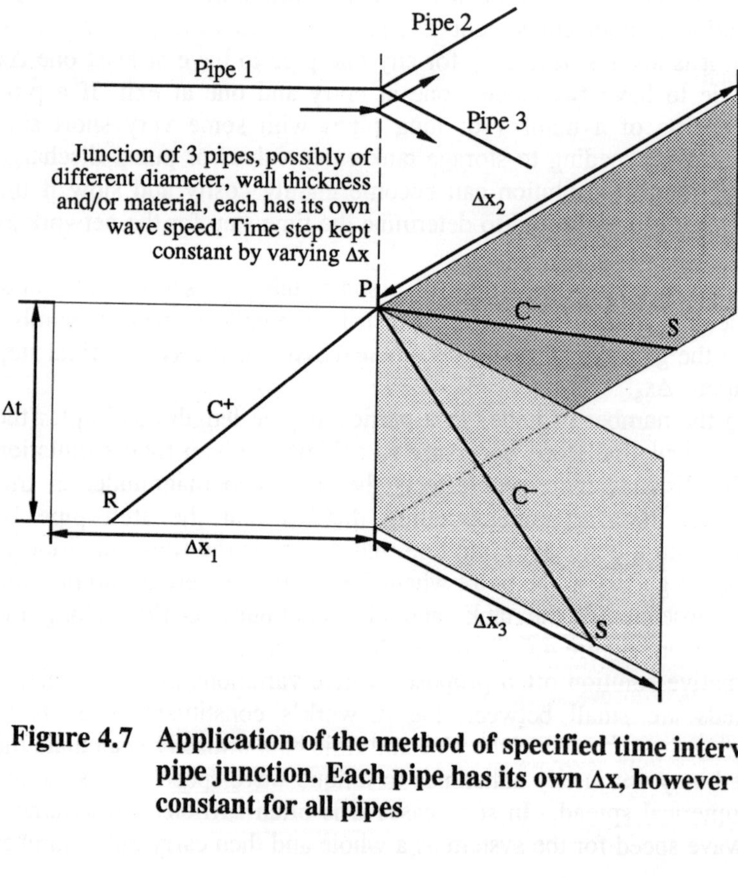

Figure 4.7 Application of the method of specified time intervals at a pipe junction. Each pipe has its own Δx, however Δt is constant for all pipes

of 'sensitivity' runs of the model to ensure that the approach does not lead to unacceptable variations in predicted pressures.

Interpolation may be seen as necessary for two reasons:

1. Variation in wave speed and flow velocity along any one pipe in the network dictates variable characteristic slopes for the nodal increments in that pipe. This implies that interpolation would be required to determine the base values at R and S so that the simulation may continue. This is often the case in gas and free surface flow transient analysis, Figure 4.4 illustrated this case.

2. Variation in wave speed between the individual pipes in a network, possibly due to changes in diameter, wall thickness or material properties, result in the highest wave speed dictating the magnitude of the time increment. In this case many pipes in the system may require analysis with a time step considerably smaller than that dictated by the Courant Criterion, resulting in the introduction of interpolation, Figure 4.7.

Both these applications of interpolation should be avoided, or at least minimised by careful choice of internodal distance increment, Δx. However this may not always be possible and alternative, more accurate interpolation methods may be necessary to replace the simplistic linear interpolation normally used and illustrated by the development of equations 4.21 to 4.29 above.

4.5 Improved interpolation methods

Figure 4.8 illustrates the time line interpolation technique applied to the second case above, ie where the wave speed is much greater than the local flow velocity. In this case each successive time increment has the same value and hence the base values at R and S, now located on the vertical time line through A and B, $\pm\Delta x$ on either side of the calculation node P respectively, may be found by linear interpolation between the known conditions at these nodes at the time coordinates bracketing the intersection points R and S.

Referring to Figure 4.8, the interpolation involves a calculation of the degree of interpolation within a time step, $\varepsilon\Delta t$, and the number of time steps 'back' to the time step that includes the intersection points R and S, defined by previous authors as the reachback number, m, Vardy (1976), Goldberg and Wylie (1983). If c is the constant wave speed for the system it follows from geometrical considerations that

$$\frac{c\Delta t}{\Delta x} = \frac{1}{m + \varepsilon} \tag{4.40}$$

and, utilising a linear interpolation, or a first degree Lagrangian Polynomial, then the interpolated base values at R and S may be expressed as

$$Z_R = Z_{A'} + (Z_{A''} - Z_{A'}) \frac{t_R - t_{A'}}{t_{A''} - t_{A'}}$$

$$Z_S = Z_{B'} + (Z_{B''} - Z_{B'}) \frac{t_S - t_{B'}}{t_{B''} - t_{B'}}$$

where Z stands for either V, c, p or H and A′, A″ are the nodes on the time line that straddles the intersection point R; similarly B′, B″ are the nodes on the time line that straddles the intersection point S. One important point worth emphasising here is that the time increment $t_P - t_R$, or $t_P - t_S$, replaces the single Δt in the friction and slope terms of the C^+ and C^- equations set out above, ie equations 4.13 and 4.14, and their respective V-H equivalents.

The time line interpolation technique for the constant wave speed case has been previously described by Vardy (1976) and Goldberg and Wylie (1983).

Application of the time line interpolation technique to the more interesting case where both V and c are comparable and where consecutive time steps may well vary as the maximum values of V and c along the pipe vary in response to changes in flow conditions, is more complex. In many respects the time line interpolation technique in this case becomes a special case of the generalised grid illustrated in Figure 4.1, with the x location defined by the vertical time lines. All points on the vertical time line A′A″ in Figure 4.9 generate a C^+ and C^- characteristic indicating the mode of transient propagation upstream or downstream from that location and forward in time. The time line interpolation therefore depends upon identifying the space-time locations that generate the C^+ and C^- that intersect at P, Figure 4.9.

The Δt applicable between P and C is based on the maximum values of c and V along the pipe at time t. Thus for all C^+ and C^- characteristics it follows that

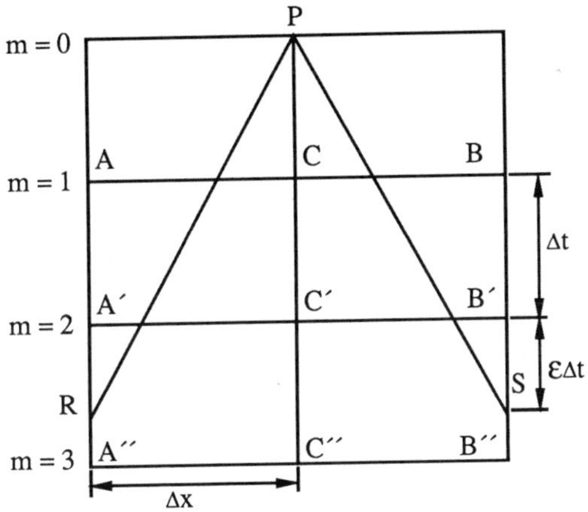

Figure 4.8 Schematic representation of time line interpolation when c>>V and constant along a pipe. The reachback number, m, relates to the interpolation factor defined by equation 4.33 (Goldberg and Wylie, 1983)

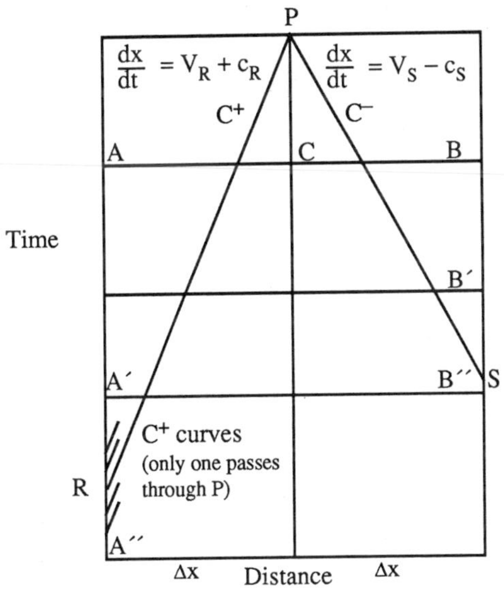

Figure 4.9 Schematic representation of time line interpolation for the general case when V and c are comparable and variable

$$\frac{\Delta t^+}{\Delta x} = \frac{t_P - t_R}{\Delta x} = \frac{1}{V_R + c_R} \tag{4.41}$$

$$\frac{\Delta t^+}{\Delta x} = \frac{t_P - t_S}{\Delta x} = \frac{1}{V_S - c_S} \tag{4.42}$$

Concentrating on the C^+ characteristic to exemplify the technique it follows that, as

$$\Delta t^+ = t_P - t_R \tag{4.43}$$

$$t_R = t_{A'} + (t_{A''} - t_{A'}) \frac{V_{A'} - V_R}{V_{A'} - V_{A''}} \tag{4.44}$$

$$t_R = t_{A'} + (t_{A''} - t_{A'}) \frac{c_{A'} - c_R}{c_{A'} - c_{A''}} \tag{4.45}$$

Due to the form of the above equations an iterative solution is required:

1. Utilise V_C and c_C to indicate likely A′A″ time increment, calculate the reachback number as

$$\Delta t^+ = t_P - t_R = t_P - \frac{\Delta x}{V_C + c_C} \tag{4.46}$$

therefore the reachback number, m, is the integer value of $\Delta t^+/\Delta t$ and

$$\varepsilon = \frac{\Delta t^+ - m\Delta t}{\Delta t}$$

2. Identify $t_{A'}$ and $t_{A''}$ and hence calculate conditions at R, and, by a similar process conditions at S, and then proceed to determine the conditions at P via the characteristic equations, noting again that the Δt term in these expressions, equations 4.13 and 4.14, is the time difference between P and R, and P and S, respectively.

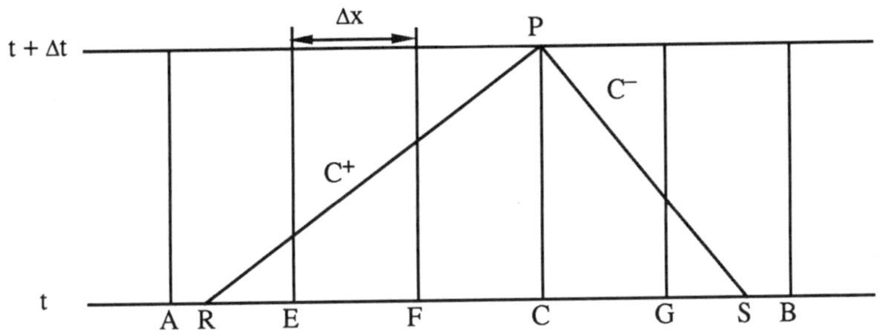

igure 4.10(a) **Illustration of the use of a decreased Δx to improve the prediction in zones of high wave speed. This improvement depends upon the base conditions being accurately known from A to B at time t (Vardy, 1976)**

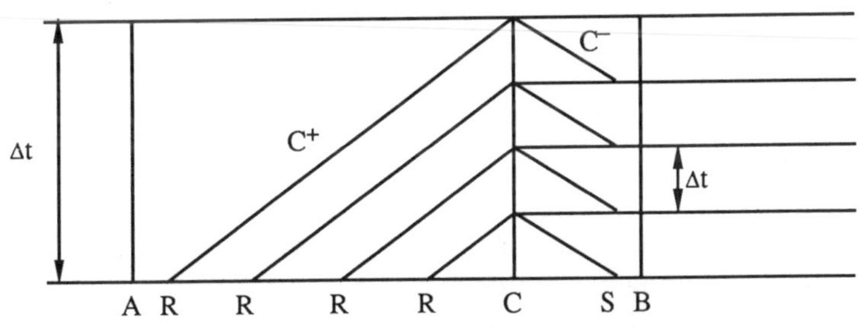

'igure 4.10(b) **Schematic representation of the interpolations required at a boundary if the time steps differ. Note the possibility of large errors at positions of R closest to C (Vardy, 1976)**

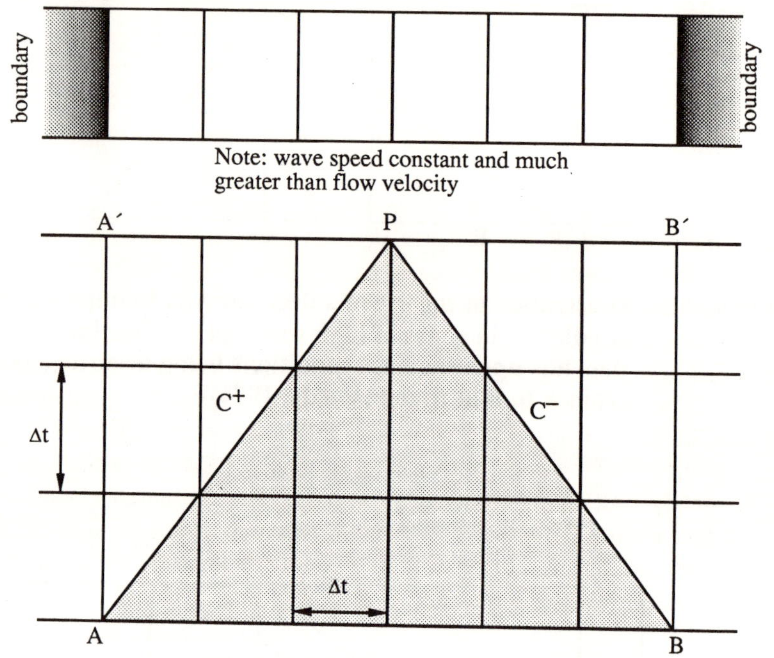

Figure 4.11 Zone APB indicates the limit of calculation within the method of specified time intervals without reference to the boundary conditions at AA´ and BB´ (Chapter 5 illustrates a range of suitable boundary conditions)

Spatial interpolation techniques that allow calculations outside the one Δx limit either side of the calculation node P have also been proposed, Vardy (1976). While these techniques are an aid in overcoming the interpolation difficulties discussed above, they do introduce complications at system boundaries. Figure 4.10 illustrates this technique and the 'time mismatch' at boundaries.

4.6 Setting up the base conditions for a transient analysis

It will already be apparent that the techniques described all depend upon the flow conditions being known at all points along all the pipes within the network at time zero. This implies that the transient calculation process must be proceeded by a steady state analysis of the network to determine these base conditions.

For full bore flow liquid, or gas flows, the steady state initial condition can be zero flow, provided that the pressures or head values along the system are known. In free surface flow situations this presents a difficulty normally overcome by assuming an initial flow presence in the channels.

The steady state analysis of networks is well documented elsewhere and is straightforward provided that accurate representations of flow frictional and separation losses are available. An interesting test of a method of characteristics transient analysis is the accuracy with which the model returns the system to its initial flow condition following the passage of a transient caused by an excursion in some boundary condition.

Once the initial conditions along the network have been determined the transient analysis may proceed, initially by determining the base conditions at all of the points identified as R and S and then calculating the time step to be used by searching for the maximum values of the local flow velocity and wave speed at each node along each pipe in the network. Reference to Figure 4.4 will indicate that at the first time step the conditions of velocity, pressure and wave speed at all internal nodes may be determined without reference to the boundary conditions, however beyond the first time step the relationships at each of the system boundaries must be incorporated. This will be discussed in Chapter 5.

In the case of time line interpolation it is obvious that the equations already developed apply to time zero, all the individual nodal point values of the velocity, pressure and wave speed variables before this time having the same values as those determined at time zero as the flow has been steady up to that time.

4.7 Frictional representation

Frictional representation remains one of the areas within transient modelling that is generally unsatisfactory. In equation 4.1 the general form of Darcy's equation was introduced in the form fV|V|/2m. The absolute sign ensures that frictional forces oppose time dependent motion and the hydraulic mean depth allows the general application of the relationship to closed conduit or free surface flows. However the friction factor f is normally determined from the Colebrook White relationship, in turn based upon a considerable body of fundamental research all undertaken under steady fully developed flow conditions. It is this that causes the difficulty in the modelling of transient frictional resistance. During the periodic reflections of a pressure transient within a pipe network the local flow will reverse in response to the passage of the transient. Any semblance of a fully developed flow condition disappears and the application of a steady flow friction factor can therefore readily be seen to be at best an approximation. While attempts have been made to address this difficulty, with some success in the laminar flow regime at the expense of model simplicity, Zielke (1968), no advance in the turbulent flow regime can be claimed, Shuy and Apelt (1983).

However it must be stressed that while frictional damping underestimation is common in all transient modelling the effect is to slightly overestimate peak pressures or column separation existence times so that the problem is not considered to be of major significance. One result of this difficulty has been the use generally of a fixed friction factor rather than introducing locally calculated friction factors dependent upon local flow Reynolds Numbers, a minor simplification to the model.

Generally frictional effects act to damp the pressure fluctuations following pressure transient propagation. However in high frictional loss systems with a low initial flow rate the regain of frictional head following a flow stoppage may exceed the transient propagated. This effect, known as 'line packing', is automatically included in the method of characteristics solution provided that the initial flow conditions along the network were correctly represented; this effect was previously demonstrated by Figure 1.5 where frictional pressure recovery was clearly predicted.

In the application of the method of characteristics to surface wave propagation in partially filled pipeflow the representation of friction becomes important. Generally the same arguments concerning the use of steady flow based friction remain, with a choice between the use of an equivalent Colebrook White friction factor or a traditional open channel approach based on the Chezy coefficient, Swaffield and Bridge (1983).

4.8 Network solutions

The presentation of the method of characteristics given in this chapter has concentrated upon the development of the finite difference equations necessary to solve for unknown flow conditions at any node along a pipe, duct or channel provided that conditions at adjacent nodes at an earlier time step were known. Figure 4.11 illustrates this zone of prediction for a simple pipe. It will be appreciated that the solution so far presented ceases at a time equal to (n $\Delta t/2$), where n is an even number of Δx increments making up the pipelength and Δt is the time step calculated in accordance with the Courant Criterion. The failure of the solution to proceed beyond this point is due to the fact that adjacent node conditions are progressively unknown as time increases due to the lack of any information as to the conditions at the upstream and downstream boundaries of the pipe. In order for the solution to proceed it will be necessary to provide this boundary information in a form compatible with the single C^+ or C^- characteristic that is always available at a pipe entry or exit.

Boundary condition representation within the method of characteristics is the most challenging aspect of the technique as it often requires a fundamental understanding of the fluid mechanics of the interface in order to generate an equation that both describes accurately the boundary conditions and also is compatible with the available characteristic.

Examples of boundary conditions will be presented in Chapter 5, however in general they may be sub-divided into three broad categories:

1. Passive boundary conditions that arise as a result of the design of the system, for example junctions of two or more pipes, constant pressure reservoirs, dead-ended pipes, open discharges or changes in pipe cross section, material or wall thickness.

2. Active boundary conditions that represent equipment connected to the system, for example valves, pumps or turbines.

3. Boundary conditions that arise as a result of the propagation of transients within the network and which are therefore not necessarily identified as being present at the design stage or by a simple examination of the system layout. In this category fall the boundary conditions needed to represent column separation, trapped air or gas release due to low system pressures. This category may also include 'moving' boundary conditions, for example the interface formed by

the representation of a train moving through a tunnel or the discontinuity between free surface and full bore flow in an initially partially filled conduit under surcharging conditions.

A further category of boundary conditions representing the interface to pressure transient control and suppression devices, such as relief valves and air chambers will be presented in Chapter 6.

In general suitable boundary equations will link either flow rate or pressure to time. In the case of some equipment boundary conditions it may be necessary to relate these variables via a monitoring of the equipment over the time period considered, for example valve position vs. time data will be linked to valve 'pressure loss' vs. position data in order to provide a suitable boundary condition to be solved with the pressure-velocity C^+ or C^- characteristic available at either pipe entry or exit. In other cases the required boundary equation will only be activated if the indicated conditions in the pipeline reach certain 'trigger' levels, for example the representation of a pressure relief valve will need to be present in the numerical model and will lie dormant until required by the solution.

Given the convention adopted in this text that distance increases in the initial flow direction it follows that a C^- characteristic will always be available at pipe entry and a C^+ characteristic will always be present at pipe exit. (It should be noted here that 'pipe entry' is assumed synonymous with inflow and a zero value of the distance measured along the pipe, this linkage is not necessary as inflow to a system can be regarded as a normal negative flow, see Chapter 7.) While the development of suitable boundary conditions provides the main area of interest for the numerical modeller utilising the method of characteristics, it has been necessary to provide a basic development of the method from the base equations of momentum and continuity and to demonstrate that the solutions generated are general in application. This has been the general objective of this chapter, together with an emphasis upon the commonality of the solution across both full bore liquid and gas flows and free surface conditions. Similarly this chapter has defined one of the main difficulties with the application of the method, namely the need to interpolate to obtain the base conditions for the characteristics, as well as outlining some possible improvements to limit the inherent rounding and dispersion errors.

The following chapters will detail suitable boundary conditions and the methodology required in order to write computer programs to model a range of transient flow conditions.

Chapter 5:

Pressure transient network boundary conditions

Chapter 4 has both outlined the solution of the momentum and continuity equations defining pressure transient propagation by means of the method of characteristics and identified the method's limitations in terms of the need to provide suitable equations to express conditions at the pipe boundaries.

Boundary equations must be defined at each pipe entry and exit and must therefore include the treatment of pipe junctions, changes in pipe material or dimensions, defined as passive boundaries, or interfaces with equipment such as pumps, valves or turbines, defined as active boundaries. It is the change in conditions at a boundary that initiates the transient to be analysed by the method of characteristics.

In addition to active and passive boundary conditions there exists a third category of boundary condition that arises as a result of the transient propagation through the pipe network, for example the constant pressure zone boundary invoked during the vapour pocket formation and column separation that may occur as a result of transient propagation or reflection, or the introduction of surge relief device boundary equations dependent upon the pressures reached.

In order to undertake the modelling of real networks it is sometimes also necessary to develop 'special' boundary conditions tailored to particular pipeline configurations in order to adequately represent the operation of the network while retaining a reasonable computing time. Examples will be presented where the modeller must utilise a knowledge of the system to generate a simulation that is both useful and computationally efficient.

5.1 Definition of boundary equations within the method of characteristics

Figure 5.1 illustrates the condition that one C^+ or C^- characteristic is always available at either pipe entry or exit. Thus an equation of the form 4.13 or 4.14 will exist at a boundary to link known conditions at time t to unknown conditions at time $t + \Delta t$. Therefore a further equation, or set of equations, is needed in terms of the velocity, pressure and possibly wave speed at the boundary in order for the solution to proceed. It is the choice and development of these boundary equations that presents the greatest interest and challenge in the application of the method of characteristics to the study of unsteady flow transient problems.

Normally a $V = \phi(p)$, or a V or $p = \phi(\text{time})$ expression is utilised at the boundary, however a range of other possibilities exist, for example in the case of valve closure the mode of closure will be important, similarly in the case of pump shut down the speed-time relationship will need to be referred to. Boundary conditions can therefore become complex in themselves, particularly if the boundary to be represented is itself in motion through the system, for example in the case of a train moving through a tunnel in an air transient application, or a moving hydraulic jump in an open channel application, both mentioned in Chapter 4. In these cases it is necessary to trace the movement of the boundary through the x-t plane and considerable care has therefore to be taken in determining the correct base conditions for the characteristic equations.

This chapter will set out a range of boundary equations, varying in degrees of complexity, that represent all three boundary condition categories defined above. For simplicity each boundary condition will be presented to the same pattern:-

1. Identification of the boundary condition.

2. Development of a suitable boundary equation to represent the physical boundary, together with a statement of any time dependency or need for any other time related data, eg valve closure mode.

3. Presentation, in as near as possible computing format, the solution of the C^+ or C^- characteristic equation with the defined boundary condition.

Equations 4.13 and 4.14 will be utilised to represent the C^+ and C^- characteristic expressions

$$V_P = K1 - K2\, p_P$$

$$V_P = K3 + K4\, p_P$$

where the definitions of K1 to K4 are given in equations 4.15 to 4.18. It is stressed that this form of the characteristic equations is in every way equivalent to the V-H expressions, the use of the relevant equations to define K1 to K4 being the only modification necessary between the two sets of defining units. (It therefore follows that pressure, p, and head, H, are interchangeable in all the equations presented in this chapter unless specifically mentioned.)

A computing style double subscript array notation will be chosen for the presentation of boundary conditions, with a particular pipe being identified as J, while a particular node will be identified as I; thus for example the pressure at the 6th node along the 4th pipe in a network to be calculated at time $t + \Delta t$ becomes p(J,I), or p(4,6).

Pipe nodes will be identified from 1 at pipe entry, ie the upstream boundary defined at $x = 0$, usually chosen to correspond to the initial flow direction, to n+1 at pipe exit. Each pipe will thus have n sections, each Δx in length, the length of Δx being chosen to comply with the Courant Criterion and the need to retain a constant Δt for the whole network.

Similarly the R,S notation adopted in Chapter 4 to indicate the base conditions for the C^+ and C^- characteristics will be maintained. However for ease of manipulation these points will be referred to in terms of the location of the point currently being calculated at time $t + \Delta t$. Thus the pressure and velocity values at R, the base of the C^+ characteristic passing through point P, ie node I in pipe J in Figure 5.1, will be identified as V_R(J,I) and p_R(J,I). The associated values of K1 to K4 will also have to be designated in an array format, becoming K1(J,I) etc. In cases where the wave speed far exceeds the local flow velocity and interpolation is not necessary it may be seen that, as mentioned in Chapter 4, values at A and R coincide, however the notation introduced above will be retained for generality in these cases.

The C^+ characteristic equation may therefore be restated as

$$V_P(J,I) = K1(J,I) - K2(J,I)\, p_P(J,I) \qquad (5.1)$$

and the C^- characteristic as

$$V_P(J,I) = K3(J,I) + K4(J,I)\, p_P(J,I) \tag{5.2}$$

where

$$K1(J,I) = V_R(J,I) + \frac{p_R(J,I)}{\rho c_R(J,I)} + g\sin\alpha_R \Delta t - \frac{f_R(J,I) V_R(J,I) |V_R(J,I)| \Delta t}{2 m_R(J,I)} \tag{5.3}$$

$$K2(J,I) = \frac{1}{\rho c_R(J,I)} \tag{5.4}$$

$$K3(J,I) = V_S(J,I) - \frac{p_S(J,I)}{\rho c_S(J,I)} + g\sin\alpha_S \Delta t - \frac{f_S(J,I) V_S(J,I) |V_S(J,I)| \Delta t}{2 m_S(J,I)} \tag{5.5}$$

$$K4(J,I) = \frac{1}{\rho c_S(J,I)} \tag{5.6}$$

The hydraulic mean depth, m, for full bore pipe flow, diameter D, is D/4 and f is the appropriate friction factor, either Reynolds Number dependent or assumed constant.

5.2 Pipe and network application of the method of characteristics

Referring to the idealised pipelength of Figure 5.1 the modelling of transient propagation may be seen to follow the steps set out below

1. Set up steady flow conditions at time zero throughout the network. This involves a knowledge of the initial flow and valve or pump settings, friction losses etc. The number of nodes per pipe are selected based on the pipelengths and wave speeds so that the Courant Criterion is maintained.

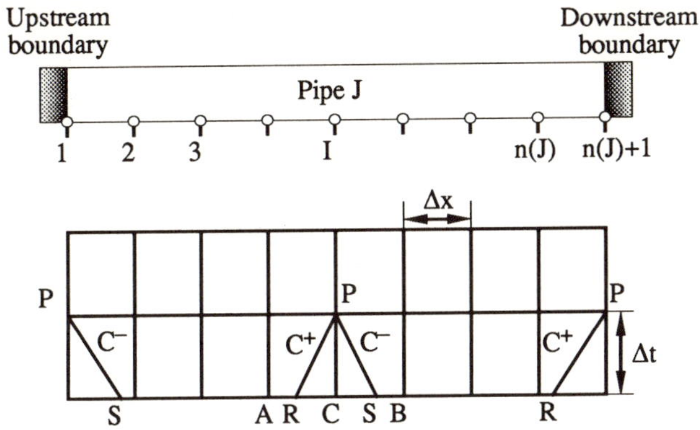

Figure 5.1 Idealised pipe J in a network, with nodes identified from I = 1 to n(J)+1. Base conditions at R and S identified by the calculation node P, ie for node I, $V_R(J, I)$, $V_S(J, I)$

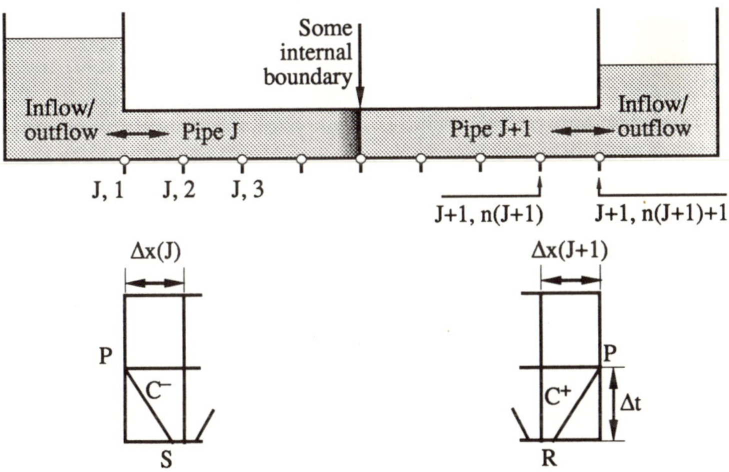

Figure 5.2 Boundary conditions provided by a constant pressure reservoir connected to the upstream or downstream pipe end

2. Allocate values to all the pressure, velocity and wave speed arrays needed at time zero. This may necessitate interpolation to give values at the various R and S base points identified, as well as at the nodal points, ie A,B,C, in Figure 5.1. Note that it follows from the notation introduced that there are no values for K1 and K2 for I = 1, ie prior to pipe entry, nor values for K3 and K4 at I = n+1, ie originating beyond pipe exit. This process is probably best achieved with a separate subroutine or procedure invoked prior to the calculations at each time step that checks for the need to interpolate, based upon relationship between the chosen and Courant Criterion time step, and then determines values for the V_R and p_R arrays accordingly.

3. Once all the base value calculations have been completed at I = 1 to I = n+1 for all the pipes J in the network the values of the unknown pressure and velocity at time $t + \Delta t$ at all the 'internal' nodes I = 2 to I = n may be calculated by the simultaneous solution of equations 5.1 and 5.2.

4. Values of the unknown pressure and velocity at time $t + \Delta t$ at the boundary nodes I = 1 and I = n+1 may now be calculated from a solution of the appropriate C^+ or C^- characteristic with the boundary equation identified. This process is best achieved by the use of specific subroutines related to each of the boundary conditions to be found in the network. This technique will also facilitate the inclusion of 'dormant' subroutines that are only activated if the conditions in the network reach certain preset levels, for example the introduction of vapour pressure as a boundary condition if pressure in the system falls to the fluid vapour pressure or the introduction of a particular surge relief device, also dependent upon local pressure levels.

5. Once all the pressure and velocity values along the pipe have been calculated at time $t + \Delta t$ the whole process from step 2 above may be repeated. In order to save storage in computing it is common to overwrite the V_R, p_R arrays with the new base conditions calculated utilising the nodal values just determined from I = 1 to n+1 for each particular pipe. This stage in the model is also appropriate for the writing to storage files of the pressure and velocity results at a preselected number of nodes for later scrutiny or graphical presentation. The size of the time step commonly required and the number of calculation nodes in a network normally precludes the

storage of data relating to every node at each time step and it is often appropriate to arrange to store nodal data at time intervals of several time steps. The techniques chosen here depend upon the individual case.

6. The process outlined above continues to cycle until the total simulation time requested is reached.

5.3 Boundary conditions

A range of boundary conditions will be presented following the methodology introduced above and utilising the appropriate C^+ or C^- characteristic, equations 5.1 and 5.2.

5.3.1 Reservoirs terminating a pipeline

The boundary formed by a reservoir may be termed a passive boundary and can be incorporated into the model based on the design and layout of the system. The presence of a constant pressure reservoir may also be used to form an artificial termination to a system if further upstream reflections are unlikely to return within the simulation time of the model.

At a constant pressure reservoir, as shown in Figure 5.2, the boundary equation is supplied by the assumption that the pressure at the pipe to reservoir interface remains unchanged as the transients propagate and are reflected within the network, an assumption already discussed in Chapter 1 in the description of the periodic pressure and velocity oscillations within a simple pipeline.

Referring to Figure 5.2

$$p_P(J,1) = p_{Res} \tag{5.7}$$

and hence

$$V_P(J,1) = K3(J,1) + K4(J,1)\, p_{Res} \tag{5.8}$$

for the upstream reservoir, when equation 5.7 is solved with the C^-characteristic, equation 5.2.

Similarly the downstream boundary may be represented as

$$p_p(J+1,n(J+1)+1) = p_{Res} \tag{5.9}$$

and hence

$$V_p(J+1,n(J+1)+1) = K1(J+1,n(J+1)+1) - K2(J+1,n(J+1)+1)\, p_{Res} \tag{5.10}$$

The equations above assume a zero separation loss at entry or exit from the reservoirs and neglect the velocity pressure, or head, loss at these boundaries. If these losses are included then the equations may be modified

$$p_p(J,1) = p_{Res} - (M+k)\, 0.5\, \rho V_p(J,1)|V_p(J,1)| \tag{5.11}$$

where M = 1 for flow out of the reservoir, and
M = 0 for flow into the reservoir

and k is the appropriate separation loss coefficient, having two values corresponding to the flow direction at the reservoir to/from pipe interface.

Equation 5.11 is solved with the C^- characteristic so that

$$V_p(J,1) = K3(J,1) + K4(J,1)(p_{Res} - (M+k)\, 0.5\, \rho V_p(J,1)|V_p(J,1)|) \tag{5.12}$$

Values of $V_P(J,1)$ are monitored at each time step to determine the time at which the value of M and the separation loss coefficient k, is to be changed from that corresponding to flow out of the reservoir to that corresponding to negative flow into the reservoir. (Positive flow is defined as that flow in the increasing distance, x, direction.) Note that the – sign together with the absolute value of $V_P(J,1)$ will ensure that the correct differential is maintained between the reservoir pressure and $p_P(J,1)$, ie with reverse flow $p_P(J,1)$ exceeds the reservoir pressure.

For the downstream reservoir the process may be repeated with the difference that the defining pressure equation 5.11 becomes

$$p_p(J+1,n(J+1)+1) = P_{Res} + (M+k)\, 0.5\, V_p(J+1,n(J+1)+1)\, |V_p(J+1,n(J+1)+1)| \tag{5.13}$$

which may then be substituted into the C^+ characteristic, equation 5.1.

In many practical applications the equations above may be simplified by using the value of the appropriate velocity calculated at the previous time step to represent the separation and velocity head terms in equations 5.12 and 5.13.

5.3.2 *Valve terminating a pipeline*

The operation of a valve terminating a pipeline is the most common source of transient propagation within a network. It is therefore an active boundary condition and will be presented next so that simple transient models may be constructed and run at this stage in the text.

Referring to Figure 5.3, modelling the transient propagation from a valve terminating a pipeline requires the solution of the available C^+ characteristic with the valve discharge coefficient at any particular valve setting. Thus the solution requires information as to the valve position, and hence discharge coefficient, at each time step.

(a) Valve positioned at the downstream end of a pipe. The C^+ characteristic at the downstream end of pipe J connected to the valve, Figure 5.3, may be expressed as

$$V_p(J,n(J)+1) = K1(J,n(J)+1) - K2(J,n(J)+1)\, p_p(J,n(J)+1)$$

hence

$$p_p(J,n(J)+1) = \frac{K1(J,n(J)+1)-V_p(J,n(J)+1)}{K2(J,n(J)+1)} \qquad (5.14)$$

Rearranging the valve loss equation

$$Q = C_d A \sqrt{\frac{2\Delta P}{\rho}}$$

where ΔP is the pressure loss across the valve, the valve discharge coefficient C_d at any valve setting may be expressed as

$$C_d = \frac{V_P(J,n(J)+1)}{V_0}\sqrt{\frac{p_o - p_{Res}}{p_P(J,n(J)+1) - p_{Res}}} \tag{5.15}$$

for flow from the pipe into the reservoir, or

$$C_d = \frac{V_P(J,n(J)+1)}{V_0}\sqrt{\frac{p_{Res} - p_0}{p_{Res} - p_P(J,n(J)+1)}} \tag{5.16}$$

for flow from the reservoir into the pipe.

In each case V_0 and p_0 refer to the valve fully open flow conditions at the pipe/valve interface. The reservoir pressure may of course be set to atmospheric, ie zero gauge, if the valve has an open discharge. The technique required to determine the appropriate valve discharge coefficient is as follows, and is shown schematically in Figure 5.4,

1. Identify the valve position corresponding to the known time, for either an opening or closing action.

2. Transfer the calculated valve position to the discharge coefficient curve and determine discharge coefficient corresponding to this position, again for either an opening or closing motion.

In practice the valve position and discharge data will probably be held in a set of arrays so that a linear interpolation routine will be necessary to extract the position based on a known time and then the discharge coefficient based on position. (In order to develop simple models it is useful to represent combined time-discharge coefficient data as a straight line.)

For flow into the reservoir, equation 5.15 may be rearranged to include $p_P(J,n(J)+1)$, by reference to equation 5.14

$$V_P(J,n(J)+1)^2 = C_d^2 V_0^2 \left(\frac{p_P(J,n(J)+1) - p_{Res}}{p_0 - p_{Res}}\right)$$

$$V_P(J,n(J)+1)^2 + K5\left[\frac{V_P(J,n(J)+1)}{K2(J,n(J)+1)}\right] - K5\left[\frac{K1(J,n(J)+1)}{K2(J,n(J)+1)} - p_{Res}\right] = 0 \tag{5.17}$$

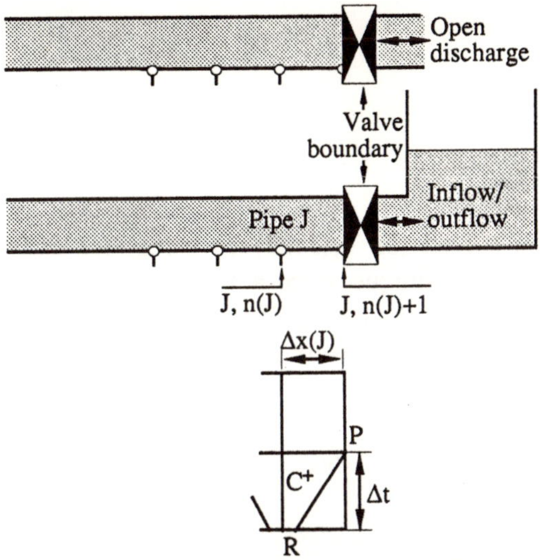

Figure 5.3 Valve boundary conditions, either open discharge to atmosphere or inflow/outflow at a reservoir

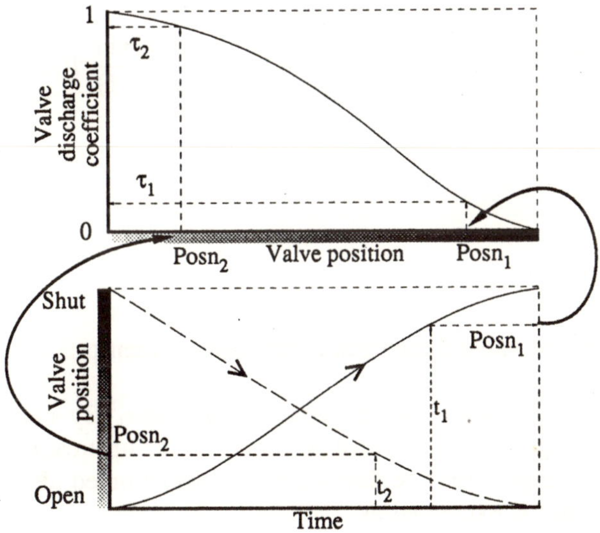

Figure 5.4 Technique used to cross reference the valve closure position to the valve discharge coefficient during valve opening or closing

where

$$K5 = \frac{C_d^2 V_0^2}{p_0 - p_{Res}}$$

Equation 5.17 is solved for $V_P(J,n(J)+1)$ using, for example a Newton Raphson or bisection technique since a very close approximation for the velocity may be obtained from extrapolation of values calculated at previous time steps. The value of the pressure at the valve boundary then becomes

$$p_P(J,n(J)+1) = \frac{V_P(J,n(J)+1)^2}{K5} + p_{Res}$$

Similarly for flow from the reservoir into the pipe, the equivalent expression may be shown to be

$$V_P(J,n(J)+1)^2 - K5\left[\frac{V_P(J,n(J)+1)}{K2(J,n(J)+1)}\right] + K5\left[\frac{K1(J,n(J)+1)}{K2(J,n(J)+1)} - p_{Res}\right] = 0 \qquad (5.18)$$

where

$$K5 = \frac{C_d^2 V_0^2}{p_{Res} - p_0}$$

Note that if the valve discharges to atmosphere then $p_{Res} = 0$.

(b) Valve positioned at the upstream end of a pipe. By a similar process the boundary equations for a valve positioned at the upstream end of a pipe, ie at node 1 in pipe J, may be developed. It must be noted that the available C^- characteristic is expressed as

$$V_P(J,1) = K3(J,1) + K4(J,1)\, p_P(J,1)$$

and the valve discharge coefficient expression becomes

$$c_d = \frac{V_p(J,1)}{V_0}\sqrt{\frac{p_{Res} - p_0}{p_{Res} - p_p(J,1)}}$$

Thus, for flow into the pipe, the unknown velocity at node 1 may be determined from

$$V_p(J,1)^2 + K5\,\frac{V_p(J,1)}{K4(J,1)} - K5\left[p_{Res} + \frac{K3(J,1)}{K4(J,1)}\right] = 0$$

where

$$K5 = \frac{C_d^2 V_0^2}{p_{Res} - p_0}$$

and the unknown pressure, at that entry node, is given by

$$p_p(J,1) = \frac{V_p(J,1) - K3(J,1)}{K4(J,1)}$$

The equivalent expressions for flow from the pipe into the reservoir is

$$V_p(J,1)^2 - K5\,\frac{V_p(J,1)}{K4(J,1)} + K5\left[p_{Res} + \frac{K3(J,1)}{K4(J,1)}\right] = 0$$

where

$$K5 = \frac{C_d^2 V_0^2}{p_0 - p_{Res}}$$

and, as before, the unknown pressure is given by

$$p_p(J,1) = \frac{V_p(J,1) - K3(J,1)}{K4(J,1)}$$

The application of the terminal valve boundary equations, together with the reservoir boundary and internal nodal calculations allows many surge problems to be investigated. Figure 5.5 (a) illustrates the application of similar equations to the analysis of pressure surges in a multitank aircraft refuelling case.

5.3.3 Boundary formed by a valve at the junction of two pipes

Figure 5.6 illustrates a valve positioned at the junction of two pipes that may have different diameters, wall thicknesses or wave speeds. The Courant Criterion is maintained by allowing each pipe to have a different Δx, but a common Δt is used.

This boundary may be termed an active boundary capable of generating transients. There are four unknowns to be determined at time $t + \Delta t$, based on known conditions at time t, namely the velocity and pressure on either side of the valve, coordinates (J,n(J)+1) and (J+1,1) respectively.

The four available equations are:

1. The C^+ characteristic at the downstream end of pipe J, from (J,n(J)) to (J,n(J)+1)

$$V_p(J,n(J)+1) = K1(J,n(J)+1) - K2(J,n(J)+1)\, p_p(J,n(J)+1) \tag{5.19}$$

2. The C^- characteristic at the upstream end of pipe J+1, from (J+1,1) to (J+1,2)

$$V_p(J+1,1) = K3(J+1,1) + K4(J+1,1)\, p_p(J+1,1) \tag{5.20}$$

3. The continuity of flow through the valve at any time, which remains true during closure

$$Q = V_p(J,n(J)+1)\, A(J) = V_p(J+1,1)\, A(J+1) \tag{5.21}$$

where A() is the cross sectional area of each pipe.

4. The valve discharge coefficient, C_d, at each time step is determined by the technique outlined in Figure 5.4.

$$Q = C_d A \sqrt{\frac{2\Delta P}{\rho}} \tag{5.22}$$

Based on an initial flow from pipe J to J+1 the valve discharge coefficient equation may be rearranged to give

$$C_d = \frac{V_P(J,n(J)+1)}{V_0} \sqrt{\frac{\Delta P_0}{p_P(J,n(J)+1) - p_P(J+1,1)}} \tag{5.23}$$

where ΔP_0 is the valve fully open pressure loss from pipe J to pipe J+1 at an initial flow velocity, V_0, in pipe J.

From equation 5.23 the pressure loss across the valve is given by

$$p_P(J,n(J)+1) - p_P(J+1,1) = \frac{C_d^2 V_0^2}{\Delta P_0} V_p(J,n(J)+1)^2 = K5 V_p(J,n(J)+1)^2$$

and by combining equations 5.19, 5.20 and 5.21 the pressure loss may be shown to be

$$p_P(J,n(J)+1) - p_P(J+1,1) = -\frac{V_p(J,n(J)+1) - K1(J,n(J)+1)}{K2(J,n(J)+1)}$$

$$-\frac{1}{K4(J+1)}\left[\frac{A(J)}{A(J+1)} V_p(J,n(J)+1) - K3(J+1,1)\right]$$

so that a quadratic may be formed in terms of $V_P(J,n(J)+1)$

$$V_p(J,n(J)+1)^2 + K6\, V_p(J,n(J)+1) - K7 = 0 \tag{5.24}$$

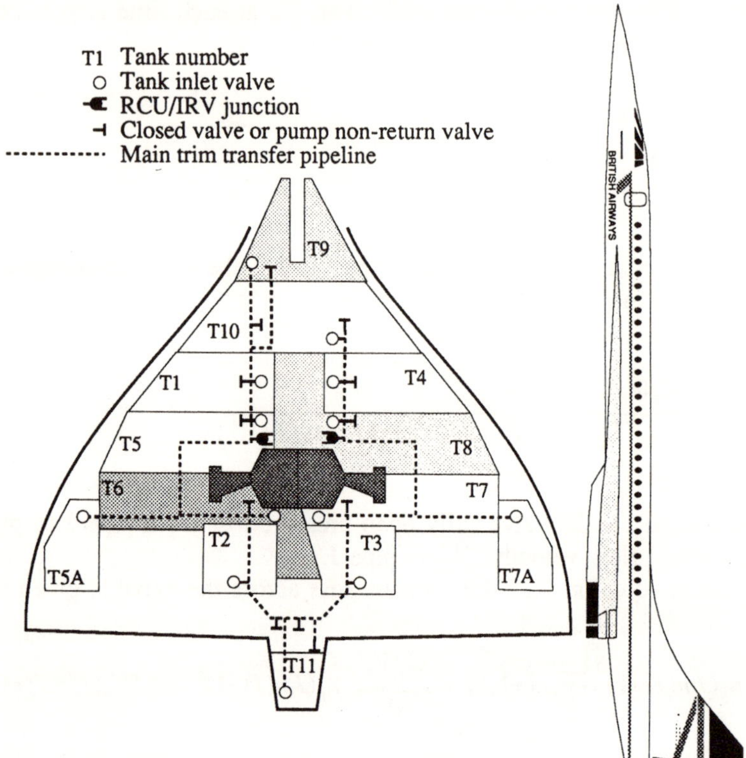

Schematic layout of an aircraft wing and fuselage fuel cells and route of the main trim transfer pipeline (mttp). Note that dead-ended pipes represent closed pipes or isolated pipes employed to link the mttp into the trim transfer or jettison systems or to connect port and starboard refuelling galleries

Figure 5.5(a) Illustration of the application of the closing valve boundary conditions to multiple valve closure during aircraft refuelling (Doyle and Swaffield (1972))

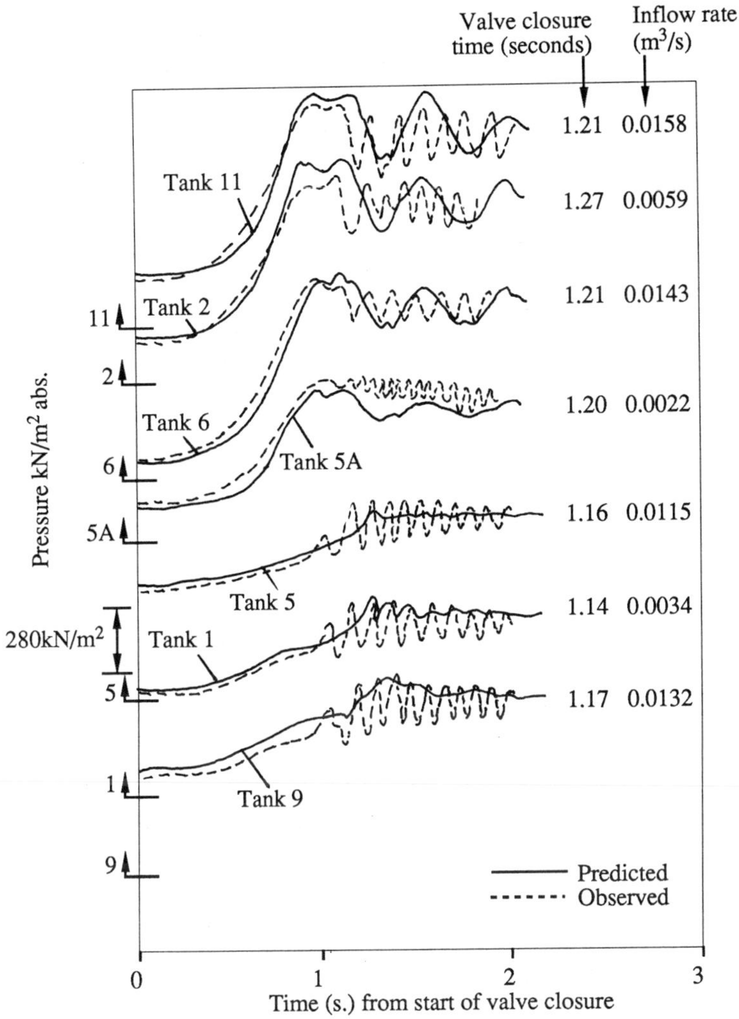

Pressure variations recorded at each tank inlet, port system, during simultaneous inlet valve closure

Figure 5.5(b) **Illustration of the application of the closing valve boundary conditions to multiple valve closure during aircraft refuelling (Doyle and Swaffield (1972))**

where

$$K6 = \frac{1}{K5}\left[\frac{1}{K2(J,n(J)+1)} + \frac{A(J)}{A(J+1)\,K4(J+1,1)}\right]$$

$$K4 = \frac{1}{K5}\left[\frac{K1(J,n(J)+1)}{K2(J,n(J)+1)} + \frac{K3(J+1,1)}{K4(J+1,1)}\right]$$

Once $V_P(J,n(J)+1)$ has been calculated the remaining unknowns may be found from equations 5.19 to 5.21.

In the case of an initial reverse flow, ie flow from pipe J+1 to pipe J at time zero an identical process may be followed if equation 5.23 is expressed as

$$C_d = \frac{V_P(J+1,1)}{V_0}\sqrt{\frac{\Delta P_0}{p_P(J+1,1) - p_P(J,n(J)+1)}} \tag{5.25}$$

where ΔP_0 and V_0 now refer to flow from pipe J+1 to pipe J. The reader is left to derive the equivalent expression to equation 5.24.

5.3.4 Boundary formed by a closed valve or dead-ended pipe

Figure 5.7 illustrates the boundary formed by a dead-ended pipe or a closed valve. In this case the available C^+ or C^- characteristic is solved with the condition that the flow velocity at the valve node is zero.

Thus for a dead end at node (J,n(J)+1)

$$V_P(J,n(J)+1) = 0 \tag{5.26}$$

and

$$p_P(J,n(J)+1) = \frac{K1(J,n(J)+1)}{K2(J,n(J)+1)} \tag{5.27}$$

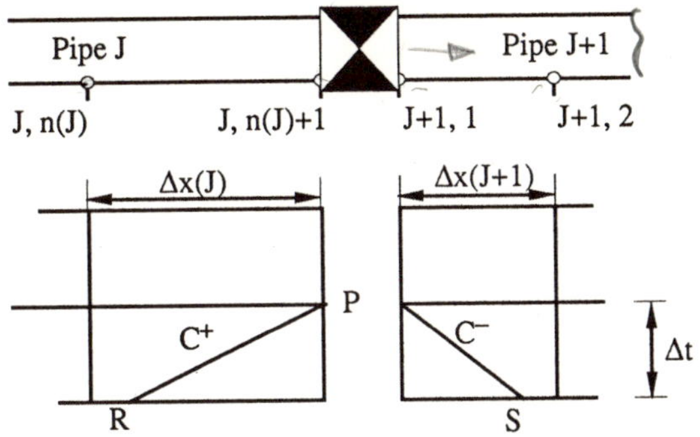

Figure 5.6 Valve located at the junction of two pipes

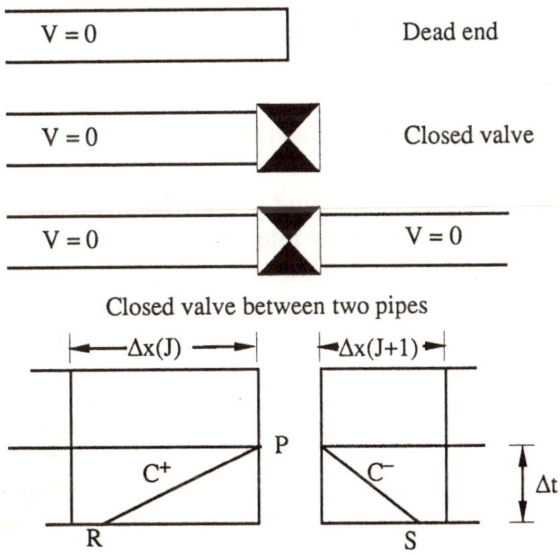

Figure 5.7 Closed valve or dead end. Boundary formed by setting flow velocity to zero. Note: only applies if pressure is above gas release or vapour pressure levels

Similarly for a dead end at node (J+1,1), eg following closure of a valve at entry to a pipe

$$V_P(J+1,1) = 0 \tag{5.28}$$

and

$$p_P(J+1,1) = -\frac{K3(J+1,1)}{K4(J+1,1)} \tag{5.29}$$

This simple boundary condition **ONLY** remains true so long as the pressure at the boundary remains above vapour or possibly gas release pressure. In cases where this criteria is not adhered to the fluid column separates from the closed valve face or pipe end and moves under the influence of the prevailing pressure gradients along the pipe and the boundary equation at the dead end is provided by the cavity pressure. This important case will be returned to in detail later as Column Separation.

5.3.5 Orifice between two pipes

This boundary may be considered passive as the details of the boundary are known purely in terms of flow rate and associated pressure loss. Figure 5.8 illustrates this special case of the inter pipe valve. The available equations are

1. C^+ characteristic at exit pipe J

$$V_P(J,n(J)+1) = K1(J,n(J)+1) - K2(J,n(J)+1)\, p_P(J,n(J)+1) \tag{5.30}$$

2. C^- characteristic at entry pipe J+1

$$V_P(J+1,1) = K3(J+1,1) + K4(J+1,1)\, p_P(J+1,1) \tag{5.31}$$

3. Continuity of flow between the two pipes

$$V_p(J,n(J)+1)\ A(J) = V_p(J+1,1)\ A(J+1) \tag{5.32}$$

4. Separation loss from pipe J to J+1 due to the local separation loss coefficient, k,

$$p_p(J,n(J)+1) = p_p(J+1,1) + 0.5\ k\rho V_p(J,n(J)+1)\,|V_p(J,n(J)+1)| \tag{5.33}$$

The use of an absolute value in equation 5.33 ensures that reverse flow is catered for in the boundary equations. Equation 5.33 forms the basis for a solvable expression

$$0.5k\rho V_p(J,n(J)+1)\,|V_p(J,n(J)+1)| + K6\ V_p(J,n(J)+1) - K7 = 0 \tag{5.34}$$

where

$$K6 = \frac{1}{K2(J,n(J)+1)} + \frac{A(J)}{K4(J+1,1)\ A(J+1)}$$

$$K7 = \frac{K1(J,n(J)+1)}{K2(J,n(J)+1)} + \frac{K3(J+1,1)}{K4(J+1,1)}$$

5.3.6 Boundary formed by the junction of two pipes

If two pipes are joined then changes in pipeline properties may result in a change in wave speed which will need to be represented in the model by this passive boundary, as shown by Figure 5.9.

Similarly changes in pipe cross section need to be modelled. Fundamentally the same as the previous boundary although it is sometimes possible, or acceptable, to neglect the local separation loss at the junction itself, initial steady state pressure losses being accommodated by distributing the junction loss through the friction terms.

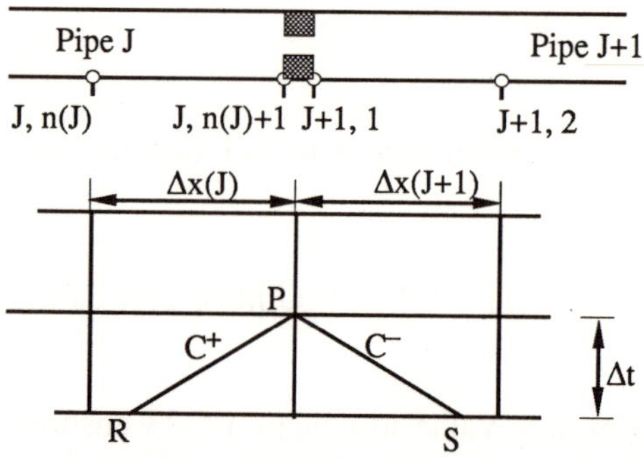

Figure 5.8 Orifice inserted in a pipe length forms two pipes in the network, although Δx may remain the same for both pipes

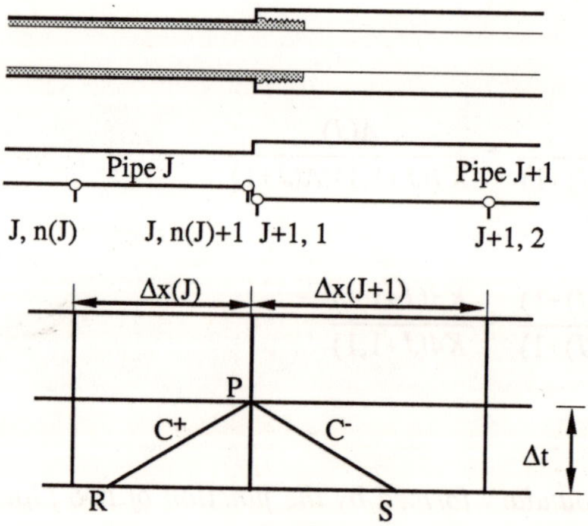

Figure 5.9 Junction of two pipes. The important change is in the wave speed, affecting Δx. Either diameter or wall thickness and/or material changes may be dealt with. Common pressure at the junction and continuity of flow provide the necessary boundary conditions

The available equations are as follows

1. C^+ characteristic at exit pipe J

$$V_p(J,n(J)+1) = K1(J,n(J)+1) - K2(J,n(J)+1)\, p_p(J,n(J)+1) \qquad (5.35)$$

2. C^- characteristic at entry pipe J+1

$$V_p(J+1,1) = K3(J+1,1) + K4(J+1,1)\, p_p(J+1,1) \qquad (5.36)$$

3. Continuity of flow between the two pipes

$$V_p(J,n(J)+1)\, A(J) = V_p(J+1,1)\, A(J+1) \qquad (5.37)$$

4. Equivalence of pressure at the boundary in the assumed absence of a junction separation loss

$$p_p(J,n(J)+1) = p_p(J+1,1) \qquad (5.38)$$

Combination of equations 5.35 to 5.38 yields

$$p_p(J+1,1) = \frac{K1(J,n(J)+1)\, A(J) - K3(J+1,1)\, A(J+1)}{K2(J,n(J)+1)\, A(J) + K4(J+1,1)\, A(J+1)} \qquad (5.39)$$

and the other terms follow by back substitution. Note that these expressions apply equally to changes in pipe bore, wall thickness or material.

In low head water distribution systems the assumption that the separation losses may be neglected may not be valid. Also the changes in the velocity head may be significant. In such cases equation 5.38 should be modified

$$p_p(J,n(J)+1) = p_p(J+1,1) + K8$$

$$K8 = 0.5\, \rho\, (\, (1+k)\, V_p(J+1,1)|V_p(J+1,1)| - V_p(J,n(J)+1)|V_p(J,n(J)+1)|\,)$$

and k is the minor loss, or separation loss, coefficient. In order to simplify the calculations it is usually acceptable to base the value of K8 on the velocity values known at the previous time step.

5.3.7 Junctions of three or more pipes

As shown in Figure 5.10 the junction of three or more pipes involves the determination of the values of 2 + 2Z unknowns at time t + Δt from known conditions at time t, namely 2 unknowns at the downstream end of pipe J and 2 at the entry to each of the Z pipes branching from pipe J.

The available equations describing this passive boundary are

1. C^+ characteristic at exit pipe J

$$V_p(J,n(J)+1) = K1(J,n(J)+1) - K2(J,n(J)+1)\, p_p(J,n(J)+1) \tag{5.40}$$

2. C^- characteristic at entry to pipes J+1 to J+Z

$$V_p(J+1,1) = K3(J+1,1) + K4(J+1,1)\, p_p(J+1,1) \tag{5.41}$$

$$V_p(J+2,1) = K3(J+2,1) + K4(J+2,1)\, p_p(J+2,1) \tag{5.42}$$

$$\ldots = \ldots$$

$$V_p(J+Z,1) = K3(J+Z,1) + K4(J+Z,1)\, p_p(J+Z,1) \tag{5.43}$$

3. Continuity of flow between the two pipes

$$V_p(J,n(J)+1)\, A(J) = V_p(J+1,1)\, A(J+1) + V_p(J+2,1)\, A(J+2) + \ldots\ldots + V_p(J+Z,1)\, A(J+Z) \tag{5.44}$$

4. Equivalence of pressure at the boundary in the assumed absence of a junction separation loss

$$p_p(J,n(J)+1) = p_p(J+1,1) = p_p(J+2,1) = \ldots\ldots = p_p(J+Z,1) \tag{5.45}$$

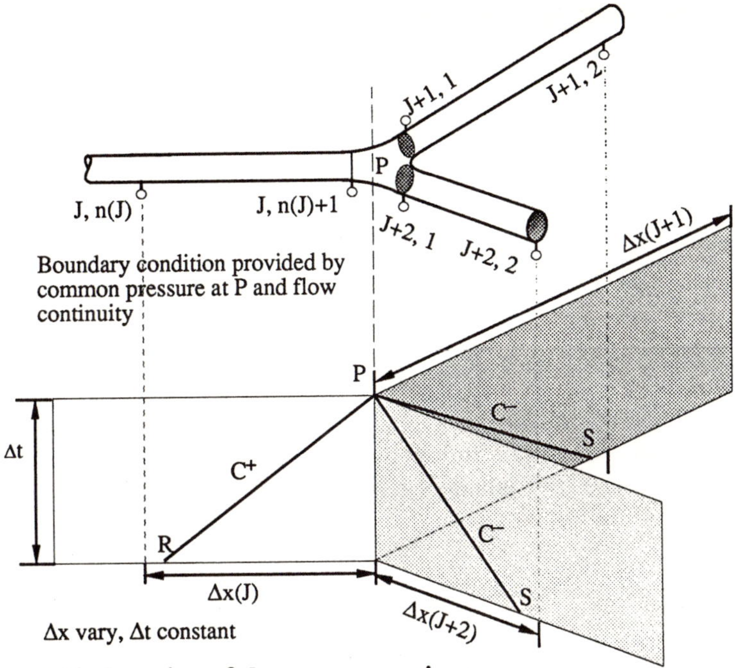

Figure 5.10 Junction of three or more pipes

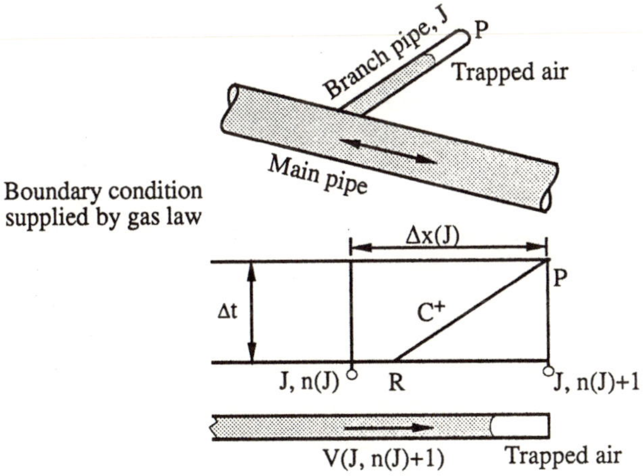

Figure 5.11 Trapped air as a boundary condition

Combination of equations 5.40 to 5.45 yields

$$p_P(J,n(J)+1) = \frac{A}{B} \tag{5.46}$$

where

$$A = K1(J,n(J)+1)\,A(J) - \sum_{M=1}^{M=Z} K3(J+M,1)\,A(J+M)$$

$$B = K2(J,n(J)+1)\,A(J) + \sum_{M=1}^{M=Z} K4(J+M,1)\,A(J+M)$$

and the other terms again follow by back substitution.

In cases where the junction separation losses must be included then equation 5.45 becomes

$$p_P(J+M,1) = p_P(J,n(J)+1) + K9(J+M) \tag{5.47}$$

for M = 1 to Z, with

$$K9(J+M) = 0.5\,\rho\,(1-k(J,J+M))\,V_P(J,n(J)+1)\,|V_P(J,n(J)+1)| - 0.5\,\rho\,V_P(J+M,1)\,|V_P(J+M,1)|$$

where k is the separation loss coefficient for flow from pipe J to pipe J+M.

It should be noted that the above analysis assumed that, at the junction, only one initial inflow pipe existed, ie only one C^+ characteristic and Z–1 C^- characteristics. In the case of looped systems, as opposed to tree networks, this will not be the case and care must therefore be taken in modelling a system to ensure that the correct characteristic is assumed for each pipe joining at the junction. The boundary equation development technique obviously remains as set out above. An example of a loop network solution for pipe and duct branches is given in Chapter 7, Section 7.6.2, where it will be seen that it is necessary to identify the pipes terminating, ie having C^+ characteristics, and originating, ie having C^- characteristics, at each junction.

For example assume that, at a junction of Z pipes, Z0 terminate and Z1 pipes originate, ie have x = 0 at the junction. Thus equation 5.40 may be written for the exit conditions of all pipes J = 1 to Z0, and equation 5.41 may be written for the entry conditions J = Z0+1 to J = Z. Equation 5.44 must however be re-written as

$$\sum_{J=1}^{J=Z0} (V_P(J,n(J)+1)\, A(J)) = \sum_{J=Z0+1}^{J=Z} (V_P(J,1)\, A(J))$$

Equation 5.45 remains true in the absence of separation losses at the junction. Equation 5.46 is also unchanged in format, however the values of the A and B terms become

$$A = \sum_{J=1}^{J=Z0} K1(J,n(J)+1)\, A(J) - \sum_{J=Z0+1}^{J=Z} K3(J,1)\, A(J)$$

$$B = \sum_{J=1}^{J=Z0} K2(J,n(J)+1)\, A(J) + \sum_{J=Z0+1}^{J=Z} K4(J,1)\, A(J)$$

5.3.8 *Trapped air at a dead end or closed valve*

Trapped air at the end of a branch pipe can occur in a range of cases and is particularly a problem if the branch is of small diameter leading to a pressure transducer. The air pocket may be treated as the terminal boundary condition for the branch and will be dealt with here both in order to provide this as one more boundary possibility and to provide an introduction to some of the techniques that will be needed in the treatment of column separation. This boundary arises as a result of flow conditions within the network and must therefore be included in a model as a boundary activated by the predicted flow conditions during transient propagation.

Figure 5.11 illustrates the situation to be modelled, the volume of gas is assumed known at some base pressure, conveniently atmospheric although the network steady flow pressure would also be suitable. The unknowns are the fluid velocity at the fluid/gas interface and the pressure in the free gas volume. The volume of gas is assumed small compared to the pipe volume represented by one computing section Δx.

The available equations are

1. C^+ characteristic at exit pipe J

$$V_P(J,n(J)+1) = K1(J,n(J)+1) - K2(J,n(J)+1)\, p_P(J,n(J)+1) \tag{5.48}$$

where

$$p_P(J,n(J)+1) = P_{gas}^{t+\Delta t}$$

2. The Gas Law expressed as

$$P_{gas}^{t+\Delta t}\left(Vol_{gas}^{t+\Delta t}\right)^n = P_{gas}^{t_0}\left(Vol_{gas}^{t_0}\right)^n \tag{5.49}$$

where $1 < n < 1.4$

The volume of the gas may be calculated at each time step from the summation

$$Vol_{gas}^{t+\Delta t} = Vol_{gas}^{t} - 0.5\, A(J)\Delta t\, (V_P(J,n(J)+1)+v_P(J,n(J)+1)) \tag{5.50}$$

where $v_P(J,n(J)+1)$ is the interface velocity at time t, ie one time step earlier.

It should be noted that the – sign in equation 5.50 ensures that the gas pocket volume decreases if the flow in the branch pipe is positive, ie towards the dead end. If n = 1 then equations 5.48 to 5.50 reduce to

$$a_1\, V_P(J,n(J)+1)^2 + a_2\, V_P(J,n(J)+1) + a_3 = 0 \tag{5.51}$$

where

$$a_1 = -\,0.5\, A(J)\Delta t$$

$$a_2 = Vol_{gas}^{t} - 0.5\, A(J)\Delta t\, (v_P(J,n(J)+1) - K1(J,n(J)+1))$$

$$a_3 = -\,K1(J,n(J)+1)(Vol_{gas}^{t} - 0.5A(J)\Delta t\, v_P(J,n(J)+1)) + K2(J,n(J)+1)\, P_{gas}^{t_0} Vol_{g}^{t}$$

If n > 1 then the solution will require an iterative approach, the bisection method being efficient in such cases.

In cases where there is air leakage from the dead-ended pipe, illustrated by Figure 5.12, as the gas pressure rises, for example in charged dry risers, the approach set out above will apply with a modification to the trapped gas volume summation, equation 5.50. The leakage of gas, q, from the pipe will depend on the equivalent orifice equation, represented by

$$P_{gas}^{time} - P_{external} = k_{orifice} \left(q_{gas}^{time} \right)^2 \tag{5.52}$$

where

$$P_{gas}^{time} = p_p(J, n(J)+1)$$

and therefore

$$Vol_{gas}^{t+\Delta t} = Vol_{gas}^{t}$$

$$- 0.5\ A(J) \Delta t\ (V_p(J,n(J)+1) + v_p(J,n(J)+1)) - 0.5 \sqrt{\frac{P_{gas}^{t+\Delta t} + P_{gas}^{t}}{k}} \tag{5.53}$$

Figure 5.13 illustrates the pressure and velocity variations expected at the interface between the trapped gas and the fluid column. The gas pressure rises sharply over the final stages of compression and the velocity of the fluid column falls to zero and then reverses as the trapped gas re-expands as a result of the reflection of the transient generated by the upstream boundary - in the case illustrated a constant pressure reservoir. Figure 5.13 demonstrates that the fluid velocity reaches a maximum value just prior to the rapid pressure rise associated with the compression of the air pocket.

Experimental observations have shown that the peak pressure generated in the trapped gas may be considerably higher than the supply reservoir pressure depending upon the rate of opening of the supply valve illustrated, Martin (1976). This example, and its extension to include gas leakage, has an important application in the study of pressure relief valve design for dry riser installations in buildings, Lawson et al (1963), Hope and Papworth (1980), and other structures requiring fire protection.

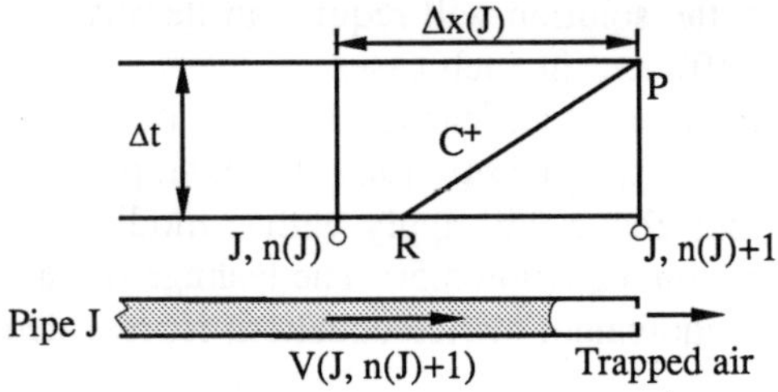

Figure 5.12 Trapped air boundary with leakage

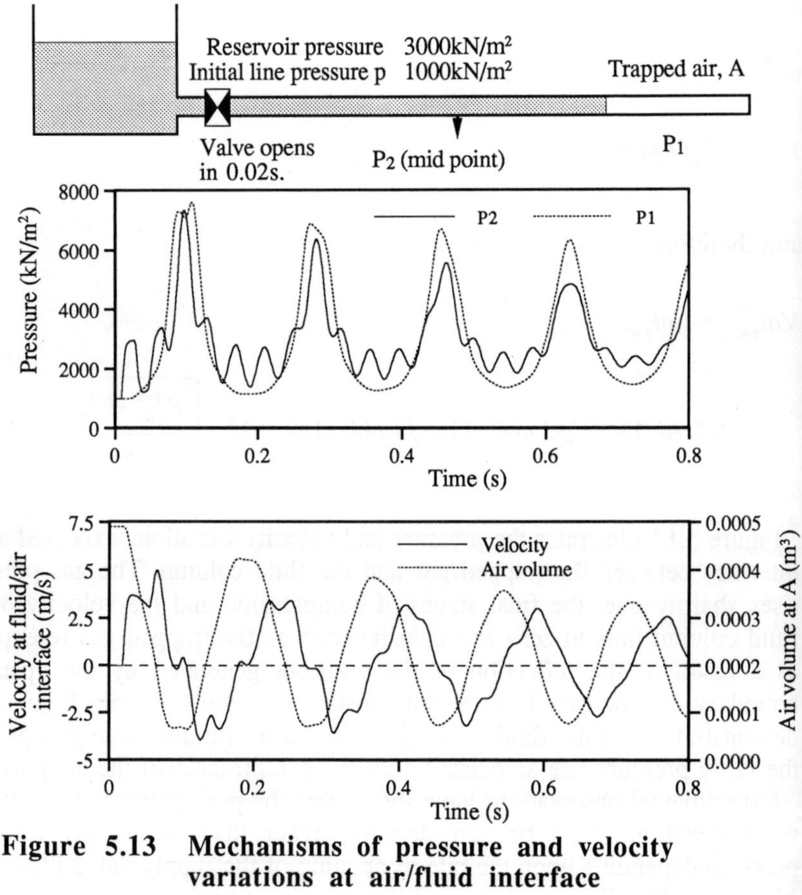

Figure 5.13 Mechanisms of pressure and velocity variations at air/fluid interface

5.3.9 *Steady pump discharge*

In many cases pumps may be considered as passive boundary conditions when the pump continues to operate at constant speed during the propagation and reflection of transients within a network. Such examples include closure of individual tank inlet valves in response to level sensors while other tank inlet valves connected to the same distribution gallery remain open. In this case the operating point of the pump will change, however it may still be represented by its steady flow discharge characteristic at the set speed. Similarly if all discharge flow is stopped by a downstream valve closure then the pump will rely on the operation of its protective non-return valve. Figure 5.14 illustrates this boundary condition for both the continuing flow and non-return valve operation conditions.

The pump discharge pressure vs. discharge flow characteristic may be represented either by a polynomial, of the form

$$\Delta P_{pump} = c_1 + c_2\, Q_{pump} + c_3\, Q_{pump}^2 + \ldots + c_n\, Q_{pump}^{n-1} \tag{5.54}$$

where

$$\Delta P_{pump} = p_P(J,1) - p_{Res} \tag{5.55}$$

or by a series of p,Q array pairs between which a linear interpolation may be performed during simultaneous solution with the C^- characteristic at entry to pipe J, expressed as

$$V_P(J,1) = K3(J,1) + K4(J,1)\, p_P(J,1) \tag{5.56}$$

Combination yields

$$p_P(J,1) - p_{Res} = c_1 + \sum_{i=2}^{i=n} c_i\, (\, A(J)\; K3(J,1) + K4(J,1) p_P(J,1)\,)^{i-1} \tag{5.57}$$

Solution by an iterative technique such as the bisection method will then yield discharge pressure and hence, from equation 5.56 the velocity at entry to the downstream pipe.

In the case of a pump located between two pipes, the boundary is defined in a manner similar to an internal valve within a pipelength, Figure 5.6. Assume that the reservoir in Figure 5.14 is replaced by a pipe J–1. The termination of this pipe may be represented by a C⁺ characteristic

$$V_P(J-1,n(J-1)+1) = K1(J-1,n(J-1)+1) - K2(J-1,n(J-1)+1)\, p_P(J-1,n(J-1)+1) \tag{5.58}$$

Similarly flow continuity must exist through the pump, thus

$$V_P(J-1,n(J-1)+1)\, A(J-1) = V_P(J,1)\, A(J) \tag{5.59}$$

Thus the reservoir pressure included in equation 5.57 may be replaced by the pressure at exit from pipe J–1, initially in terms of the velocity at exit and then in terms of the flow velocity in pipe J, via the continuity of flow equation. Manipulation results in an equation in terms of the downstream pressure and known conditions on either side of the pump one time step earlier

$$\begin{aligned} \Delta P_{pump} &= p_P(J,1) - p_P(J-1,n(J-1)+1) \\ &= p_P(J,1) - \frac{1}{k_2}\left(- \frac{V_P(J,1)\, A(J)}{A(J-1)} + k_1 \right) \\ &= \frac{1}{k_2}\left(- \frac{A(J)}{A(J-1)}\left(k_3 + k_4\, p_P(J,1) \right) + k_1 \right) \end{aligned} \tag{5.60}$$

where

$$k_1 = K1(J-1,n(J-1)+1)$$

$$k_2 = K2(J-1,n(J-1)+1)$$

$$k_3 = K3(J,1)$$

$$k_4 = K4(J,1)$$

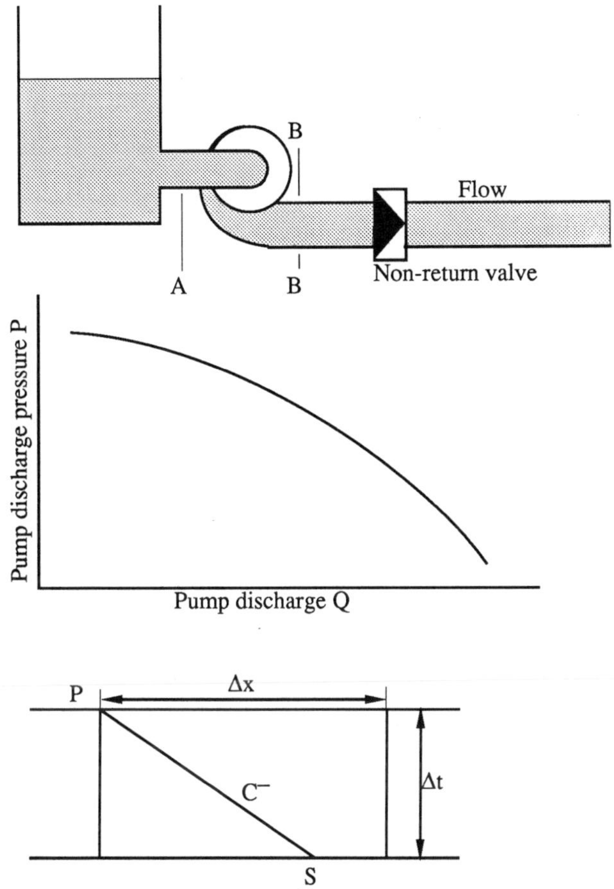

Figure 5.14 Pump and non-return valve boundary. Pump assumed to remain at the steady state speed as the transients propagate within the system

In the case of a pump between two pipes therefore the remaining three unknowns may be determined by back substitution once the discharge pressure is known.

Closure of a downstream valve will cause a transient to arrive at the pump non-return valve, position P in Figure 5.14. If this transient exceeds the pump max discharge pressure the flow will attempt to reverse through the pump. The inclusion of a non-return valve on the pump discharge prevents this.

In terms of the boundary equations available, closure of the non-return valve may be simulated by the solution of the available C^- characteristic, equation 5.56, with the no flow equation

$$V_P(J,1) = 0 \qquad (5.61)$$

thus

$$p_P(J,1) = -\frac{K3(J,1)}{K4(J,1)} \qquad (5.62)$$

If there is an upstream pipe, J–1 as above, then the no through flow condition applies via the continuity equation, 5.59, and the pressure generated on the pump suction is

$$p_P(J-1,n(J-1)+1) = \frac{K1(J-1,n(J-1)+1)}{K2(J-1,n(J-1)+1)} \qquad (5.63)$$

5.4 Variable pump discharge and speed

The starting or stopping of pumps is one of the most common boundary conditions which generates transients in a network and under such conditions the pump is termed an active boundary. During normal start-up and shut-down transient flows are propagated through the network, and must be regarded as normal occurrences.

Transients caused by emergency pump operation, such as sudden power failure to the pump motor, are usually very severe and their effects on the network should be investigated during the design phase.

Power failure initiates a reduction in the pump speed since the inertia of the pump/motor combination is usually small compared to the inertia of the fluid in the network. Therefore the pump discharge decreases rapidly causing negative transient pressure waves to travel downstream into the pump discharge pipe network, and positive transient pressure waves to travel upstream in the pump suction pipe.

It is normal practice to install a non-slam check valve (in addition to the isolating valve) on the discharge, which usually prevents any reverse flow through the pump.

If there is no check valve, the flow through the pump reverses followed by a reversal of the pump rotation, at which time the pump is operating as a turbine. Pump performance characteristics covering the zones of normal pumping, energy dissipation (ie positive speed, negative flow) and turbining are required to simulate the transient response of the pump under such circumstances. A detailed discussion of all the possible zones of operation is presented by Chaudhry (1987).

In the general case of a pump located between two pipes, the boundary is defined in a manner similar to an internal valve as shown in Figure 5.6. Since the performance characteristics of pumps are usually expressed in terms of head and power input as a function of discharge, it is appropriate to use the C^+ and C^- characteristic equations, in terms of velocity and piezometric head, developed in Chapter 4 as equations 4.19 and 4.20. Therefore equations and pump performance characteristic data for this active boundary condition may be expressed as:

1. The C^+ characteristic at the downstream end of pipe J

$$V_p(J,n(J)+1) = K1(J,n(J)+1)-K2(J,n(J)+1)H_p(J,n(J)+1) \tag{5.64}$$

2. The C^- characteristic at the upstream end of pipe J + 1

$$V_p(J+1,1) = K3(J+1,1)+K4(J+1,1)\ H_p(J+1,1) \tag{5.65}$$

3. Continuity of discharge at any time. Assuming the pump boundary consists of n_p identical pumps operating in parallel, each discharging a flow of Q_p, then

$$V_p(J,n(J)+1)\ A(J) = n_p\ Q_p = V_p(J+1,1)\ A(J+1) \tag{5.66}$$

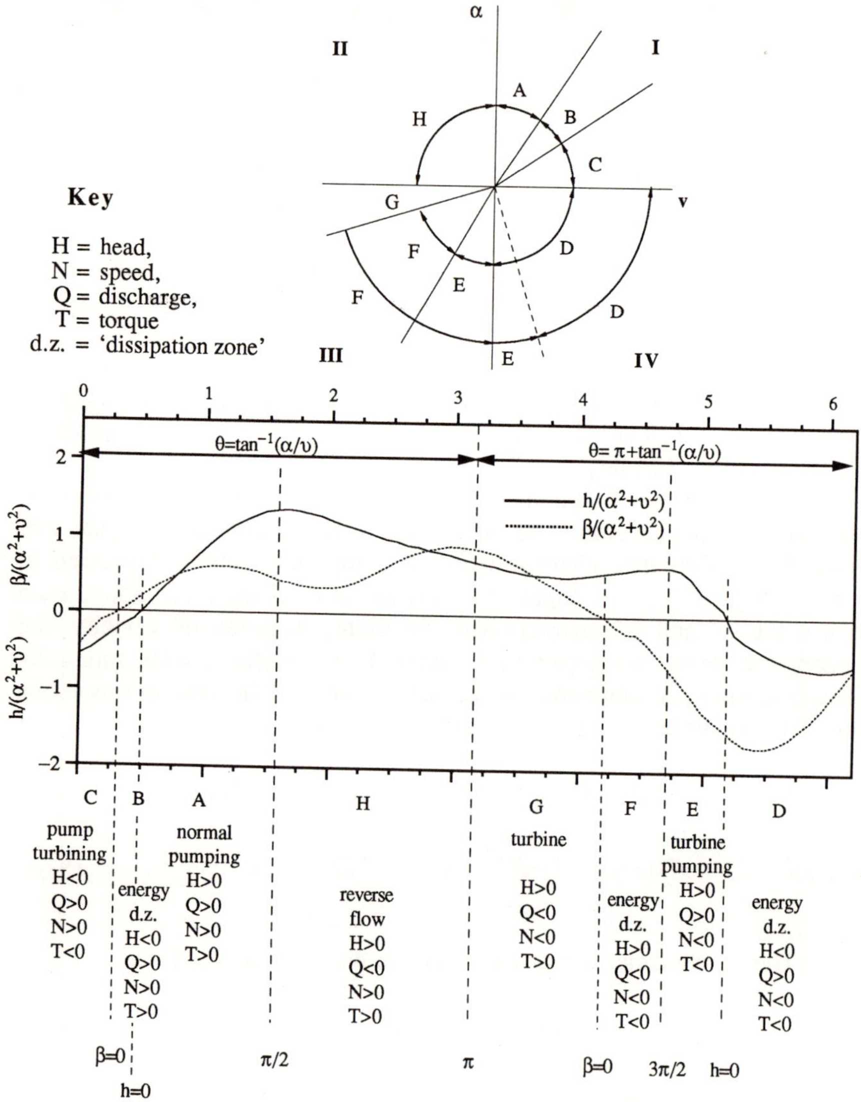

Figure 5.15(a) **Suter representation of pump performance characteristic (specific speed N_s = 0.46) for all possible operation zones**

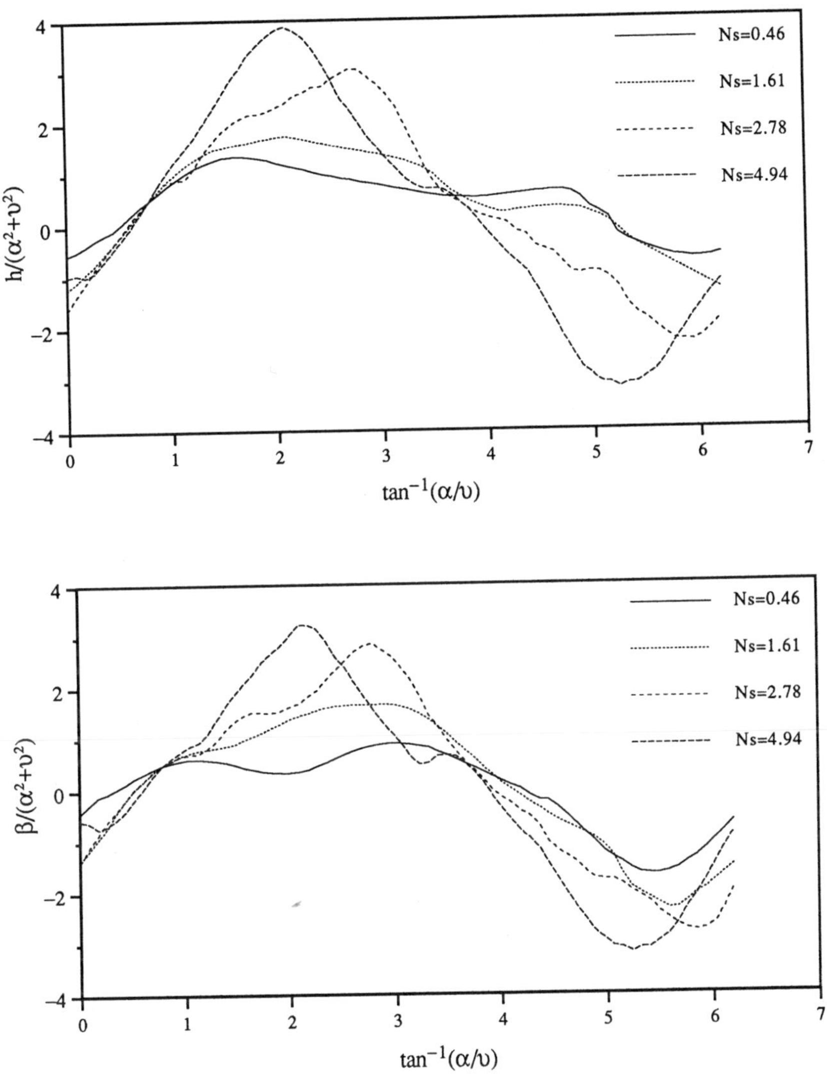

Figure 5.15(b) Suter representation of pump performance characteristics for range of specific speeds

4. The head H_p across the pump is defined as the change in total energy and hence must include both the piezometric and velocity head

$$H_p = H_P(J+1,1) - H_P(J,n(J)+1) + \Delta VH^{t+\Delta t} \tag{5.67}$$

where

$$\Delta VH^{t+\Delta t} = \frac{1}{2g}\left(V_P^{t+\Delta t}(J+1,1)^2 - V_P^{t+\Delta t}(J,n(J)+1)^2\right)$$

5. Pump performance characteristics. The method of representing the pump performance characteristics is based on that proposed by Marchal, Flesch and Suter (1965) and commonly referred to as the Suter Method. The prototype pump characteristics are obtained from model test data using homologous relationships, which in turn enables the characteristics to be transformed into the following non-dimensional forms

$$\frac{h}{\alpha^2 + \upsilon^2} \quad \textit{as a function of} \quad \tan^{-1}\left(\frac{\alpha}{\upsilon}\right) = \theta \tag{5.68}$$

$$\frac{\beta}{\alpha^2 + \upsilon^2} \quad \textit{as a function of} \quad \tan^{-1}\left(\frac{\alpha}{\upsilon}\right) = \theta \tag{5.69}$$

where

$$h = \frac{H}{H_R}, \ \upsilon = \frac{Q}{Q_R} \ \alpha = \frac{N}{N_R}, \ \beta = \frac{T}{T_R}$$

and subscript R refers to a reference point on the pump characteristic. Figure 5.15 shows an example of a pump performance characteristic, for all possible zones of operating, presented in a form based on the Suter Method.

These curves may be stored in a data bank as a series of points such that linear interpolation is appropriate between adjacent points. Values of α_p, υ_p, h_p, and β_p at the end of the time interval are required and the relevant section of the pump performance characteristic is obtained as follows

a) Extrapolate values of α and υ calculated at previous time intervals to obtain estimated values α_e and υ_e. Both α and υ undergo a smooth change during any transient flow, unlike h, and therefore α_e and υ_e are a very close approximation for α_p and υ_p. Calculate θ_e from

$$\theta_e = \tan^{-1}\left(\frac{\alpha_e}{\upsilon_e}\right)$$

b) Determining the stored data points (from the specified data bank) on either side of θ_e and the corresponding values of

$$\frac{h}{\alpha^2 + \upsilon^2} \text{ and } \frac{\beta}{\alpha^2 + \upsilon^2}$$

The coefficients for the straight line passing through these points may then be calculated and assuming that the operating point corresponding to the required values of α_p, υ_p, h_p and β_p lie on these lines then

$$\frac{h_p}{\alpha_p^2 + \upsilon_p^2} = a_1 + a_2 \tan^{-1}\left(\frac{\alpha_p}{\upsilon_p}\right) \tag{5.70}$$

$$\frac{\beta_p}{\alpha_p^2 + \upsilon_p^2} = a_3 + a_4 \tan^{-1}\left(\frac{\alpha_p}{\upsilon_p}\right) \tag{5.71}$$

6. The differential equation for the rotating masses. During the transient flow the out-of-balance torque, T, on the pump is related to the moment of inertia I, Nm^2, of the rotating masses (impeller, motor, water) and the deceleration $d\omega/dt$ by

$$T = -I\frac{d\omega}{dt} = -I\frac{2\pi}{60}\frac{dN}{dt} \tag{5.72}$$

Calculation procedure - The objective is to determine α_p and υ_p. The moment of inertia of the rotating masses is usually expressed as:

$$GD^2 \ (N.m^2) = 4gI$$

Since

$$\beta = \frac{T}{T_R} \text{ and } \alpha = \frac{N}{N_R}$$

equation 5.72 becomes

$$\beta = -\frac{GD^2}{4g}\frac{2\pi}{60}\frac{N_R}{T_R}\frac{d\alpha}{dt} \tag{5.73}$$

Since the value of Δt used in the computer simulation of the transient flow is very short (usually less than 0.1s), an average value for β may be assumed. Also a first order approximation for the deceleration is appropriate, therefore equation 5.73 becomes

$$-\left(\frac{\beta + \beta_p}{2}\right)\frac{4g60T_R}{GD^2 2\pi N_R} = \frac{\alpha_p - \alpha}{\Delta t}$$

or

$$\alpha_p + X1\beta_p = X2 \tag{5.74}$$

where

$$X1 = \frac{60gT_R\Delta t}{GD^2\pi N_R}$$

and

$$X2 = \alpha - \beta X1$$

The C^+ and C^- characteristic equations, 5.64, and 5.65, are expressed in terms of discharge as follows

$$Q_p(J,n(J)+1) = X3 + X4\ H_p(J,n(J)+1) \qquad (5.75)$$

$$Q_p(J+1,1) = X5 + X6\ H_p(J+1,1) \qquad (5.76)$$

where

$$X3 = A(J)\ K1(J,n(J)+1)$$

$$X4 = -A(J)\ K2(J,n(J)+1)$$

$$X5 = A(J+1)\ K3(J+1,1)$$

$$X6 = A(J+1)\ K4(J+1,1)$$

Multiplying equations 5.75 and 5.76 by X6 and X4 respectively, subtracting, and combining with equations 5.66 and 5.67 yields

$$n_p Q_p(X4-X6) = (X4\ X5 - X3\ X6) + X6\ X4(H_p - \Delta VH^{t+\Delta t}) \qquad (5.77)$$

In order to simplify the calculations the value of the velocity head difference at time t may be used instead of the unknown value at time t + Δt. Therefore equation 5.77 is expressed as

$$X7\ Q_p = X8 + X9\ H_p \qquad (5.78)$$

where

$$X7 = n_p(X4 - X6)$$

$$X8 = X4\,X5 - X3\,X6 - X9\,\Delta VH'$$

$$X9 = X6\,X4$$

Substituting the unknown non-dimensional ratios υ_p and h_p equation 5.78 produces the following expression for h_p

$$h_p = X10\,\upsilon_p + X11 \tag{5.79}$$

where

$$X10 = \frac{X7\,Q_R}{X9\,H_R}$$

$$X11 = -\frac{X8}{X9\,H_R}$$

Combining equations 5.79 and 5.70 results in the following relationship between the two unknowns α_p and υ_p

$$f_1 = a_1\left(\alpha_p^2+\upsilon_p^2\right) + a_2\left(\alpha_p^2+\upsilon_p^2\right)\tan^{-1}\left(\frac{\alpha_p}{\upsilon_p}\right) - X10\upsilon_p - X11 = 0 \tag{5.80}$$

Combining equations 5.74 and 5.71 results in a second relationship for the two unknowns

$$f_2 = a_3\left(\alpha_p^2+\upsilon_p^2\right) + a_4\left(\alpha_p^2+\upsilon_p^2\right)\tan^{-1}\left(\frac{\alpha_p}{\upsilon_p}\right) + \frac{\alpha_p}{X1} - \frac{X2}{X1} = 0 \tag{5.81}$$

Equations 5.80 and 5.81 form a pair of nonlinear equations in the two unknowns α_p and υ_p, and are solved using an iterative Newton-Raphson technique. If $\alpha_p(1)$ and $\upsilon_p(1)$ represent the initial estimates for α_p and υ_p

which satisfy equations 5.80 and 5.81, then better estimates are given by

$$\alpha_p(2) = \alpha_p(1) + \delta\alpha_p \tag{5.82}$$

$$\upsilon_p(2) = \upsilon_p(1) + \delta\upsilon_p \tag{5.83}$$

where

$$\delta\alpha_p = \frac{f_2 \dfrac{\partial f_1}{\partial \upsilon_p} - f_1 \dfrac{\partial f_2}{\partial \upsilon_p}}{\dfrac{\partial f_1}{\partial \alpha_p}\dfrac{\partial f_2}{\partial \upsilon_p} - \dfrac{\partial f_1}{\partial \upsilon_p}\dfrac{\partial f_2}{\partial \alpha_p}} \tag{5.84}$$

and

$$\delta\upsilon_p = \frac{f_2 \dfrac{\partial f_1}{\partial \alpha_p} - f_1 \dfrac{\partial f_2}{\partial \alpha_p}}{\dfrac{\partial f_1}{\partial \upsilon_p}\dfrac{\partial f_2}{\partial \alpha_p} - \dfrac{\partial f_1}{\partial \alpha_p}\dfrac{\partial f_2}{\partial \upsilon_p}} \tag{5.85}$$

In the above equations the value of the functions f_1 and f_2, and their derivatives, are evaluated using $\alpha_p(1)$ and $\upsilon_p(1)$.

The required differentiations of f_1 and f_2 are

$$\frac{\partial}{\partial \alpha_p}\left[\tan^{-1}\left(\frac{\alpha_p}{\upsilon_p}\right)\right] = \frac{\upsilon_p}{\upsilon_p^2 + \alpha_p^2}$$

$$\frac{\partial}{\partial \upsilon_p}\left[\tan^{-1}\left(\frac{\alpha_p}{\upsilon_p}\right)\right] = \frac{-\upsilon_p}{\upsilon_p^2 + \alpha_p^2}$$

therefore

$$\frac{\partial f_1}{\partial \alpha_p} = 2a_1\alpha_p + a_2\upsilon_p + 2a_2\alpha_p \tan^{-1}\left(\frac{\alpha_p}{\upsilon_p}\right)$$

$$\frac{\partial f_1}{\partial \upsilon_p} = 2a_1\upsilon_p - a_2\alpha_p + 2a_2\upsilon_p \tan^{-1}\left(\frac{\alpha_p}{\upsilon_p}\right) - X10$$

$$\frac{\partial f_2}{\partial \alpha_p} = 2a_3\alpha_p + a_4\upsilon_p + 2a_4\alpha_p \tan^{-1}\left(\frac{\alpha_p}{\upsilon_p}\right) + \frac{1}{X1}$$

$$\frac{\partial f_2}{\partial \upsilon_p} = 2a_3\upsilon_p - a_4\alpha_p + 2a_4\upsilon_p \tan^{-1}\left(\frac{\alpha_p}{\upsilon_p}\right)$$

Convergence of the solution - The required value of the unknowns α_p and υ_p are obtained from equations 5.82 and 5.83 if

$$|\delta\alpha_p| \quad \text{and} \quad |\upsilon_p| \leq \varepsilon \tag{5.86}$$

where ε is set at some small value (0.001 say). If either of the inequalities stated in equation (5.86) are not satisfied then

$$\alpha_p(1) = \alpha_p(2) \quad \text{and} \quad \upsilon_p(1) = \upsilon_p(2)$$

the coefficients a_1, a_2, a_3, a_4 of equations 5.70 and 5.71 are re-established, and the calculation procedure repeated until the required convergence of the solution is achieved. The value of all the remaining unknowns are obtained by appropriate substitution in the above equations, from equation 5.70

$$h_p = (\alpha_p^2+\upsilon_p^2)\left(a_1 + a_2 \tan^{-1}\left(\frac{\alpha_p}{\upsilon_p}\right)\right)$$

and, from equation 5.71

$$\beta_p = (\alpha_p^2 + \upsilon_p^2)\left(a_3 + a_4 \tan^{-1}\left(\frac{\alpha_p}{\upsilon_p}\right)\right)$$

Then, by definition

$$Q_p = \upsilon_p \, Q_R \quad \text{and} \quad N_p = \alpha_p \, N_R$$

and, from equation 5.66

$$Q_P(J,n(J)+1) = Q_P(J+1,1) = n_p \, Q_p$$

therefore

$$V_P(J,n(J)+1) = \frac{Q_P(J,n(J)+1)}{A(J)}$$

and

$$V_P(J+1,1) = \frac{Q_P(J+1,1)}{A(J+1)}$$

hence, from equation 5.75

$$H_P(J,n(J)+1) = \frac{Q_P(J,n(J)+1) - X3}{X4}$$

and, from equation 5.76

$$H_P(J+1,1) = \frac{Q_P(J+1,1) - X5}{X6}$$

No suction pipe - In some pumping installations the suction pipe is either absent or short enough to be neglected. In such cases the above calculation procedure can still be used by modifying the parameters X7, X8 and X9 as follows. The water elevation in the suction reservoir is assumed to remain constant during the transient flow, therefore $H_P(J,n(J)+1)$ in equation 5.69 is replaced by H_{suc} = constant and substituting for $H_P(J+1,1)$ from equation 5.76 yields equation 5.78 where

$$X7 = n_p$$

$$X8 = X5 + X6(H_{suc} - \Delta VH')$$

$$X9 = X6$$

5.5 Hydraulic turbine

The boundary conditions for a hydraulic turbine is similar to that of a pump but the following factors complicate the simulation:

1. In addition to the four quantities of dynamic head H, discharge Q, torque T and speed N defining a pump characteristic, an additional parameter has to be introduced for any turbine in order to define the position, a_o, of the guide vanes (wicket gates). For a Kaplan turbine the blade angle β_o must also be taken into account. For a reversible pump-turbine the performance characteristics cover the four quadrants obtained by the combinations of ±Q and ±N.

2. In some installations the pressure rise upstream of the turbine generated during a load rejection is controlled by opening a relief valve connected to the spiral casing at the same time as the guide vanes are closed. The discharge characteristics of the relief valve must be incorporated in the boundary condition and the continuity equation becomes

$$Q_P(J,n(J)+1) - Q_{rv} = Q_t = Q_P(J+1,1) \quad (5.87)$$

where Q_{rv} is the relief valve discharge and Q_t the turbine discharge.

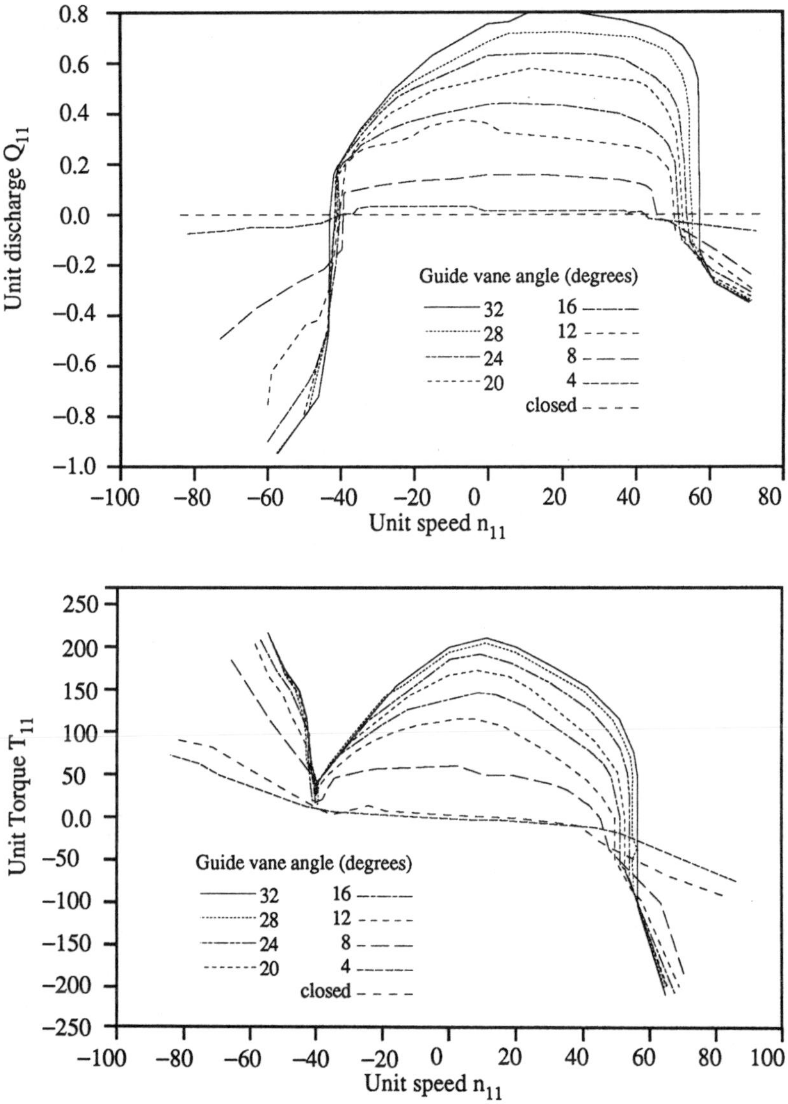

Figure 5.16 Four-quadrant performance characteristics of reversible pump-turbine

3. If the turbine is not connected to a large electrical network, the turbine speed will not remain at the synchronous speed during a load change. The differential equations describing the control function of the turbine governor (hence the movement of the guide vanes) must be incorporated, introducing a number of additional variables.

A complete description of the method of solution for a hydraulic turbine boundary condition is outside the intended scope of this book. Detailed information may be obtained from the following Boldy (1976, 1983, 1989), Pejovic (1989), Chaudhry (1987), Martin (1988). The performance characteristics of a reversible pump-turbine are discussed below, defining the additional parameters which must be considered.

5.5.1 Pump-turbine performance characteristics

For the purpose of transient simulations the pump-turbine performance characteristics are usually represented by two sets of four-quadrant curves which define

1. the variation of unit discharge Q_{11} with unit speed n_{11},

2. the variation of unit torque T_{11}, with unit speed,

for the range of guide vane openings a_o.

Figure 5.16 shows a typical set of such performance characteristics. The unit parameters are defined by

Unit speed

$$n_{11} = \frac{ND}{\sqrt{H}} \tag{5.88}$$

Unit discharge

$$Q_{11} = \frac{Q_t}{D^2\sqrt{H}} \tag{5.89}$$

Unit torque

$$T_{11} = \frac{T}{D^3 H} \tag{5.90}$$

where N is the turbine speed (rpm)
D is the turbine runner diameter (m)
H is the net head across the turbine (m)
Q_t is the turbine discharge (m^3s^{-1})
T is the torque on the turbine (Nm)

The turbine performance characteristics provide two relationships which can be expressed as

$$Q_{11} = \phi_1 \ (n_{11}, a_o) \tag{5.91}$$

$$T_{11} = \phi_2 \ (n_{11}, a_o) \tag{5.92}$$

but has also introduced three new variables in the turbine, namely speed N, torque T, and guide vane position a_o. Therefore two more relationships are required in order to solve the turbine boundary condition. Equation 5.72, relating the torque to the deceleration of a pump, applies to a turbine and can be expressed as

$$T = -I \frac{d\omega}{dt} = -\frac{GD^2}{4g} \frac{2\pi}{60} \frac{dN}{dt} \tag{5.93}$$

and the final relationship, defining the guide vane closure as a function of time, is usually specified as data

$$a_o = \phi_3(\text{time}) \tag{5.94}$$

Therefore the turbine boundary provides five relationships, equations 5.87, 5.91, 5.92, 5.93 and 5.94, which together with the C^+ and C^- characteristic equations provide seven equations enabling the following seven unknowns to be determined

$V_P(J,n(J) + 1)$ and $H_P(J,n(J) + 1)$
$V_P(J + 1,1)$ and $H_P(J + 1,1)$
Turbine speed, N,
Turbine torque, T,
Guide vane position, a_o

5.6 Summary of boundary conditions

The active and passive boundary conditions presented, together with the general guidance notes on the development of transient models, should be sufficient to allow the reader to write simple programs to model transient propagation. The main solution techniques presented have been mainly simultaneous solutions of linear equations with some quadratics. The iterative solutions necessary in the pump and trapped air cases are easily accomplished by standard methods, the bisection technique presented in Appendix III is usually sufficient.

This chapter will continue with a treatment of column separation as a separate series of boundary conditions that arise as a result of the propagation of transients through the pipe network.

5.7 Column separation and gas release

The propagation of pressure transients through a pipe system sets up a normally periodic pressure and velocity variation at any particular point on the network. Generally the period of the pressure oscillation depends upon the wave speed appropriate to the fluid-pipe combination and the distance between 'reflectors', ie boundaries such as reservoir, pipe junctions, pumps or dead-ended pipes.

Two important transient phenomena disrupt this orderly view of the subject, both connected with the occurrence of 'negative', ie below atmospheric, pressures. The first is the appearance of dissolved gas in the flow as the pressure falls below the gas release level, often atmospheric pressure, and the second is the formation of vapour pockets in the flow when the pressure falls to the fluid vapour pressure. For some volatile fluids these two phenomena are often linked.

Chapter 1 outlined the importance of air content in the determination of wave speed. It will have been noted that a small air content can considerably affect the wave speed and hence affects the application of the method of characteristics through the Courant Criterion and the slope of the characteristic curves in the x-t plane. Considerable research effort has been applied to this subject as evidenced by the range of available research papers, Enever (1972), Weyler et al (1971), Tullis, Streeter and Wylie (1976), Wiggert and Sundquist (1979), Fox and McGarry (1983).

In many practical applications the true air release situation is unknown and a sensitivity analysis approach has been proposed where the particular network analysis is re-run with a range of free gas affected wave speeds to determine the sensitivity of the pressure transient predictions to this variable. It is quite incorrect to assume that as the wave speed falls the transient pressures propagated will also decrease. As stressed in Chapter 1 the transient generated by, for example, a closing valve depends on its closure rate expressed in terms of the local pipe period not in terms of absolute time. A reduction in wave speed lengthens the pipe period and hence effectively 'reduces' the valve closure time, in pipe period units, introducing the possibility of generating a higher transient pressure.

Vapour pocket formation can, in contrast to the air release phenomena, be included in the transient analysis presented with some accuracy. Again this topic has been the subject of considerable study since the introduction of the method of characteristics and digital computing as a combination opened up this research area in the 1960s.

The mechanism of column separation is inherently simple. If the local pressure at a point in the pipe network falls to vapour pressure then a vapour pocket forms, normally assumed to completely fill the pipe cross section at that location, and the fluid column breaks into two, upstream and downstream of the cavity. These two columns then move independently of each other under the influence of the prevailing pressure gradients between the vapour cavity, assumed to remain at the fluid vapour pressure and at the original formation site, and whatever terminal pressures the network possesses.

In practice the phenomenon is complicated by the formation of partial cavities overlaying the flow, the generation of gas brought out of solution by the reduced pressure levels, the migration of the cavity etc. However, it has been shown that the simple model is remarkably accurate in predicting cavity formation and the after effects of cavity collapse.

The formation of vapour cavities, while posing a pipe implosion failure, are generally a cause for concern in terms of the pressures generated on their collapse. As the separated fluid columns move independently within the pipeline they are eventually brought to rest and are then accelerated back towards the site of the cavity formation by these same pressure gradients. The collapse of the cavity as the columns rejoin is effectively an instantaneous flow stoppage and therefore generates the full Joukowsky pressure rise based on the final column velocity prior to cavity collapse. When it is appreciated that cavity formation in low head systems can result from 'slow' valve closures, ie closures in longer than a pipe period, the potential hazard posed by column separation becomes clear, as does the justification for the research effort expended in this area.

Cavities can be formed within the length of a pipe or at a pipe boundary. In the latter case the fluid column separates from the pipe end or valve face, forming one separated column as opposed to the two discussed above.

Figure 5.17 illustrates the range of situations in which column separation may occur.

Downstream of either a closing valve or a pump during planned or failure shutdown the pressure falls rapidly as the flow is reduced. If the system is at a relatively low head and if the flow reduction is rapid, then column separation can occur either during valve closure or pump shutdown. In these cases it is likely that gas release will also occur.

On the upstream face of a closing valve column separation may occur following total valve closure as a result of the arrival at the closed valve of the first negative pressure transient reflection from an upstream reflecting boundary, such as a reservoir. As discussed in Chapter 1 the incoming transient is re-reflected at the closed valve with no change in sign so that a 'double' pressure drop is experienced and vapour levels may be easily reached even if the valve initial closure was slow.

In each of the cases cited above the pressures generated on cavity collapse can exceed the safe limits for the system and thus accurate predictions are required at the design stage to identify the problem and to either allow it to be 'designed out' or protected against by the specification of surge relief devices. Column separation boundary condition definitions within the method of characteristics therefore clearly fall into the third category, namely boundaries formed as a result of pressure variations within the system.

The earliest work specifically dealing with column separation was due to Le Conte (1937) who considered column separation upstream of a closed valve and concluded that accurate prediction of the interface velocity was necessary to determine the cavity collapse pressure rise. Bergeron (1935) included the monitoring of cavity volume in a graphical representation of this type of column separation, assuming a full bore cavity. The relatively large time steps necessary in the graphical method limited the accuracy.

While the upstream column separation event is the easiest to study, and the closest to the simple model described above, Bunt (1953) and Binnie and Thackrah (1951) studied separation downstream of a closing valve, in the latter case as a means of simulating the effect of pump shutdown, finding that severe pressure surges accompanied cavity collapse and that the phenomena was repetitive. Gayed and Kamel (1959) found that, for the case studied by Binnie and Thackrah, a secondary series of pressure transients was identifiable propagating between the cavity, a constant pressure zone, and the downstream reservoir commensurate with the presence of two −1 reflection coefficient boundaries. This analysis fits with later predictions and observations.

Column separation in practice was reported by Apelt (1956), Richards (1956) and Brown (1968), while Duc (1959) provided photographic evidence of the cavity growth and collapse cycle following a pump failure. Following the introduction of the method of characteristics analysis considerable interest was shown in the prediction of column separation, both for the upstream/downstream of the valve case, which is naturally very suitable for laboratory and computational study, but also in the case of pump trip initiated separation.

Swaffield (1970), Weyler (1971), Safwat et al (1972) and Martin (1983) all studied the upstream separation problem with reasonable agreement being achieved between the pressure generated on first cavity collapse and its prediction via the method of characteristics. Li and Walsh (1964) had considered column separation downstream of a closing valve with no allowance for possible gas release and concluded that the dissipation within the model severely underestimated that measured following the collapse of the first cavity. Swaffield (1972, 1972[a]) and Driels (1973), reporting on studies of downstream separation in a kerosene pipeline incorporated air release utilising Henry's Law as a first approximation. Swaffield (1972[a]) demonstrated that once out of solution free gas remains in the region of the cavity formation and accounts for the increased energy dissipation observed by other investigators. Kranenberg (1974) also identified the need to include air release in the models used for water column separation downstream of shutdown pumps.

Thus the basis for a representation of column separation within the method of characteristics exists and the relevant boundary equations for valve and pump initiated separation will be presented.

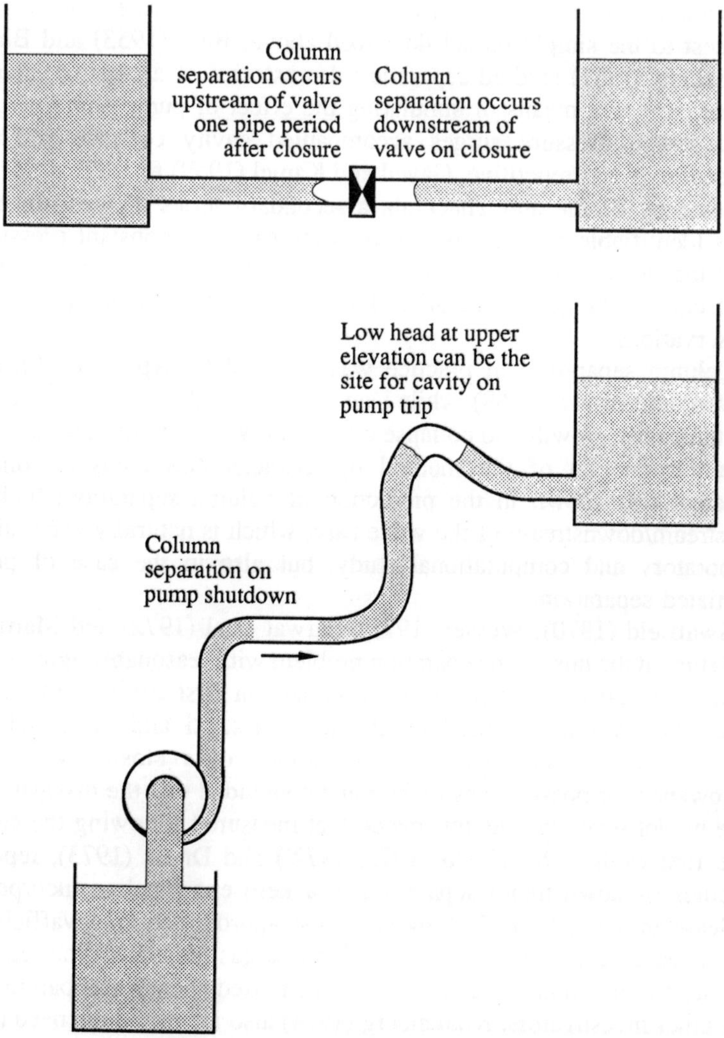

Figure 5.17 Range of column separation cases

5.7.1 *Column separation upstream of a closing valve or at a dead-ended pipe within a complex pipe network*

The boundary equations appropriate for a valve termination at the downstream discharge of a pipe have been developed in Section 5.4.2. The boundary during closure was provided by the valve discharge coefficient, while following closure the boundary was provided by the zero velocity condition at the closed valve, or at a dead end.

It is this second boundary condition that must be replaced in order to model the occurrence of column separation upstream of the closed valve. Figure 5.18 illustrates the condition during the existence of the cavity, while Figure 5.19 illustrates the likely pressure oscillations at the closed valve.

As mentioned above it is the reflected wave from the upstream reservoir, Figure 5.19, arriving at the closed valve that initiates the column separation by dropping the pressure to fluid vapour pressure. In order for this to occur the line pressure initially must have been less than p_{final} above the fluid vapour pressure, where p_{final} is the surge pressure, above initial line pressure, reached at the valve at the instant of final closure, independent of the time taken to close the valve. The column then separates from the closed valve, forming a full bore cavity that grows and collapses as the fluid column is first retarded and then accelerated back towards the closed valve by the pressure gradient that exists between the cavity, at vapour pressure, and the upstream reservoir. Figure 5.20 illustrates this mechanism.

Therefore the available boundary equations during cavity existence are the C^+ characteristic

$$V_p(J,n(J)+1) = K1(J,n(J)+1) - K2(J,n(J)+1)\, p_p(J,n(J)+1) \tag{5.95}$$

and the cavity pressure condition

$$p_p(J,n(J)+1) = P_{vap} \tag{5.96}$$

The duration of this boundary depends on the duration of the cavity. It is therefore necessary to monitor the cavity growth within the model to determine the time at which the boundary reverts to the no flow condition

$$V_p(J,n(J)+1) = 0 \tag{5.97}$$

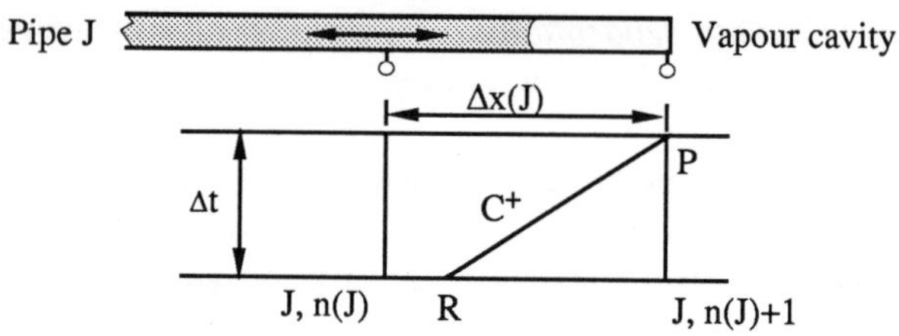

Figure 5.18 Boundary formed by the formation of a vapour cavity

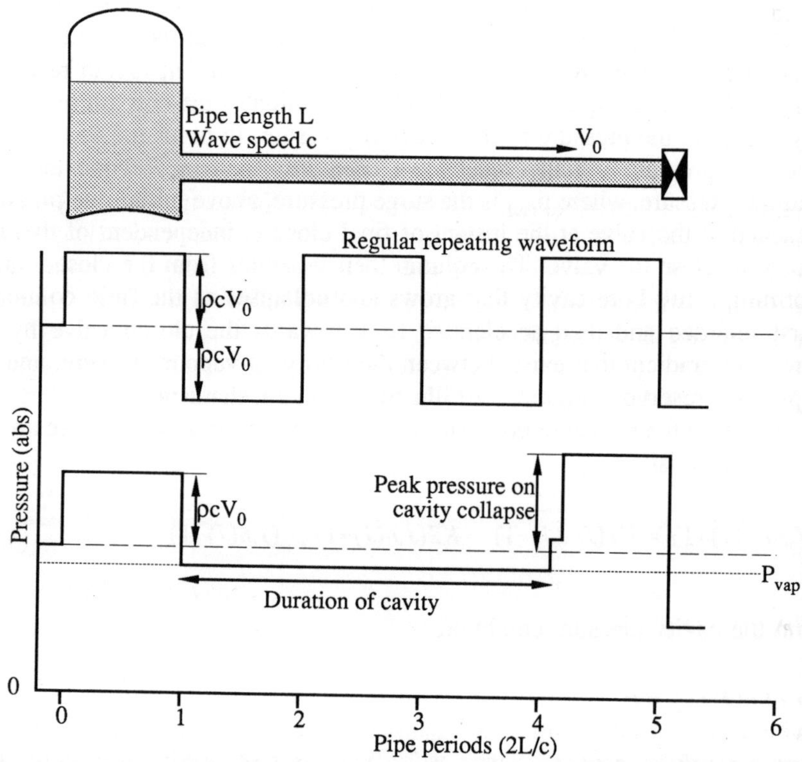

Figure 5.19 Comparison between pressure variation upstream of the closed valve with and without separation. Instantaneous valve closure, frictional damping neglected

The cavity volume, assumed full bore may be monitored at each time step by an expression

$$Vol_{cav}^{t+\Delta t} = Vol_{cav}^{t} - 0.5\, A(J)\, \Delta t\, (V_p(J,n(J)+1)+v_p(J,n(J)+1)) \qquad (5.98)$$

Note that the – sign ensures that flow away from the closed valve increases the cavity volume, a similar approach to that adopted for the trapped air boundary. The velocity term, v, indicates the value of the interface velocity at the start of the time step to be monitored.

Figure 5.21 illustrates the form of the resulting pressure variations predicted at the closed valve. It should be noted that the presence of the cavity disrupts the repeating waveform normally associated with this simple case. Once the transient dissipates, ie the pressure reductions at the closed valve are insufficient to generate a reopening cavity, these periodic pressure oscillations are re-established.

In terms of the reflection coefficient models presented in Chapter 2, the pressure variations at the valve during cavity existence may be explained by the transmission and reflection of transients between two –1 reflecting boundaries, the closed valve boundary reflection coefficient reverting to +1 on cavity collapse. Once the cavity has collapsed the prediction of pressure at the closed valve returns to

$$p_P(J,n(J)+1) = \frac{K1(J,n(J)+1)}{K2(J,n(J)+1)} \qquad (5.99)$$

If friction is neglected and the pipe is assumed to be horizontal, then

$$p_P(J,n(J)+1) = \rho\, c_R(J,n(J)+1)\, \left(V_R(J,n(J)+1) + \frac{p_R(J,n(J)+1)}{\rho\, c_R(J,n(J)+1)} \right)$$

which, as Δx tends to zero and $V_R(J,n(J)+1)$ tends to the flow final approach velocity prior to cavity collapse, reduces to the Joukowsky (1904) expression for pressure rise on an instantaneous flow stoppage

$$\Delta P = \rho\, c\, V_{final} \qquad (5.100)$$

Figure 5.22 illustrates the level of agreement possible between this model and observations of cavity collapse pressures.

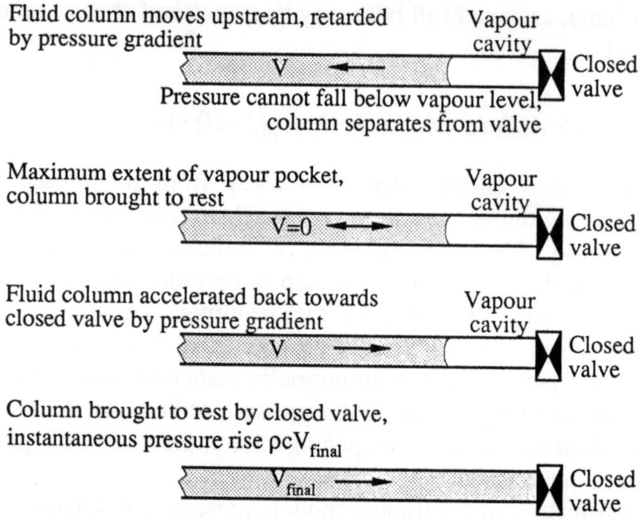

Figure 5.20 Column separation upstream of a closed valve

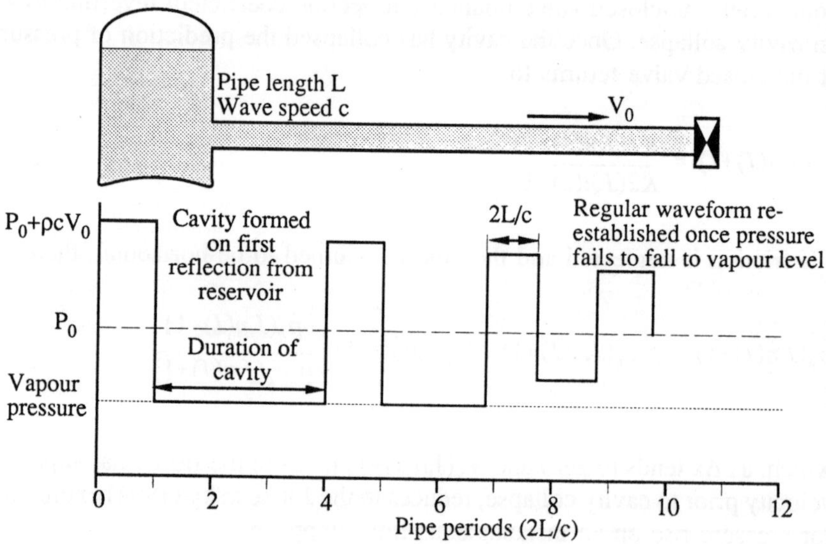

Figure 5.21 Pressure variations at a closed valve before, during and following column separation and vapour pocket formation.

Note: Actual pressure variations have been simplified, see Figure 5.31, for identification of secondary wave effects

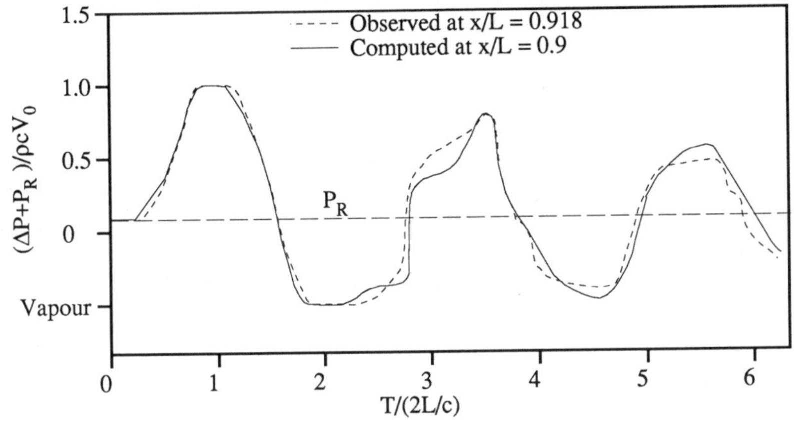

$V_0 = 0.29\text{ms}^{-1}$ $P_R = 28.0\text{ kNm}^{-2}$ $T_C = 0.030\text{s}$ $L = 14\text{m}$, $D = 0.05\text{m}$

Figure 5.22 Comparison between observed and predicted pressure variations upstream of a closing valve, with column separation at the valve

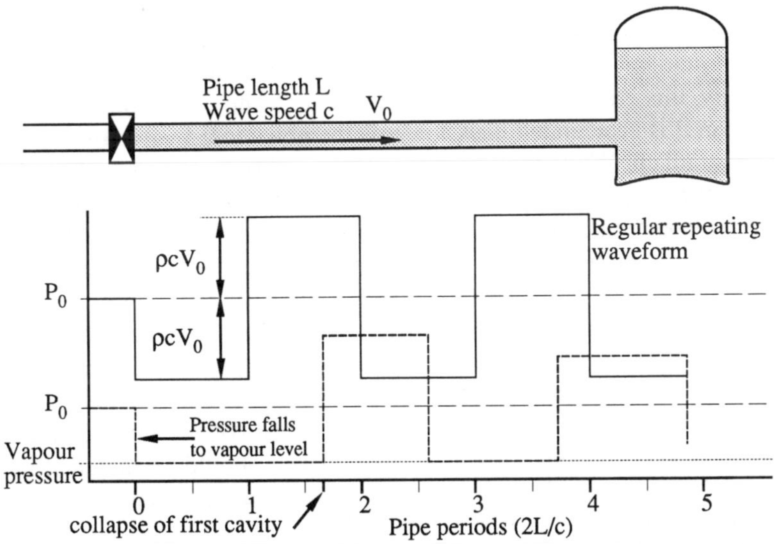

Figure 5.23 Comparison of pressure variation downstream of a closed valve with and without column separation. Note the instantaneous valve closure, frictional damping neglected, compared with Figure 5.19

5.7.2 *Column separation downstream of a closing valve, including gas release effects*

Figure 5.23 illustrates the occurrence of column separation downstream of a closing valve and the comparison between this and the pressure transient variations that would be expected in the absence of column separation.

In contrast to the upstream case discussed above, column separation on the downstream face of a closing valve occurs as a result of the first transmitted transient and can occur while the valve is still closing. As the valve closes so the pressure on its downstream face falls. In the extreme case of an instantaneous closure then the downstream pressure falls by the Joukowsky pressure, based on the initial fully open valve throughflow, Figure 5.23. If the line pressure exceeds the fluid vapour pressure by an amount less than this, in the instantaneous closure case, then a vapour cavity is formed.

The complication arises as the cavity may be spread downstream by the continuing flow through the valve and therefore a more representative boundary condition must be sought while the valve is still in motion.

Figure 5.24 illustrates the mechanism of column separation downstream of a closing valve. This general description also applies to the case of pump shutdown. The separated column moves downstream and is retarded by the pressure gradient between the cavity and the downstream reservoir. Eventually this, together with fluid friction, brings the column to rest, whereupon it is accelerated back towards the now closed valve by this same pressure gradient. While the effect of friction will be to attenuate the column velocity, the pressure rise generated as the cavity collapses may still be considerable and much greater than any prediction based on the closure time of the valve as the action of cavity collapse is effectively an instantaneous flow stoppage. Thus a slow initial valve closure taking several pipe periods, may generate positive pressure commensurate with an instantaneous closure.

The release of dissolved gas into the cavity region downstream of the closing valve cannot be neglected for fluids with a high gas content. The agitation provided by the closing action of the valve will be a major contributory factor in releasing gas. These effects should be included in the model as the total gas released will effect the accurate prediction of the later peak pressures beyond the first cavity collapse.

Figure 5.25 illustrates a comparison between the expected pressure transient records at the closed valve with and without gas release. The effect of gas release is to remove the instantaneous pressure rise on cavity collapse as the gas remains out of solution as the pressure rises, and remains out over several pressure oscillations, Swaffield (1972[a]), and the resulting traces then resemble the trapped gas case already discussed, Section 5.3.8. It must however not be assumed that the gas in some way forms a cushion to reduce the resurge pressure rise as the interaction of the returning fluid column with a compressible boundary can result in extremely high, if not instantaneous, pressure rises.

In the vapour only case the cavity will reopen one pipe period later, calculated on the basis of the wave speed in, and length of, the downstream pipe. Cavities will then open and close with decreasing duration of column separation until sufficient energy has been dissipated for the minimum pressure at the closed valve to remain above vapour pressure. Beyond this the pressure transient oscillations at the valve then take up the periodic repeating waveform expected, Chapter 1.

In the gas release case, the gaseous volume remaining out of solution continues to expand and contract until the transients in the pipe die away completely. Figure 5.26 illustrates this decay process together with the variation of the linked column velocity, shown as an absolute value. It will be seen that the column comes to rest at both the minimum and maximum gas pressures and attains its maximum velocity just prior to the rapid rise in pressure associated with the final stages of the gas compression, the equivalent to cavity collapse in the vapour only model.

5.7.3 *Vapour only column separation downstream of a closing valve*

The boundary conditions for a valve linking two pipes have been presented in Section 5.3.3. These relationships are used up to the time when the pressure, $p_P(J+1,1)$ in the notation adopted, falls to vapour level. If this level is not reached then on valve closure the zero velocity boundary, equation 5.28, is utilised.

If the pressure on the downstream face falls to the fluid vapour pressure during valve closure the boundary conditions at the valve become:

1. The valve discharge coefficient rewritten as

$$C_d = \frac{V_P(J,n(J)+1)}{V_0}\sqrt{\frac{p_P(J,n(J)+1)-P_{vap}}{\Delta P_0}} \tag{5.101}$$

where ΔP_0 and V_0 are the valve fully open pressure loss, pipes J to J+1, and through flow velocity referred to pipe J.

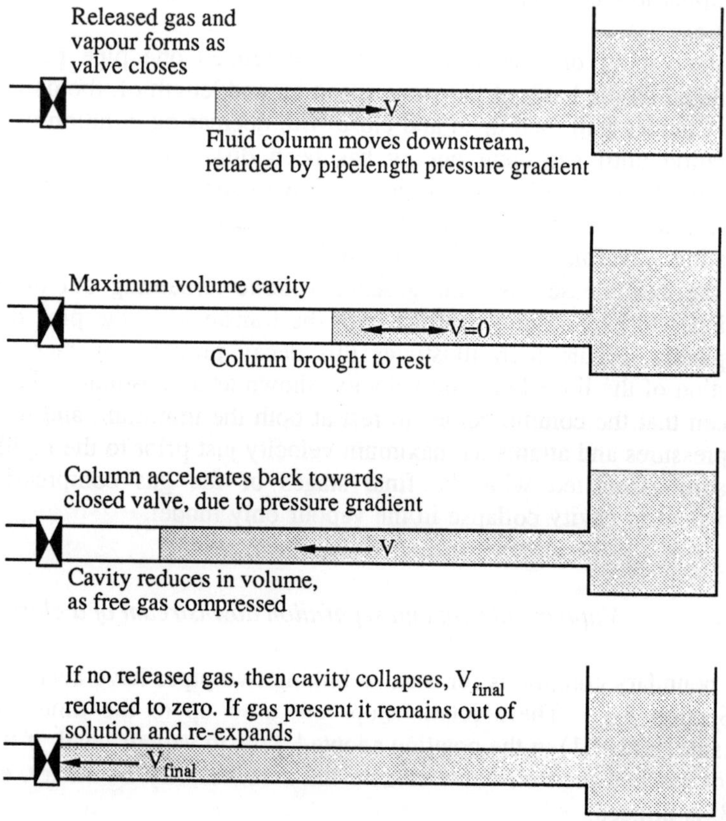

Figure 5.24 Column separation downstream of a closing valve, with gas release to the cavity

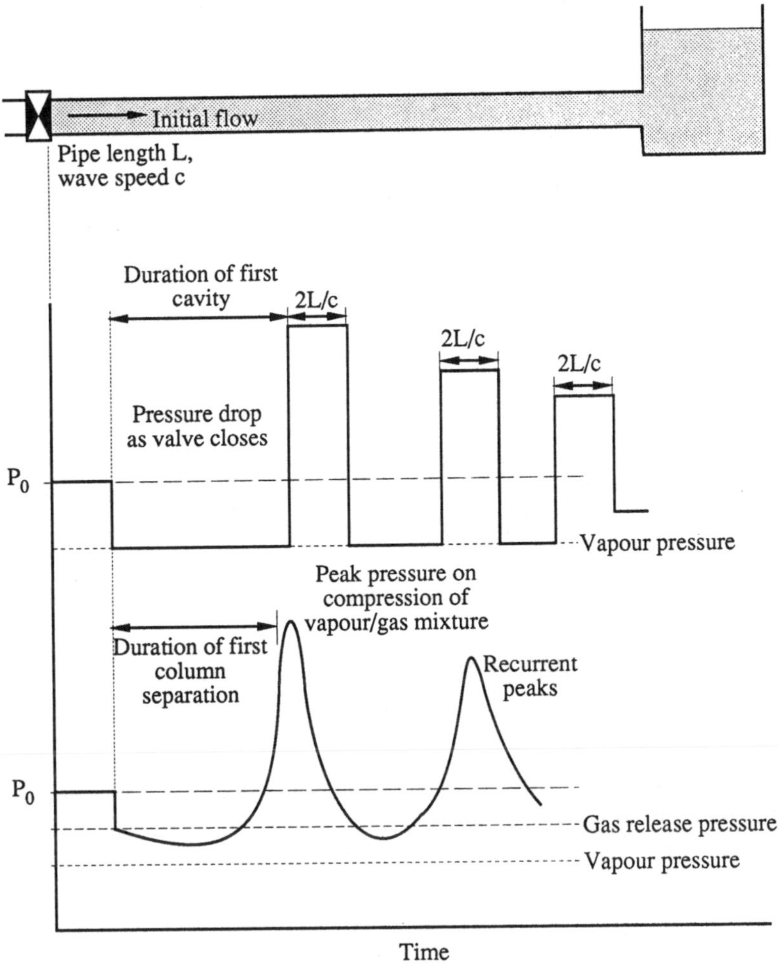

Figure 5.25 Comparison of pressure variations downstream of a closing valve, with and without gas release into the vapour pocket

2. The pressure downstream becomes the fluid vapour pressure

$$p_P(J+1,1) = P_{vap} \tag{5.102}$$

3. The C^+ characteristic from (J,n(J)) to (J,n(J)+1)

$$V_P(J,n(J)+1) = K1(J,n(J)+1) - K2(J,n(J)+1)\, p_P(J,n(J)+1) \tag{5.103}$$

4. The C^- characteristic from (J+1,1) to (J+1,2)

$$V_P(J+1,1) = K3(J+1,1) + K4(J+1,1)\, p_P(J+1,1) \tag{5.104}$$

If a cavity forms during valve closure the vapour is assumed to be concentrated at the valve. The length of the cavity is assumed small compared with Δx so that a valid C^- characteristic exists between sections (J+1,1) and (J+1,2).

The volume of the growing cavity may be monitored during valve closure

$$Vol_{cav}^{t+\Delta t} = Vol_{cav}^{t} + 0.5\, A(J+1)\, \Delta t\, (V_P(J+1,1)+v_P(J+1,1))$$

$$- 0.5\, A(J)\, \Delta t\, (V_P(J,n(J)+1)+v_P(J,n(J)+1)) \tag{5.105}$$

by assuming that a full bore cavity is formed between the upstream face of the separated fluid column continuing towards the downstream reservoir and the downstream face of the fluid column that continues to pass through the valve while the cavity is forming, Figure 5.27.

Following valve closure the volume of the cavity will continue to grow, during this period the volume of the cavity is obtained from a consideration of conditions on the downstream face of the closed valve

$$Vol_{cav}^{t+\Delta t} = Vol_{cav}^{t} + 0.5\, A(J+1)\, \Delta t\, (V_P(J+1,1)+v_P(J+1,1)) \tag{5.106}$$

where positive downstream flow increases the cavity volume. The collapse of the cavity is indicated by equation 5.106 having a negative or zero value at the end of a time step. The boundary at the closed valve then becomes

$$V_P(J+1,1) = 0 \tag{5.107}$$

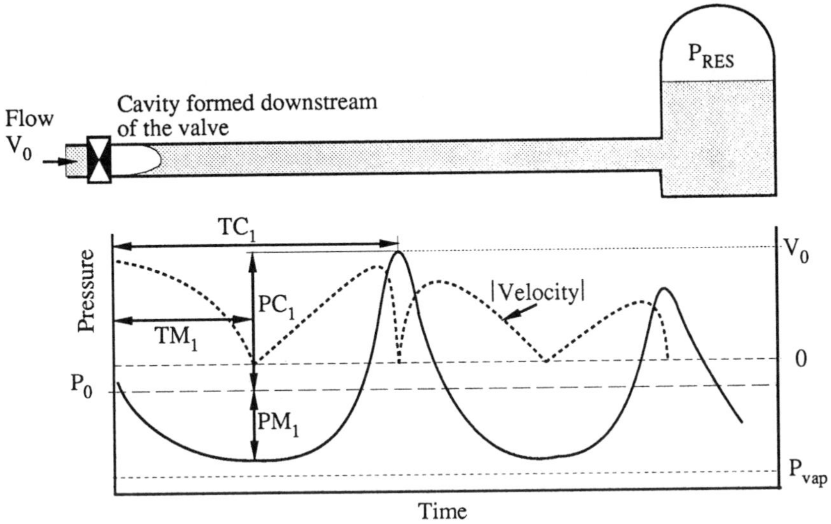

TC_1 cavity collapse time

TM_1 minimum pressure time

PC_1 cavity collapse pressure

PM_1 minimum pressure during growth

Figure 5.26 Column separation on the downstream side of a closing valve

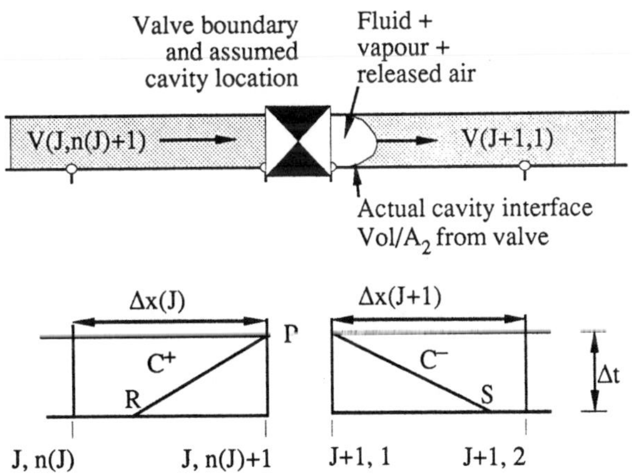

Figure 5.27 Separation downstream of a closing valve

and the pressure at the downstream face of the valve becomes

$$p_P(J+1,1) = -\frac{K3(J+1,1)}{K4(J+1,1)} \tag{5.108}$$

Subsequent cavities open and are monitored by equation 5.106. The process repeats until sufficient energy has been dissipated.

5.7.4 Cavity boundary conditions including released air

If the vapour cavity at the downstream face of the valve is considered to contain freed gas, as shown in Figure 5.27, then the pressure boundary condition becomes

$$p_P(J+1,1) = P_{vap} + P_{gas} \tag{5.109}$$

where p_{gas} is the partial pressure of the released gas within the total volume of the cavity.

The value of the gas partial pressure may be determined as

$$P_{gas} = P_{atm}\left(\frac{Vol_{gas}^{t+\Delta t}}{Vol_{cav}^{t+\Delta t}}\right)^{n} \tag{5.110}$$

where n is the polytropic coefficient of expansion, Vol_{gas} is the volume of released gas measured at n.t.p. and Vol_{cav} is the cavity volume calculated from equation 5.105 during valve closure. The volume of gas released may be calculated from Henry's Law, Derry et al (1952). If K is the Bunsen solubility coefficient, defined as the volume of gas that will dissolve in one unit volume of fluid under a partial pressure of one atmosphere, then the quantity of gas released by a unit volume of fluid at a pressure $p_P(J+1,1)$ is

$$Vol_{gas}^{t+\Delta t} = K\left[\frac{P_{atm} - p_P(J+1,1)}{P_{atm}}\right] \tag{5.111}$$

where P_{atm} is atmospheric pressure and $p_P(J+1,1) \leq p_{atm}$.

The volume of fluid giving up its dissolved gas may be calculated as that volume that passed through the valve while the downstream pressure was below atmospheric; already utilised in equation 5.105

$$Vol_{fluid}^{t+\Delta t} = Vol_{fluid}^{t} + 0.5\, A(J)\, \Delta t\, (V_p(J,n(J)+1)+v_p(J,n(J)+1)) \tag{5.112}$$

and the released gas volume generated by this fluid volume, by reference to equation 5.111, becomes

$$Vol_{gas}^{t+\Delta t} = Vol_{fluid}^{t+\Delta t}\, K \left[\frac{P_{atm} - P_{average\ \Delta t}}{P_{atm}} \right] \tag{5.113}$$

based on the average pressure on the valve's downstream face during that time step and the fluid volume available to release its dissolved gas.

During valve closure the equations above, together with the C^+ and C^-characteristics may be used to solve for the pressure and velocity on either side of the valve. If n = 1 then this results in a quartic in $p_P(J+1,1)$ that is readily solvable. This value of n would seem reasonable as the gas volume is small compared to the likely fluid volume so that the fluid acts as a heat sink during the transient propagation. If n > 1 then the equations may be solved by an iterative technique.

Following valve closure, and within the opening phase of the cavity, pressure and velocity on the cavity-fluid interface continue to be calculated from equations 5.104 and 5.110. Gas is assumed to be released from the fluid throughout the opening phase of the cavity, the fluid volume involved in gas release is assumed to be the final volume Vol_{fluid} on valve closure indicated by equation 5.112.

During the closing phase of the cavity no gas is allowed back into solution, an assumption backed by observations. The cavity will reopen a number of times before sufficient energy is dissipated.

For a valve discharging directly from a tank to a pipeline, the analysis discussed above will continue to apply with the obvious simplification that

$$p_P(J,n(J)+1) = P_{supply\ reservoir} \tag{5.114}$$

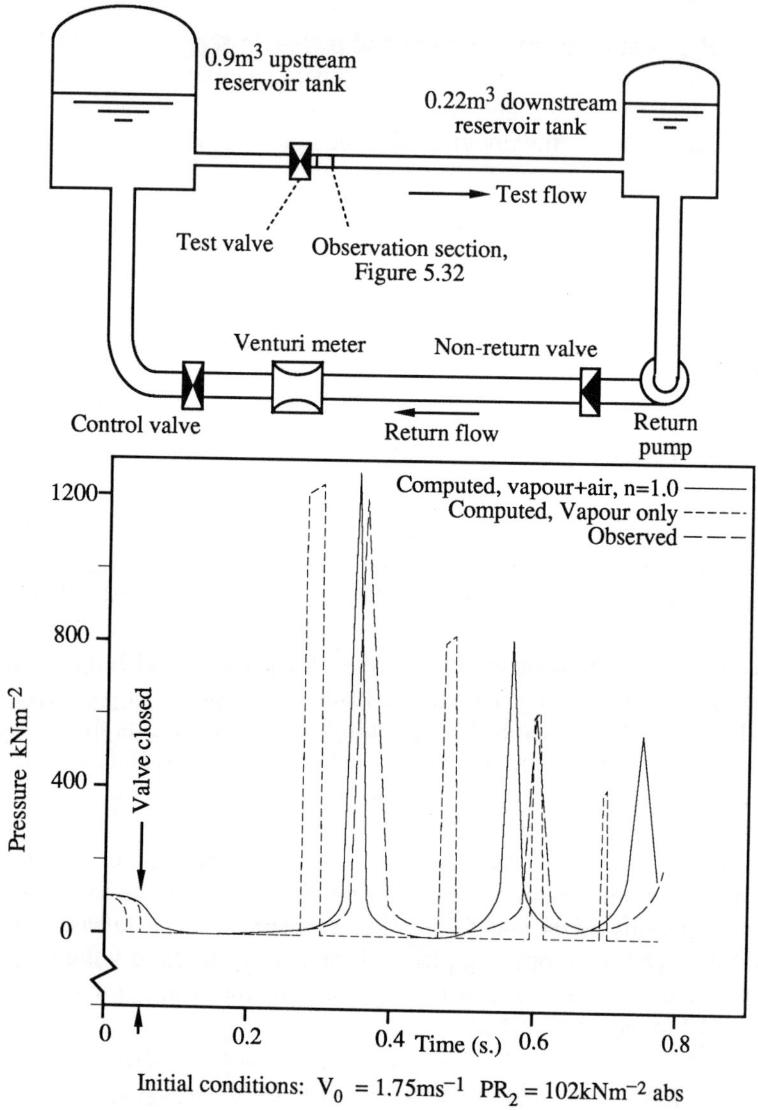

Figure 5.28 Comparison of the vapour only and vapour/released gas boundary conditions for column separation downstream of a closing valve (Swaffield 1972)

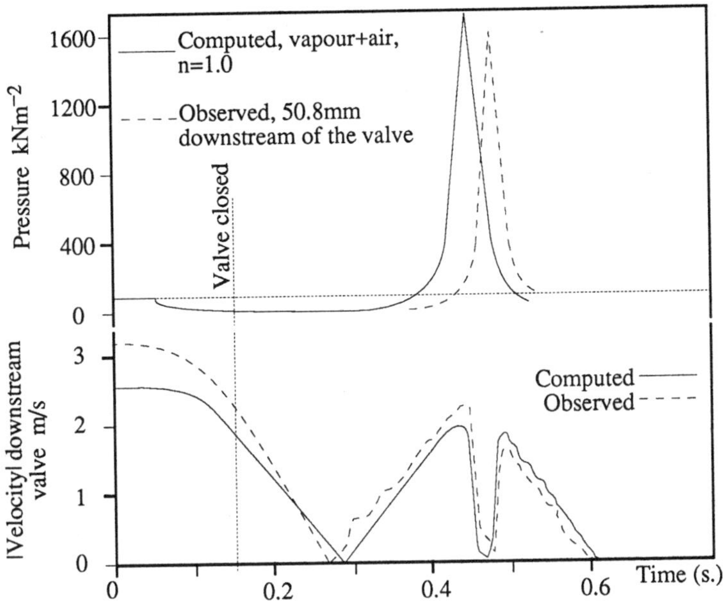

Initial conditions: V_0 = 2.67ms^{-1} PR_2 = 120Nm^{-2} abs TC = 0.14s.

Figure 5.29 Predicted and measured centre line separated fluid column velocity downstream of a closed valve (Swaffield 1972)

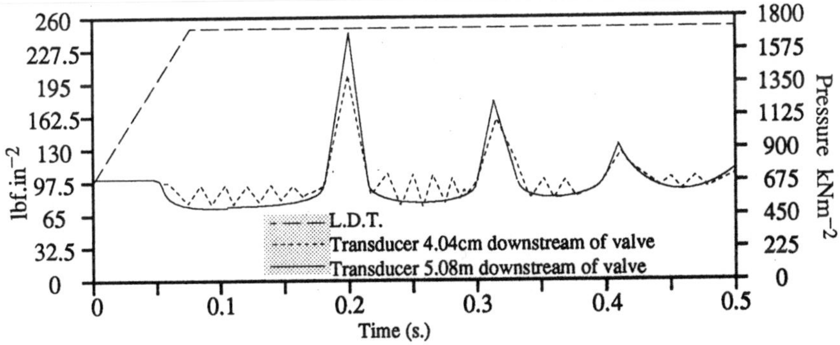

V_0 = 5.75fts^{-1} (1.75ms^{-1}) PR_2 = 32.2lbf.in^{-2} abs (222kNm^{-2})

Figure 5.30 Pressure traces on a valve downstream face and midway to reservoir for pipeline shown in Figure 5.28, illustrating the secondary wave propagation

In this case therefore it would be necessary to continue to calculate the valve throughflow to a downstream pressure lying between gas release pressure and fluid vapour pressure while the valve continued to close in order to be able to determine the Vol_{fluid} and Vol_{gas} terms in equations 5.110 and 5.111.

Figure 5.28 illustrates the degree of agreement achieved in the prediction of the downstream column separation effect, both with and without gas release. The case illustrated utilised aviation kerosene as the working fluid and resulted in a high gas content for the cavity. It will be seen that the inclusion of the gas has a considerable effect on the predicted pressure variations downstream of the valve. This is in marked contrast to the separation upstream of the closed valve, Figure 5.22, where no gas release effects were identified. This difference may be ascribed to the fact that it is the flow agitation caused by passage through a closing valve that initiates the gas release from the fluid in the second case. Figure 5.29 illustrates the measurement of the separated fluid column velocity, obtained using a centre line mounted hot film anemometer probe, and indicates that the form of the separated column velocity variations with time conform to those expected, Figure 5.24, and also agree with the variations described for the trapped air case, Section 5.3.8.

Downstream of the cavity pressure transients continue to propagate and reflect through the separated fluid column between the cavity and the downstream reservoir. These two boundaries may be seen as two −1 reflectors, Chapter 2, as both are effectively constant pressure zones during the existence of the column separation. Discussion may be found in the literature concerning the appropriate wave speed for the fluid column downstream of the cavity, due to the possible effect of free gas liberated at the valve and not concentrated in the cavity area, Kranenberg (1974), Safwat and de Kluyver (1972), Swaffield (1972, 1972[a]).

Figure 5.30 reproduces the observed pressure variations at the closed valve and midway to the downstream reservoir under column separation conditions, Swaffield (1972[a]). It will be seen that the two −1 reflectors result in the detection at a mid pipe length pressure transducer of an oscillation caused by the propagation and subsequent re-reflection of a wave equal in magnitude to the pressure drop from initial line pressure to cavity minimum pressure that left the cavity at the instant that the column separates. The observed frequency of this wave was 44-48 cycles/s, while the frequency corresponding to a fluid only wave speed for the downstream pipe was 47.3 cycles/s. The agreement clearly indicates that the assumption that gas release effects may be 'lumped' at the cavity location was justified.

The magnitude of these secondary oscillations depend upon the rate of initial pressure reduction on valve closure and increase following the collapse of the first cavity. This reinforces the point that cavity collapse is effectively identical to an instantaneous flow stoppage.

The presence of this secondary pressure transient also allows the form of the pressure traces commonly recorded at, and following column separation to be understood. Figure 5.31 illustrates schematically the sequence of pressure waves for an instantaneous valve closure. The secondary wave reflects between two −1 reflectors until the cavity closes, whereupon it perceives on its next arrival at the closed valve a +1 reflector, ie a dead end. This produces a pressure rise wave equal in magnitude to twice the pressure difference between initial line pressure and cavity minimum pressure. As the secondary pressure wave is independent of the duration of the cavity it arrives at the collapsed cavity at any instant during the first pipe period following cavity collapse so that the secondary pressure rise it generates at the closed valve can appear on pressure traces at any time during this first post cavity pipe period.

Subsequently each re-opening of the cavity generates its own secondary wave which continues to propagate along with the all those produced by each preceding cavity, resulting in the, at first sight, confusing traces generated by column separation. The one important point that arises from this is that the maximum absolute pressure level reached within the first pipe period following cavity collapse, either upstream or downstream of a closed valve is

$$P_{max} = P_{vap} + \rho c V_{final} + 2\,(P_{t=0} - P_{vap}) \qquad (5.115)$$

Swaffield (1972[a]) presented photographic evidence of the development of the vapour/released gas pocket at the valve downstream face, filmed at 1000 frames/s, during valve closure and subsequent cavity growth and decay. A few frames are reproduced in Figure 5.32, illustrating the growth and decay of the released gas filled cavity on the downstream side of a valve during and following its closure. The original film confirmed that gas release commences during the later stages of valve closure and that the released gas remains out of solution following the final decay of the pressure transients.

With the valve initially open, at time = 0.0 s, gas is seen to be released from the fuel as the pressure on the valve downstream face falls below atmospheric during closure; completed at time = 0.1 s. By time = 0.25 s the cavity/released gas volume has reached its maximum extent, collapsing at

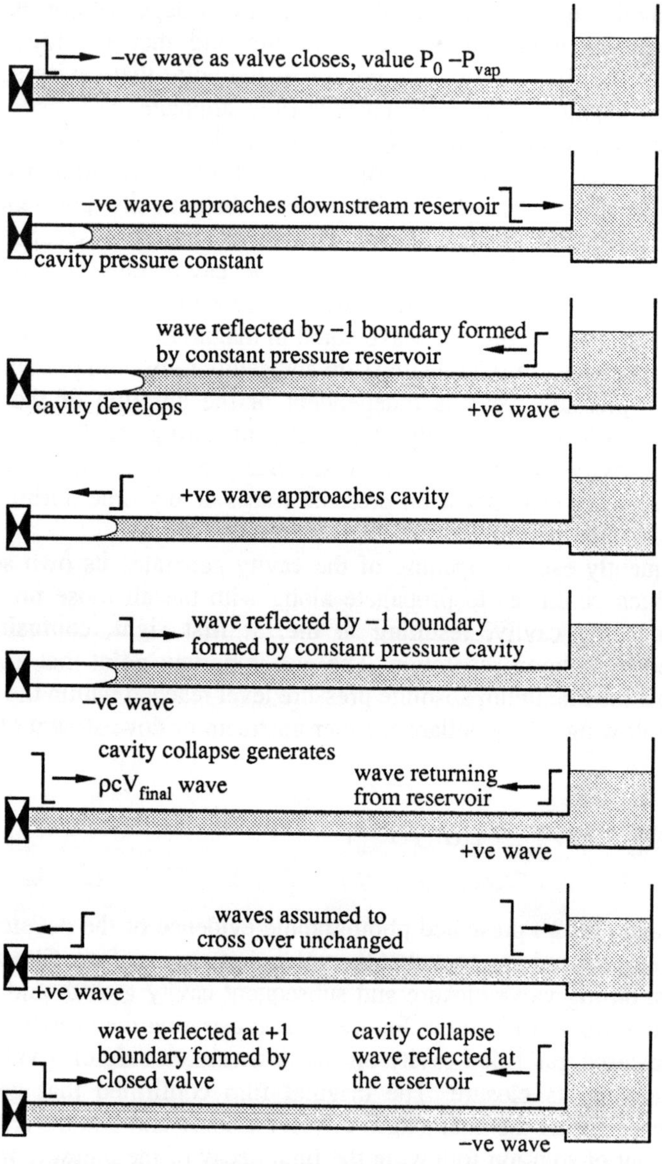

Figure 5.31(a) General form of pressure variations following valve closure and during column separation

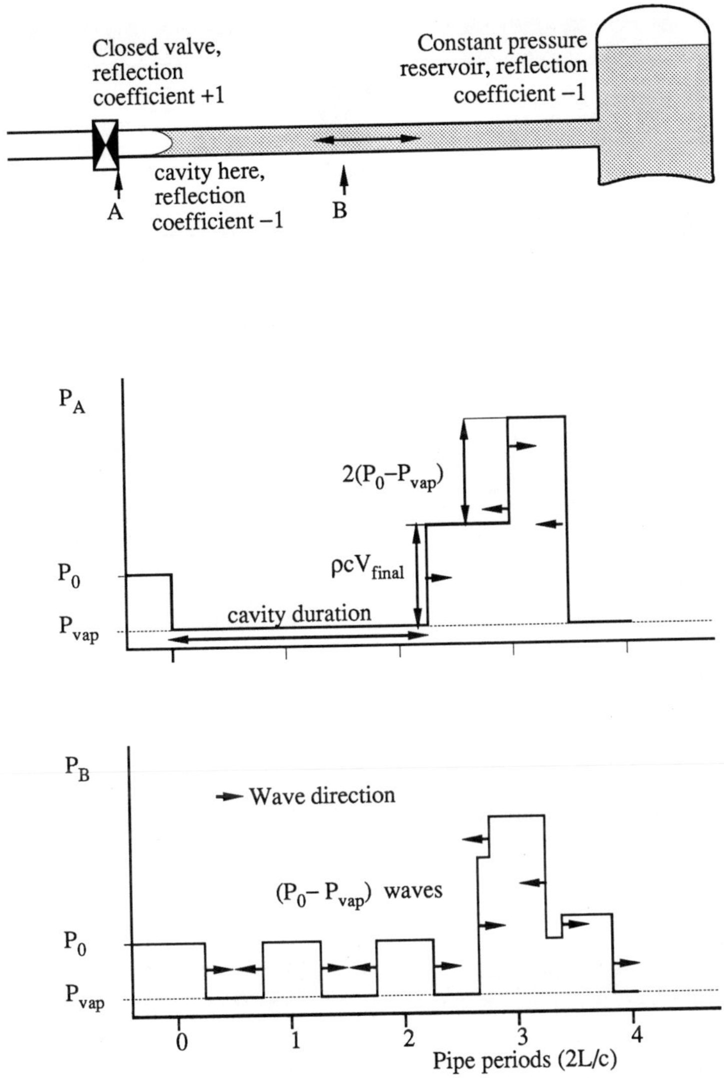

Figure 5.31(b) General form of pressure variations following valve closure and during column separation

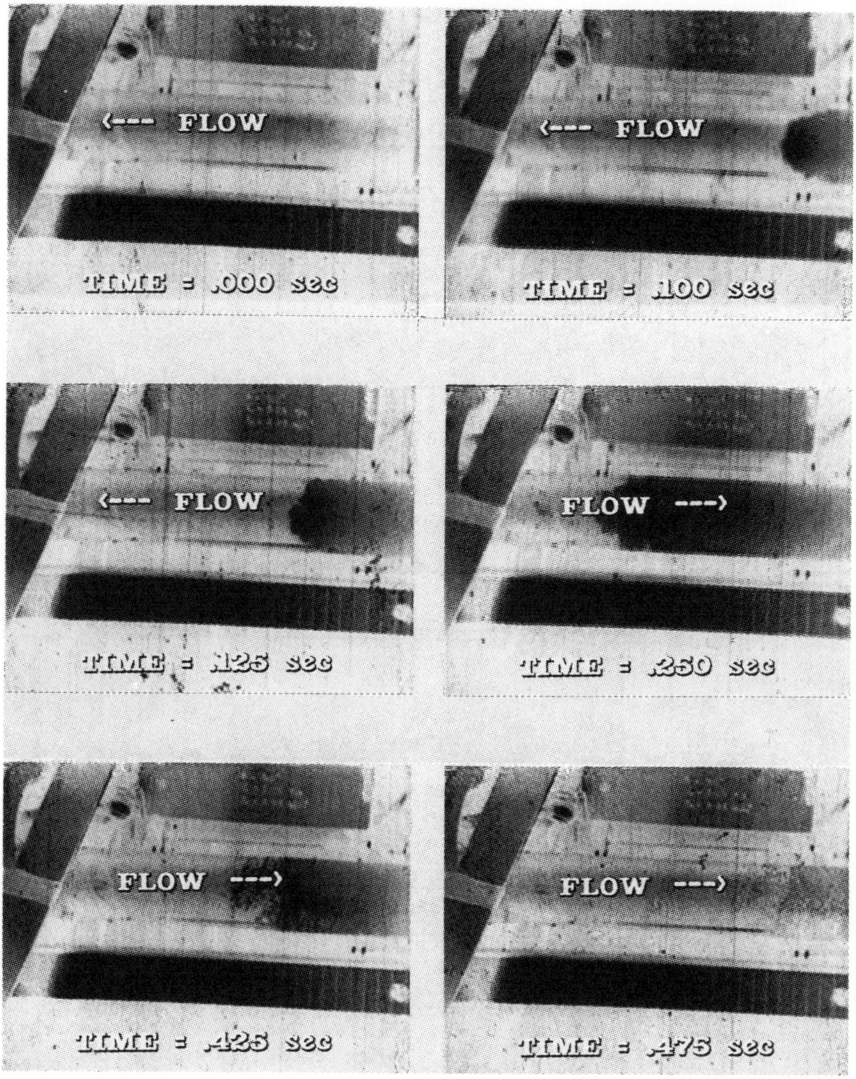

Figure 5.32(a) **Column separation and air release filmed at 1000 frames per second downstream of a closing valve in a 50mm diameter glass and aluminium aviation kerosene pipeline, Swaffield (1972[a])**

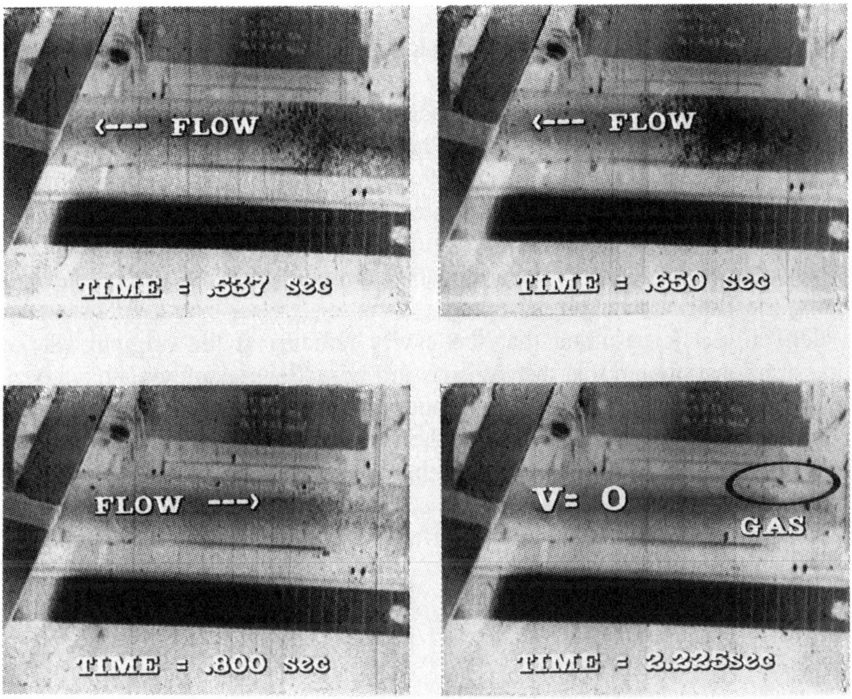

Figure 5.32(b) **Column separation and air release filmed at 1000 frames per second downstream of a closing valve in a 50mm diameter glass and aluminium aviation kerosene pipeline, Swaffield (1972[a])**

the end of the first cycle by time = 0.475 s. Subsequent cavity/released gas volume expansions and contractions repeat until the fuel column finally comes to rest by time = 2.25 s. The gas released from the fuel during the pressure transient event remains out of solution and may be identified on the crown of the glass pipe at time = 2.25 s.

5.7.5 Column separation within a pipe length

Column separation may occur as a negative transient passes through a low head region of pipe, for example at the top of a gradient or at a high point in the system, for example at condensers in power plant cooling circuits, Safwat (1976). Figure 5.33 illustrates the mechanism, in this case the analysis of the pipeline is then split into two sections, upstream and downstream of the cavity location, the common pressure at the interface being the fluid vapour pressure.

Generally it is assumed that the cavity remains at the original vapour pressure location and that the cavity is full bore. Several attempts have been made to introduce free surface conditions beneath the vapour pocket, Marsden and Fox (1976), however the simpler assumptions presented here still represent the usual approach to the problem.

The two C^+ and C^- characteristics still exist on either side of the cavity, although they now indicate two interface velocities

1. Upstream of the cavity

$$V_P(J,I)_{up} = K1(J,I) - K2(J,I)\, p_P(J,I) \tag{5.116}$$

2. Downstream of the cavity

$$V_P(J,I)_{down} = K3(J,I) + K4(J,I)\, p_P(J,I) \tag{5.117}$$

The volume of the cavity may be monitored as

$$Vol_{cav}^{t+\Delta t} = Vol_{cav}^{t} + 0.5\, A(J)\, \Delta t\, (V_P(J,I)+v_P(J,I))_{up}$$
$$- 0.5\, A(J)\, \Delta t\, (V_P(J,I)+v_P(J,I))_{down} \tag{5.118}$$

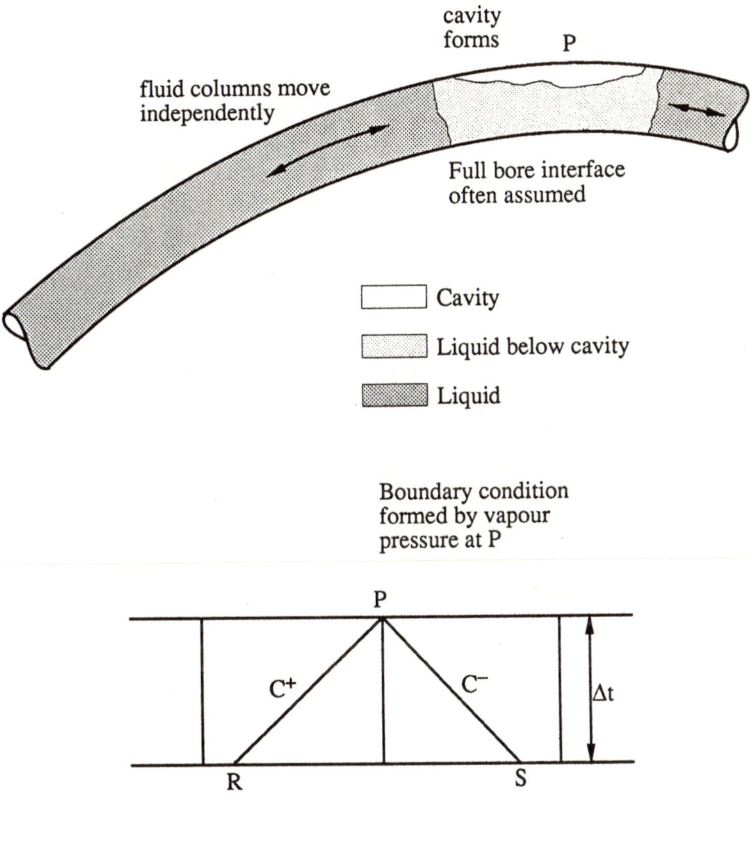

Figure 5.33 Column separation at a high point in the pipeline profile

where the sign chosen for the upstream and downstream velocities ensure that the cavity grows as the columns move apart.

The collapse of the cavity is indicated by a zero or first negative value of equation 5.118. The collapse pressure and the subsequent velocity of the combined column may be determined directly by the solution of equations 5.116 and 5.117.

In this case, and also in the valve located column separation examples, it will be found that the method of characteristics model will predict pressures at or below vapour pressure over several nodes adjacent to the cavity. It has been found acceptable to set these nodal pressures to vapour pressure and to regard the fluid as fluid at vapour pressure once a main cavity location has been identified, rather than introduce a separate 'internal' cavity at each affected node, Safwat and de Kluyver (1972), Swaffield (1970, 1972).

5.8 Development of individual boundary conditions

The boundary equations discussed have so far been common to a wide range of piping installations or have been concerned with identifiable conditions arising from flow conditions during transient propagation. In many cases it is necessary to 'customise' boundary equations to fit a particular case. One such example was introduced in Figure 4.6 where the development of a special boundary to include a valve at the junction of three pipes in a fluid distribution gallery would have considerably shortened the analysis by allowing the pipe section lengths in the gallery to be lengthened and not made dependent, via the Courant Criterion, on the short stub pipes leading to a series of tanks.

Figure 5.34 illustrates this example. The available boundary equations at the junction incorporating the valve are

1. C^+ characteristic for the upstream pipe J

$$V_P(J,n(J)+1) = K1(J,n(J)+1) - K2(J,n(J)+1)\, p_P(J,n(J)+1)$$

2. C^- characteristic for the downstream pipe J+1

$$V_P(J+1,1) = K3(J+1,1) + K4(J+1,1)\, p_P(J,n(J)+1)$$

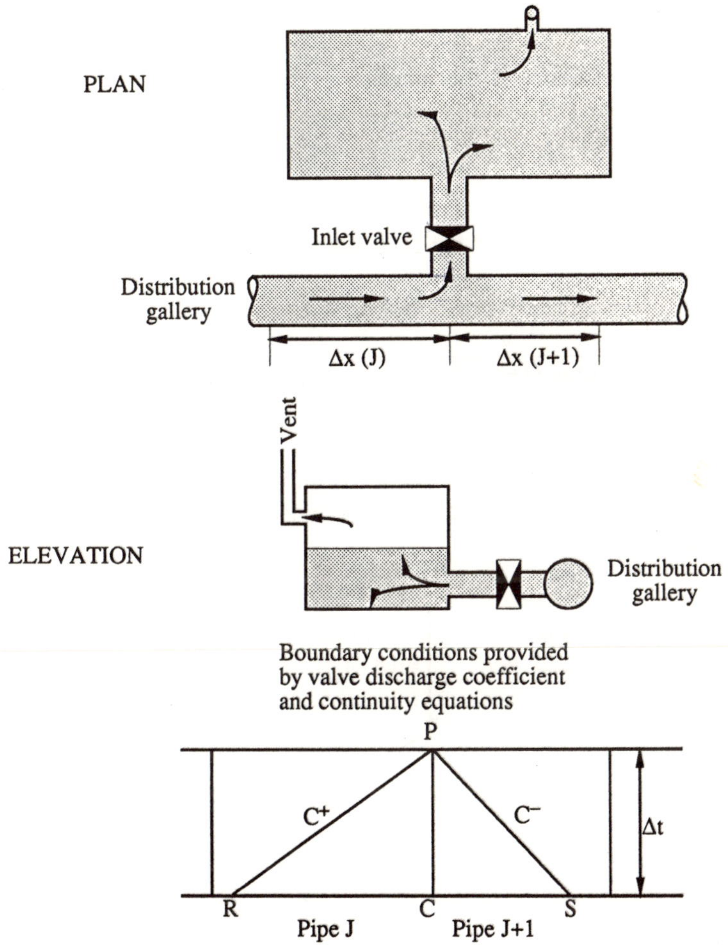

Figure 5.34 Development of a *special case* boundary condition, see Figure 4.6

3. Continuity of flow

$$V_p(J,n(J)+1)\ A(J) = V_p(J+1,1)\ A(J+1) + (AV)_{tank\ inlet}$$

4. Pressure equivalence at the junction

$$p_p(J,n(J)+1) = p_p(J+1,1)$$

5. Discharge coefficient for the tank inlet valve

$$C_d = \frac{V_{tank\ inlet}}{V_0}\sqrt{\frac{\Delta P_0}{p_p(J,n(J)+1) - P_{tank}}}$$

where Δp_0 is the pressure drop through the valve necessary to generate an inflow velocity V_0 through the valve-tank connection pipe.

Combining terms yields

$$A(J)\left[k_1 - k_2\frac{V^2_{tank\ inlet}\,\Delta P_0}{C_d^2 V_0^2} + P_{tank}\right] = (AV)_{tank}$$

$$+ A(J+1)\left[k_3 + k_4\frac{V^2_{tank\ inlet}\,\Delta P_0}{C_d^2 V_0^2} + P_{tank}\right]$$

where

$k_1 = K1(J,n(J)+1)$

$k_2 = K2(J,n(J)+1)$

$k_3 = K3(J+1,1)$

$k_4 = K4(J+1,1)$

resulting in a quadratic solvable for the tank inlet velocity, or flow rate as the connection pipe diameter would be known.

The importance of this approach may be appreciated as the time step, Δt, is linked to the inter node distance, Δx, by the wave speed and local flow velocity through the Courant Criterion. If the Δx value is restricted then the time step can become prohibitively short, for example a doubling in the value of Δx results in a quartered value of Δt.

A complication in this example would involve the fact that the tank pressure would change with time as it filled. This could be accommodated by summing the inflow and determining at each time step the fluid depth over the inlet, hence giving the required reservoir head. In practice the venting of the tank might also be a consideration as this would produce an above atmospheric pressure in the air above the fluid. This effect could be accommodated by monitoring numerically the necessary displacement air flow and checking against the resistance of the venting path. An iterative solution would then yield tank air pressure for use in the determination of the P_{tank} term above.

This example is given to illustrate the relative ease with which specialised boundary conditions may be accommodated within the method of characteristics.

Chapter 6:

Surge control and suppression

Pressure transient problems within pipe and duct systems may be characterised as due to either changes in the flow conditions at the boundaries of the system, changes that generate transients as a means of propagating the information on the change throughout the network, or to local interactions between these transients and local pipeline features. The first category obviously includes any change in pump demand or valve adjustment while the latter encompasses such problems as column separation at high points in the local pipeline profile.

In general the severity of any transient is dependent upon the rate of change of the conditions at its source, for example the rate of valve closure, or opening, determines for any given pipeline the magnitude of the resulting transient. A well understood analysis of transient propagation may allow potentially damaging transients to be avoided by good design, however this is not always possible for a range of reasons and in these cases it may be necessary to incorporate surge control and suppression devices.

The overall generic objective of surge control devices must therefore be the reduction in the rate of change of boundary flow conditions, for example air chambers allow the maintenance of flow conditions while pumps run down or valves close, inwards relief valves maintain pressure levels downstream of either pumps or closing valves. To achieve this end surge control devices must be positioned as close as possible to the source of the transient.

It would be tempting to propose that all pressure surge problems can be suppressed by control devices. Unfortunately this is not the case for a variety of reasons, including cost, probability of transient generation, inadequate design information and system operating requirements. In the large pumping main systems surge relief devices may be practical, however in other systems, equally prone to pressure surge, operating conditions and design criteria preclude the more common surge control devices. An example of the latter case being aircraft fuel systems where system weight and the need to provide rapid changes in engine fuel supply combine to present particular design problems.

This chapter will introduce a range of devices and techniques to achieve these objectives, in each case developing a suitable boundary model to allow the effects of the control device to be incorporated into the method of characteristics analysis presented in previous chapters.

6.1 Control of valve operation generated transients

As outlined in Chapter 1 valve operation is the most common cause of pressure transient propagation. In general, valve generated transients are severe due to the possibility of a rapid change in boundary conditions, defined in terms of the relevant pipe period rather than absolute time. Valve closures in less than one pipe period generate the Joukowsky pressure rise, or fall, and are termed 'fast' while valve closures in longer than one pipe period are 'slow'. Thus the simplest form of surge control might be seen as an increase in the valve closure time.

Figure 6.1 illustrates the change in generated pressure rise on simple valve closure for a range of closure times. While this would appear to be the simplest solution it does not address the whole problem as the form of the valve discharge coefficient vs. position relationship determines the 'effective' valve closure time. For example many valve designs result in flow modification over a relatively minor segment of the valve characteristic, thus reducing markedly the valve closure time. Two, or more, stage valve closures are an effective means of reducing valve generated surge. Figure 6.2 illustrates the effect on the surge generated of closing a valve to 80% shut in 20% of the overall closure time, the final 20% being accomplished in the remaining 80% of the overall closure time.

The modelling of controlled valve closures, including multi-stage operation, has been covered by Sections 5.3.2 and 5.3.3 as the form of the valve position vs. time curve may be used to describe the mode of valve closure, or opening required, Figure 5.4. Similarly the simple system program

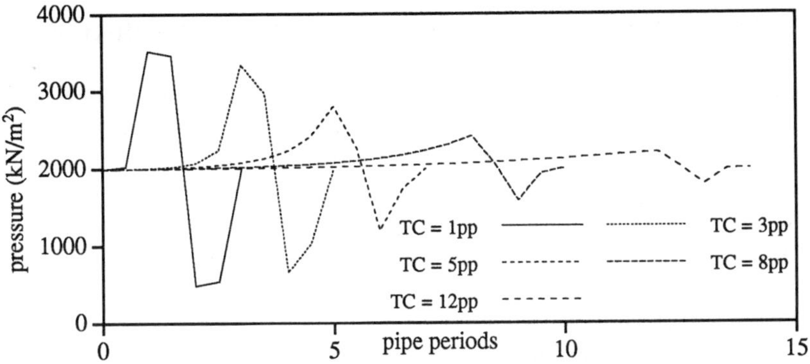

Figure 6.1 Effect of valve closure time, note that the peak pressure surge reduces with increased valve closure time once it exceeds one pipe period

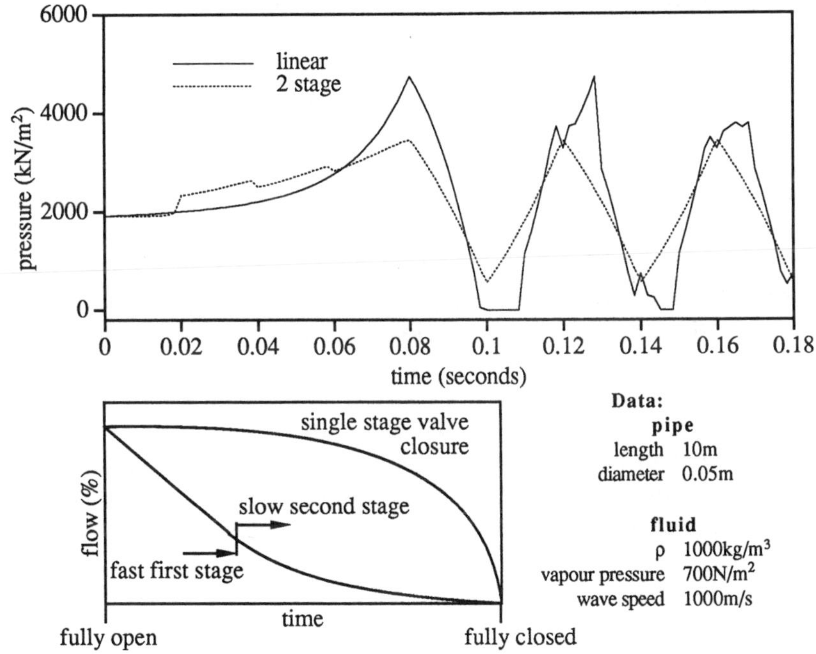

Figure 6.2 Pressure reduction possible with 2-stage valve closures. Note that the column separation present as a result of a single stage closure in this example is suppressed by 2-stage valve motion.

presented in Appendix II, may be utilised to demonstrate the effects of valve motion control.

The use of multi-stage valve closure may be seen as a step in the development of the total control of transient propagation by linking the valve motion to the propagation and reflection of transients within a particular system. Known as valve stroking, this technique may be traced to early work by Knapp (1937) and was developed fully by Streeter (1963, 1967). The technique will not be developed in detail here, however it does provide a basis for the elimination of transient propagation in particular cases provided sufficient information is available. The technique has also been applied to gas pipelines, Stoner (1968), open channel surges, Wylie (1969) and fluids with a high dissolved gas content, Driels (1974, 1975). The full potential of valve stroking will await the introduction of on-line computer control and monitoring of pipeline flow conditions, however even at its most basic level this technique may be seen to hold considerable promise for the future control of transients.

6.2 Outwards relief valves as surge controllers

Possibly the simplest means of reducing the pressure surge generated on valve closure or vapour cavity collapse is to vent the pressurised fluid. Figure 6.3 illustrates a typical outwards relief valve that opens when the pressure in the pipeline reaches a preset value. The effect of this venting is to provide an alternative fluid flow path, thereby reducing the rate of change of flow conditions, which is, as mentioned above, a prerequisite of any surge control device. Figure 6.4 illustrates the effect of introducing such a device downstream of a closing valve as a means of suppressing the cavity collapse pressure surge, reduction of approximately 80% being observed.

The presence of an outwards relief valve may be included in a method of characteristics model. Figure 6.5 illustrates such a device mounted at a dead-ended pipe or at a closed valve. Following valve closure

$$V_p(J,n(I)+1) = 0.0 \tag{6.1}$$

while

$$p_P(J,n(J)+1) = K1(J,n(J)+1) \;/\; K2(J,n(J)+1) \qquad \textit{if} < p_{orv} \tag{6.2}$$

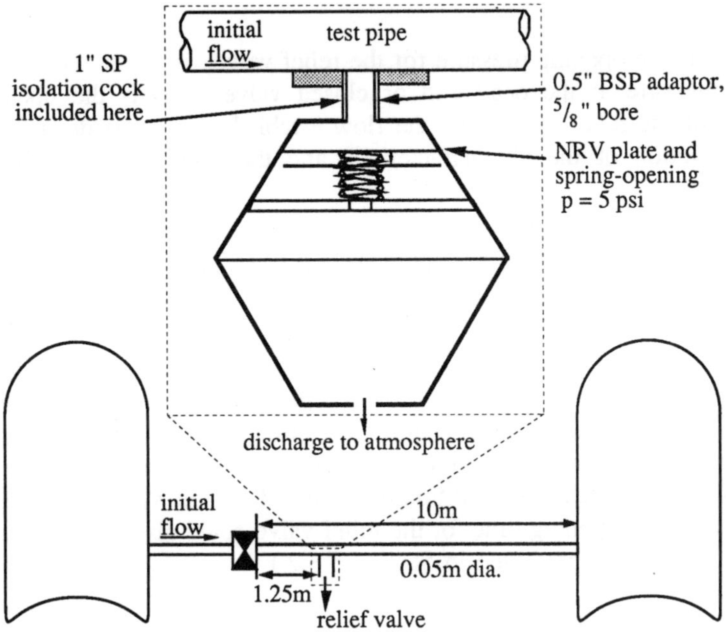

Figure 6.3 Relief valve design and test position

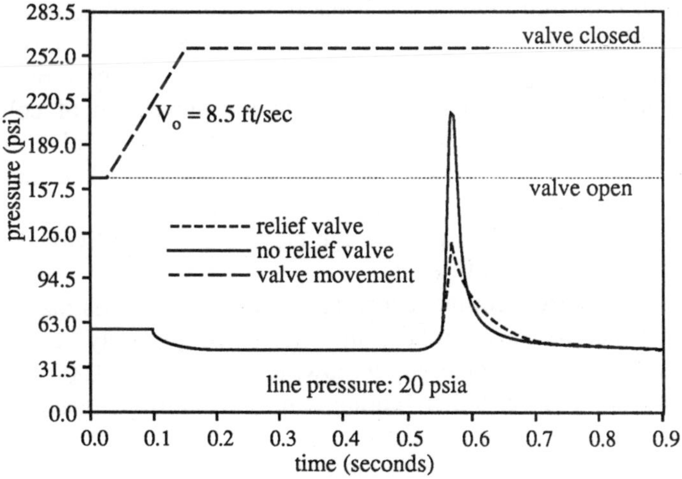

Figure 6.4 ORV effect on peak surge pressure following cavity collapse downstream of closing valve, Figure 6.3

where p_{orv} is the opening pressure for the relief valve.

If the pressure at the dead end or closed valve exceeds the opening pressure for the relief valve then the flow established may be determined from a solution of the C^+ characteristic and the relief valve discharge characteristic

$$V_P(J,n(J)+1) = K1(J,n(J)+1) - K2(J,n(J)+1)\, p_P(J,n(J)+1) \tag{6.3}$$

$$V_P(J,n(J)+1)\, A(J) = q_0 \left((p_P(J,n(J)+1) - ATM) / \Delta p_0 \right)^{0.5} \tag{6.4}$$

where q_0 and p_0 are the relief valve fully open throughflow and pressure drop, the valve is assumed to discharge to atmosphere although a contained discharge to a pressurised zone could obviously be included. The effect of introducing this boundary condition is included in the system program, Appendix II. (An opening time for the relief valve could also be introduced using the techniques presented in Section 5.3.2.)

6.3 Inwards relief valves to alleviate column separation

The natural corollary to the venting of surge overpressures is to allow the ingress of air or fluid into a below atmospheric section of pipeline to reduce the volume and duration of column separation, or to prevent local line pressures actually falling low enough to allow vapour cavity formation. Figure 6.6 illustrates this approach applied to the prevention of column separation on the downstream side of a main refuelling inlet valve on an aircraft fuel system. In such systems safety considerations dictate that the main inlet valve closes automatically, and quickly, if there is a power failure in the aircraft itself that would prevent the closure of any individual tank valves on reaching preset 'full' levels.

In an aircraft fuel system component weight is an added criteria that limits the choice of relief device. Similarly the 'not to be routinely' exceeded pressure surge level is traditionally quite low, in the region of 1000 kN/m^2, or 150 psi, thus relief valves either venting to or drawing from adjacent fuel tanks provide suitable control solutions.

The necessary equations to allow the modelling of the boundary formed by the main inlet valve, distribution gallery and the inwards relief valve may be expressed as follows, if it is assumed that the main external fuel supply to the aircraft may be represented by an effectively constant pressure source, this assumption is justified by the use of flexible hose connections to bowser

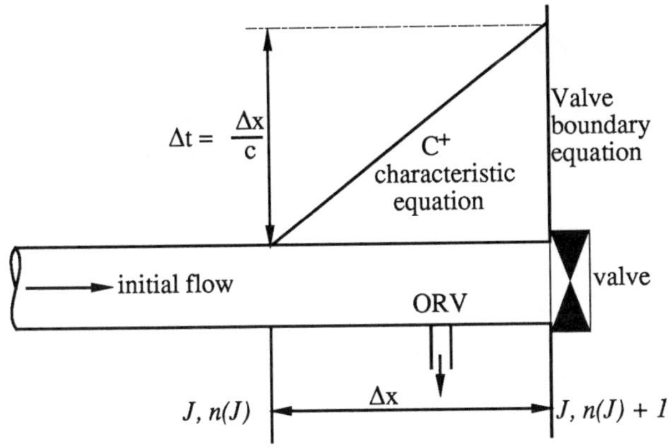

Figure 6.5 Outwards Relief Valve (ORV) positioned at closed valve, opens at preset pressure

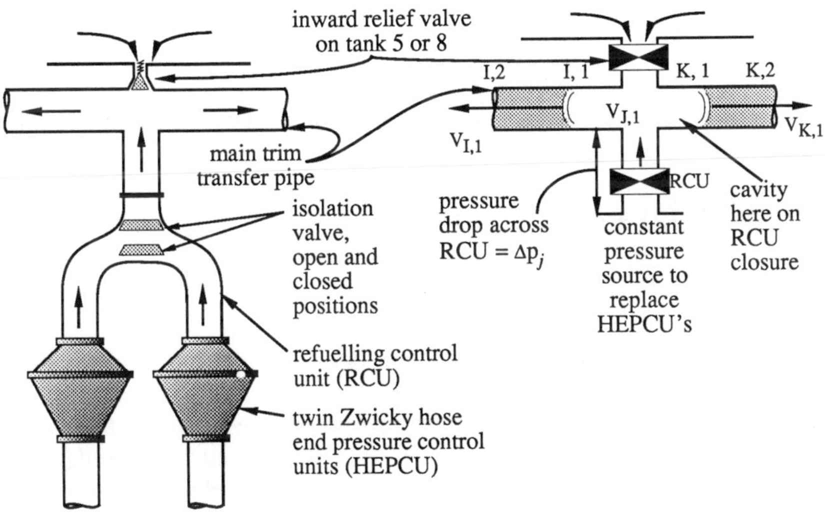

Figure 6.6 Comparison between the twin hose supply to the Concorde RCU and the layout assumed in the analysis. Modelling of main refuelling inlet valve, incorporating IRV

control vehicles and long airfield distribution networks.

1. The valve discharge coefficient for the main inlet valve, assumed to be mounted directly onto the distribution gallery as shown

$$C_d = (V_P(J,1)\ A(J) \ /\ q_0) \ (p_0 \ /\ (p_P(J,1) - p_R))^{0.5} \tag{6.5}$$

where q_0 and p_0 are the inlet valve throughflow and associated pressure drop while p_R is the assumed constant supply pressure.

2. The continuity of flow at the junction

$$V_P(J,1)\ A(J) = V_P(K,1)\ A(K) + V_P(I,1)\ A(I) \tag{6.6}$$

3. The common pressure at the junction assuming that local separation losses are subsumed into the pipe friction terms

$$p_P(J,1) = p_P(K,1) = p_P(I,1) \tag{6.7}$$

4. The C^- characteristics appropriate for entry to the distribution gallery pipes, I and K, as defined in Chapter 5

$$V_P(I,1) = K3(I,1) + K4(I,1)\ p_P(I,1) \tag{6.8}$$

$$V_P(K,I) = K3(K,1) + K4(K,1)\ p_P(K,1) \tag{6.9}$$

Combination of equations 6.6 through 6.9 while the main inlet valve is open yields a quadratic in inflow velocity $V_P(J,1)$ in terms of known conditions a time step earlier and the assumed valve closure-time curve, Figure 5.4. If the valve is to remain open throughout the simulation of closure of tank valves further down the system then the value $\tau = 1.0$ is utilised in equation 6.5 throughout. If the flow reverses at the inlet then the non-return valves incorporated into the refuelling unit close and equation 6.5 is replaced by

$$V_P(J,1) = 0.0 \tag{6.10}$$

If column separation and gas release occur downstream of the main inlet valve during closure then the continuity equation 6.6 is replaced by a relation governing the pressure within the forming cavity

$$p_P(K,1) = p_P(I,1) = p_{vap} + GasVol\, p_{atm} / Vol \qquad (6.11)$$

as developed in Section 5.3.8, where GasVol is the volume of gas released from the fuel at normal temperature and pressure (NTP), Vol is the cavity volume, p_{atm} is atmospheric pressure and the polytropic coefficient is unity.

During closure of the inlet valve the cavity volume may be defined as

$$Vol(t+\Delta t) = Vol(t)$$

$$+ 0.5\Delta t(A(K)(V(K,1)+v(K,1)) + A(I)(V(I,1)+v(I,1)) - A(J)(V(J,1)+v(J,1)))$$

$$- q_R\, \Delta t \qquad (6.12)$$

where q_R is the inwards relief flow if $p_P(I,1)$ falls below the relief valve opening pressure, p_{irv}, and the 'V','v' velocity terms refer to conditions known at P at the end and start of the time step being considered.

The relief flow may be expressed as

$$q_R = q_0\, (\,(p_P(I,1) - p_{tank}) / p_0\,)^{0.5} \quad if\ p_P(I,1) < p_{irv} \qquad (6.13)$$

or

$$q_R = 0.0 \quad if\ p_P(I,1) > p_{irv} \qquad (6.14)$$

where q_0 and p_0 are the relief valve throughflow and associated pressure loss and p_{tank} is the ambient pressure in the tank supplying the relief fuel or gas inflow, note the supply tank may be empty.

The quantity of gas released from the fuel passing through the inlet valve once the downstream pressure falls below gas release level, taken for kerosene as atmospheric pressure, may be expressed as

$$GasVol(t+\Delta t) = FuelVol(t+\Delta t)\, Acon + AirVol(t+\Delta t)_{irv} \qquad (6.15)$$

where Acon is the % of dissolved gas released at any pressure by Henry's Law, ie the amount of gas dissolved is proportional to its partial pressure in the vapour phase and the Bunsen Solubility coefficient K.

$$Acon = (p_{atm} - p_{av}) K / p_{atm} \tag{6.16}$$

where p_{av} is the mean value of $p_P(I,1)$ across a time step.

K may be defined as the volume of gas at 0° C and 760 mm Hg pressure which is absorbed by a unit volume of the fuel. AirVol is the volume of air that would have passed into the cavity volume from the relief tank if it was empty of fuel when the relief valve opened.

The value of FuelVol may be determined as

$$FuelVol(t+\Delta t) = FuelVol(t) + 0.5\, A(J)\Delta t\, (V(J,1)+v(J,1)) + q_R\, \Delta t \tag{6.17}$$

Combination allows the determination of the pressure and velocity downstream of the cavity, with or without relief valve inflow of air or gas releasing fuel.

Following the closure of the main inlet valve the equations above continue to apply with the $V_P(J,1)$ term zero. The relief valve inflow continues until the pressure rises above p_{irv} when these terms are set to zero. No gas is allowed back into solution and this released and trapped gas volume becomes the boundary condition at the junction following the first post valve closure peak pressure.

The inclusion of an inwards relief valve downstream of a closing valve is included in the system program, Appendix II, to enable the reader to investigate these effects further.

Figure 6.7 illustrates a comparison between measured and observed pressure traces for such an inwards relief valve in an aircraft fuel system, analysed using the technique presented above. Surge pressure relief is clearly provided, as is the prevention of column separation, Doyle and Swaffield (1972). Inwards relief valves are safe in this type of application as there is no danger of contamination. This would not be the case in other applications, such as water supply pipelines. A further disadvantage of inwards relief valves is the possibility of entraining large quantities of air in the flow if the inwards relief valve allows air rather than fluid to enter the system. The subsequent removal of this air may be difficult and may also provide problems on subsequent pump startup if not fully removed, Joseph and Hammill (1972), Sheer et al (1973).

6.4 Accumulators as surge alleviators

The use of air chambers to relieve column separation transients is common, however small accumulators may also be utilised to control pressure surge on valve closure in certain intractable design cases. One such example involves the specialist area of in-flight refuelling, particularly of military aircraft, where the need to disengage the receiving aircraft from the tanker under emergency conditions leads to one of the few examples of designed near instantaneous valve closure. Figure 6.8 illustrates schematically the equipment layout, often poded beneath the tanker wing and connected to the tanker's main fuel tanks via pump suction piping and to the receiving aircraft via reinforced flexible hose, often of relatively small diameter. The transfer to the receiving aircraft's tanks is via a fast action valve positioned within the drogue that closes automatically when the receiving aircraft's probe is withdrawn. Thus the system is the classic reservoir-constant speed pump-pipe-fast action valve-pipe-reservoir that will generate the full Joukowsky pressure rise when contact is broken. The obvious and simplistic solution to the resulting surge is to allow fuel leakage on disconnection, however operational specifications forbid this. In many cases the transient is either limited by limiting the maximum fuel transfer rate, obviously not desirable, or by the introduction of alleviators in the tanker aircraft at the junction of hose and tanker pipework.

The available equations to describe the boundary at the accumulator, Figure 6.9, are:

1. The C^- characteristic for the downstream hose

$$V_p(J+1,1) = K3(J+1,1) + K4(J+1,1)\, p_p(J+1,1) \qquad (6.18)$$

2. The C^+ characteristic for the upstream tanker pipework

$$V_p(J,n(J)+1) = K1(J,n(J)+1) - K2(J,n(J)+1)\, p_p(J,n(J)+1) \qquad (6.19)$$

3. The gas law applied to the gas in the accumulator

$$p_t = p_{t=0} \left(AirVol_{t=0} \,/\, Airvol_t \right)^n \qquad (6.20)$$

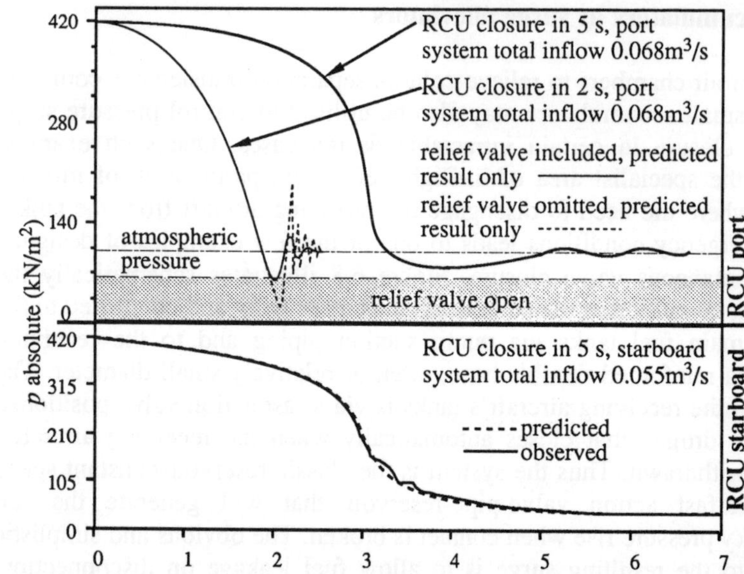

Figure 6.7 Inwards relief valve effect on refuelling surge condition. Predicted and observed pressure variations downstream of the Concorde RCU during and after its closure in an all-tanks-refuelling case

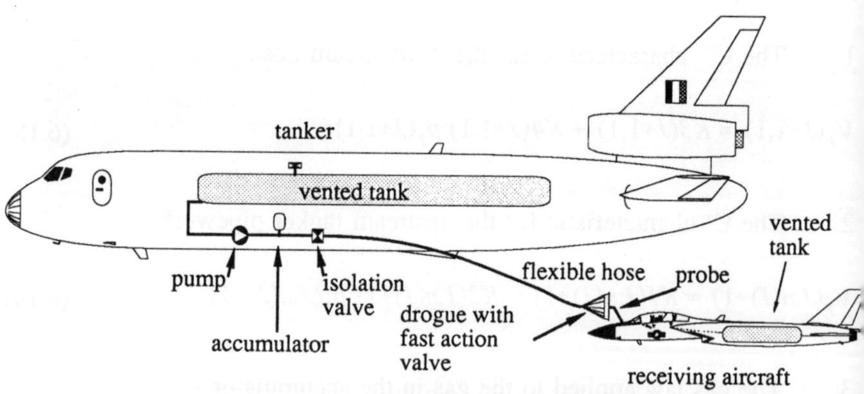

Figure 6.8 Schematic of in-flight refuelling system

Figure 6.8(a) **In-flight refuelling presents one of the few examples of near instantaneous valve closure as a transient generator if emergency disconnection is necessary. Photographs courtesy of Flight Refuelling Ltd**

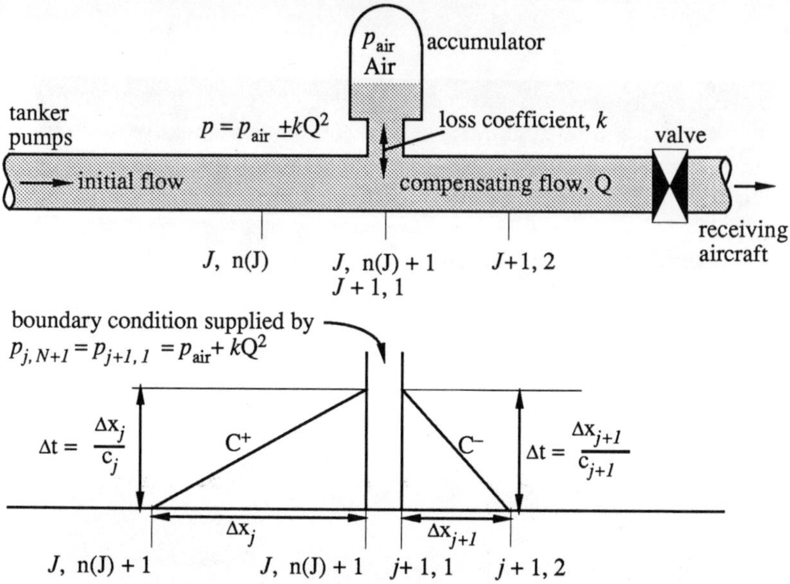

Figure 6.9 Boundary conditions at accumulator in tanker aircraft pod, in-flight refuelling case

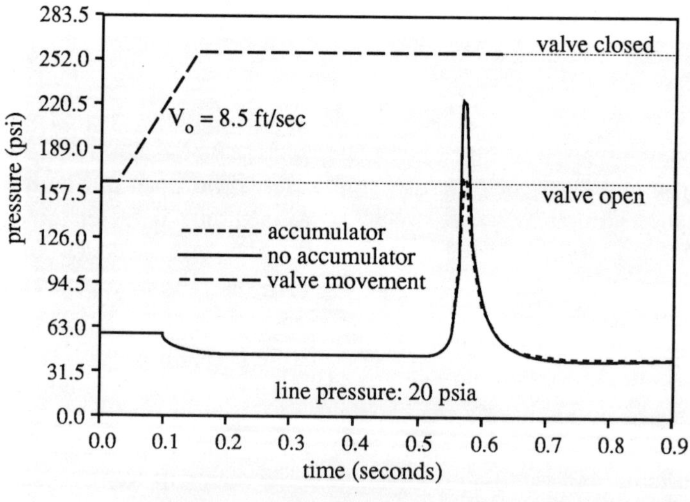

Figure 6.10 1.14 Litre accumulator mounted 1.25m downstream of closing valve to reduce surge on cavity collapse, identical piping arrangement to Figure 6.3

where p is the gas pressure and AirVol is the volume in the accumulator at that pressure, both initially at time zero when the accumulator is charged to the line pressure and during the propagation of transients. The polytropic coefficient of expansion, n, is often assumed to be 1.2.

4. The pressure drop across the accumulator inlet

$$p_P(J+1,1) = p_t + k\,Q\,|Q| \tag{6.21}$$

where Q is the flow into the accumulator taken as positive.

5. Pressure continuity at the junction

$$p_P(J,n(J)+1) = p_P(J+1,1) \tag{6.22}$$

6. Flow continuity

$$V_P(J,n(J)+1)\,A(J) = V_P(J+1,1)\,A(J+1) + Q \tag{6.23}$$

Combination of these equations allows an iterative solution for the pressure at the junction to be developed.

Figure 6.10 demonstrates the degree of surge alleviation achieved with a 1.14 litre accumulator positioned to reduce the pressure surge following cavity collapse in a kerosene pipeline, similar test conditions to Figure 6.3, the performance limited by the use of 150 mm of 6 mm diameter tubing to connect the accumulator to the test pipe. The effectiveness of the accumulator may be seen to improve as the transient rise time increases.

Alternative surge alleviators have been investigated for applications similar to that described above. The utilisation of distributed air along a short length of pipe was proposed by Grundy and Fox (1980) as a means of reducing the local wave speed by increasing the elasticity of the local pipe section. The surge alleviator consisted of a double skinned pipe, the inner skin perforate, and the inter skin void filled with a compressible material. Model tests and method of characteristics analysis indicated that the device could be suitable for transient protection in short pipelines where upstream reflections can return to lower the pressure levels at the attenuator before its storage capacity is exhausted.

6.5 Air vessels as a means of surge control

An air vessel is basically the same as an accumulator, except that air (instead of gas) inside the vessel is usually in direct contact with the fluid in the pipeline.

The most common use of an air vessel is as a pressure surge control device to limit the minimum transient pressure head in a pipeline system downstream of a pump during the transient flow regime following a pump shutdown. In this case, the air vessel is placed downstream of the pumps and of any check valves.

The operation of an air vessel is well documented; Thorley (1991) presents design charts, produced by Graze and Horlacher (1986), that enable a preliminary estimate of the initial air volume and total volume of the air vessel to be made.

Based upon a knowledge of the physical characteristics of the pipeline to be protected, including the pump delivery pressure; the appropriate wave speed and initial flow velocity, and hence the Joukowsky instantaneous flow stoppage pressure surge value; the frictional losses to be expected along the pipeline and the designer specified allowable maximum and minimum pressure heads in the pipeline, the Graze and Horlacher charts allow an estimate of the initial air volume required in the air vessel.

The total air vessel volume may then be estimated by reference to the expansion of this initial air volume as its pressure falls to the minimum acceptable. It is necessary to emphasise that this is a first estimate of the air volume required and it is normal design practice to add in a factor of safety of up to 25%. In the estimation of the gas expansion, as shown below, an appropriate value of the polytropic coefficient of expansion must be selected, see also equation 6.20 relating to accumulators. Values in the range 1 to 1.4 are normally utilised, Thorley (1991) recommends 1.3 to 1.4 if little or no heat transfer is expected, values closer to 1.0 if significant heat transfer occurs, effectively rendering the process isothermal. Similar considerations were introduced in the treatment of released gas during column separation, Section 5.7.4 and equation 5.110.

The air vessel capacities suggested by the Graze and Horlacher charts form the basis for further computer simulations utilising the techniques introduced in this text. The detailed air vessel boundary conditions are developed below and their application illustrated in the case of a pumped main to be protected against transients generated by pump failure. It is important that these simulations be run for a sufficient time to allow the minimum pressures generated with an air vessel in place to be identified.

6.5.1 *Representation of air vessel boundary conditions*

In the general case of an air vessel located between two pipes, the boundary is defined in a manner similar to the boundary conditions presented in Chapter 5. Since air vessels are principally used in water distribution systems, it is appropriate to use the C^+ and C^- characteristics in terms of velocity and piezometric head, developed in Chapter 4, equations 4.19 and 4.20. Therefore the equations for this active boundary condition may be expressed as follows with reference to Figure 6.11(a).

1. The C^+ characteristics at the downstream end of pipe J

$$V_P(J,n(J)+1) = K1(J,n(J)+1) - K2(J,n(J)+1)\, H_P(J,n(J)+1) \tag{6.24}$$

2. The C^- characteristic at the upstream end of pipe J+1

$$V_P(J+1,1) = K3(J+1,1) + K4(J+1,1)\, H_P(J+1,1) \tag{6.25}$$

3. Continuity of discharge at any time

$$Q_P(J,n(J)+1) = Q_P(J+1,1) + Q_{AV}^{t+\Delta t} \tag{6.26}$$

where Q_{AV} is the discharge into the air vessel at the end of any time interval, $t+\Delta t$.

4. The air volume inside the air vessel is assumed to satisfy the polytropic gas law expressed as

$$(H_D^{t+\Delta t} - Z_w^{t+\Delta t} + H_b)\,(\text{Airvol}^{\,t+\Delta t})^{1.2} = C_0 \tag{6.27}$$

where, at time $t+\Delta t$, H_D is the piezemetric head of the air inside the vessel, Z_w is the water elevation, and H_b is the barometric pressure head; the constant, C_0, is defined by steady state conditions.

5. The head at the junction and the head loss across the orifice entry/exit of the air vessel may be expressed as

$$H_P(J,n(J)+1) = H_P(J+1,1) = H_D^{t+\Delta t} + \frac{k\, Q_{AV}^{t+\Delta t} \mid Q_{AV}^{t+\Delta t} \mid}{2g\, A_R^2} \tag{6.28}$$

where k is the air vessel entry/exit orifice head loss coefficient, a function of the flow direction and specified as data, A_R is the cross sectional area of the 'riser pipe' connecting the air vessel to the pipeline.

6. The change in the air volume during the time interval is given by

$$d\ \text{airvol}^{\Delta t} = \text{Airvol}^{\,t} - \text{Airvol}^{\,t+\Delta t} = 0.5\ \Delta t\ (Q_{AV}^{t+\Delta t} + Q_{AV}^{t}) \tag{6.29}$$

7. The elevation of the water surface in the air vessel is given by

$$Z_w^{t+\Delta t} = Z_w^t + \frac{d\ \text{Airvol}^{\Delta t}}{A_{AV}} \tag{6.30}$$

where A_{AV} is the cross sectional area of the air vessel at the water elevation Z_w^t.

The C^+ and C^- characteristic equations 6.24 and 6.25 are expressed in terms of discharge as follows

$$Q_P(J,n(J)+1) = x3 - x4\ H_P(J,n(J)+1) \tag{6.31}$$

$$Q_P(J+1,1) = x5 + x6\ H_P(J+1,1) \tag{6.32}$$

where

$$x3 = A(J)\ K1(J,n(J)+1)$$

$$x4 = A(J)\ K2(J,n(J)+1)$$

$$x5 = A(J+1)\ K3(J+1,1)$$

$$x6 = A(J+1)\ K4(J+1,1)$$

Substituting equations 6.31 and 6.32 into the continuity equation 6.26 and rearranging yields

$$H_P(J+1,1) = Y3 + Y4\ Q_{AV}^{t+\Delta t} \tag{6.33}$$

where

$$Y3 = (x5 - x3)\ /\ (x4 - x6)$$

$$Y4 = 1\ /\ (x4 - x6)$$

and

$$H_P(J,n(J)+1) = H_P(J+1,1)$$

Substituting equation 6.33 into equation 6.28 yields

$$H_D^{t+\Delta t} = Y3 + Y4\ Q_{AV}^{t+\Delta t} - \frac{k\ Q_{AV}^{t+\Delta t}\ |\ Q_{AV}^{t+\Delta t}\ |}{2g\ A_R^2} \tag{6.34}$$

Substituting equations 6.34, 6.29 and 6.30 into equation 6.27 yields the following expression for Q_{AV} at time t+Δt

$$\left[Y5 + Y6\ Q_{AV}^{t+\Delta t} + Y7\ (Q_{AV}^{t+\Delta t})^2\right]\left[Y8 + Y9\ Q_{AV}^{t+\Delta t}\right]^{1.2} - C_0 = 0 \tag{6.35}$$

where

$$Y5 = Y3 - Z_w^t - 0.5\ \Delta t\ Q_{AV}^t / A_{AV} + H_b$$

$$Y6 = Y4 - 0.5\ \Delta t / A_{AV}$$

$$Y7 = -\ 0.5\ k / (gA_R^2)$$

$$Y8 = Airvol^t - 0.5\ \Delta t\ Q_{AV}^t$$

$$Y9 = -\ 0.5\ \Delta t$$

and k, the orifice loss coefficient, is positive for inflow and negative for outflow; the respective values are included in the boundary data and the solution monitors predicted flow conditions at each time step in order to determine the correct value to be utilised.

Equation 6.35 is solved for the unknown $Q_{AV}^{t+\Delta t}$ using the Newton-Raphson technique which is stable since a very good estimate for Q_{AV} at time t+Δt may be obtained by extrapolating values of Q_{AV} calculated at previous time intervals.

The calculated value of Q_{AV} at time t+Δt may then substituted into the relevant equations in order to calculate the remaining time-dependent variables for the air vessel boundary condition.

6.5.2 *Application of the air vessel boundary conditions*

Figure 6.11(b) illustrates the pipeline profile for a typical pumped main together with the hydraulic gradeline corresponding to a steady state discharge of 3.52 m^3/s.

The Graze and Horlacher charts mentioned in Section 6.5.2 indicate that an air vessel with an initial air volume of approximately 28.0 m^3 would be required to limit the minimum static pressure head generated at the junction of pipes 8 and 9 the during transient flow following pump trip to −2.0 m. A series of computer simulations of pump trip incorporating an air vessel at

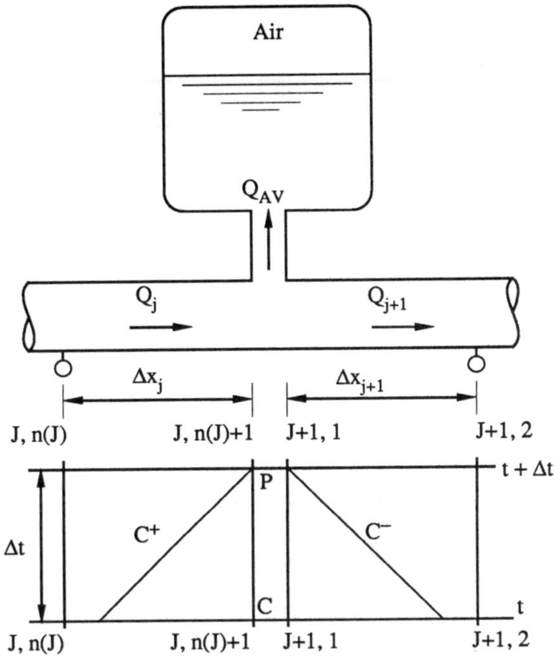

Figure 6.11(a) Air vessel boundary condition

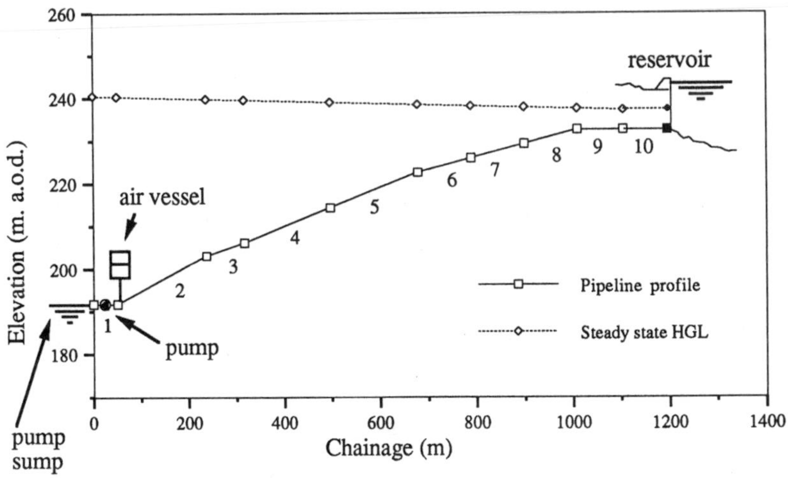

Figure 6.11(b) Variation of static pressure head against time for air vessel with initial air volume of 28m³

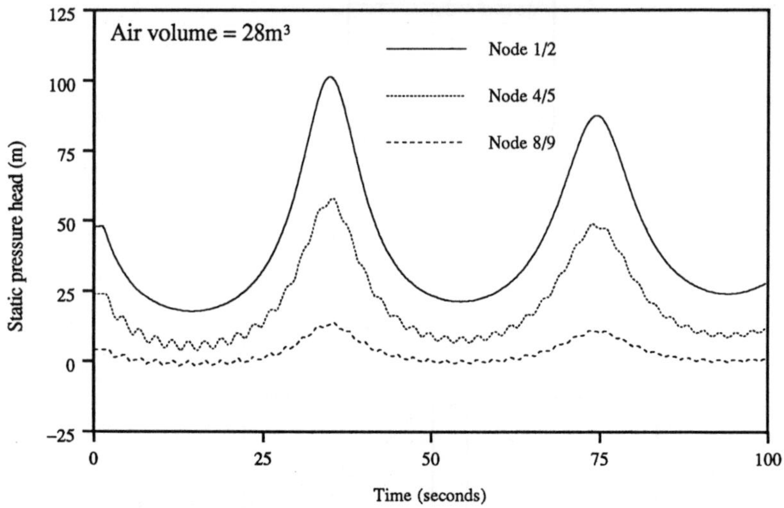

Figure 6.11(c) Variation of static pressure head against time for air vessel with initial air volume of 28m³

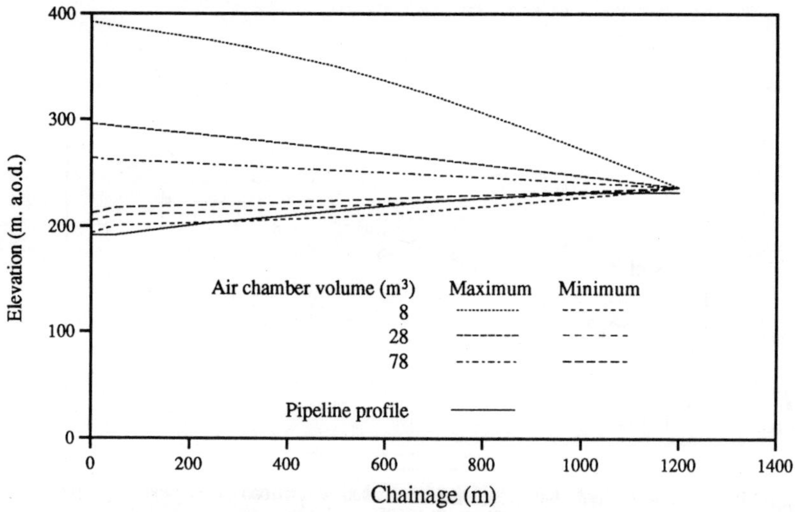

Figure 6.11(d) Maximum and minimum static pressure head envelopes for range of initial air volumes in air vessel

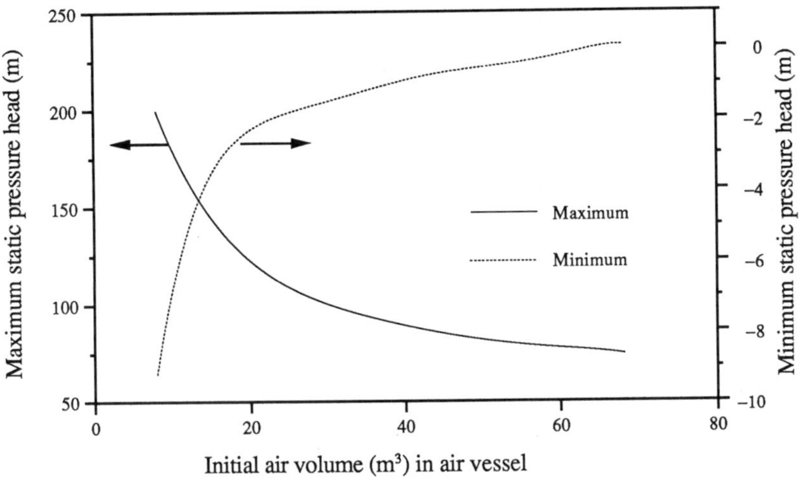

Figure 6.11(e) Maximum and minimum static pressure heads as a function of initial air volume in air vessel

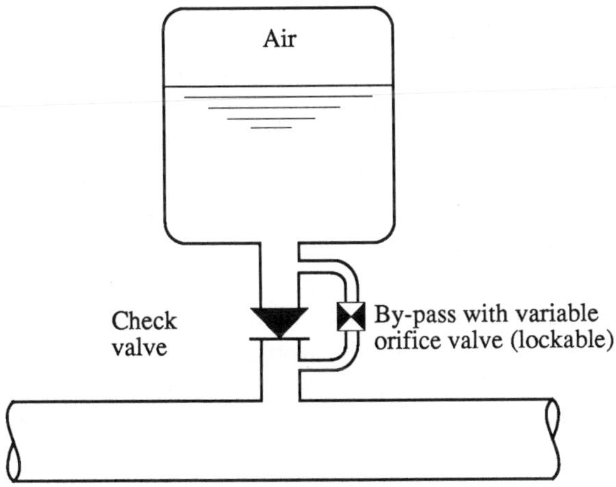

Figure 6.11(f) Air vessel throttled by-pass arrangement

the junction of pipes 1 and 2 illustrate the effect of initial air vessel volume on the subsequent transient pressure history. Figure 6.11(c) illustrates the pressure head vs. time history at selected points along the pipeline, while Figure 6.11(d) illustrates the maximum and minimum pressure head envelopes along the whole pipeline for three specific air vessel volumes, including the Graze and Horlacher recommended value. Figure 6.11(e) shows the predicted maximum and minimum transient static pressure head as a function of initial vessel air volume. These results indicate that the minimum transient static pressure head generated in the pipeline may be controlled by the introduction of an air vessel and that utilisation of the recommended volume of 28 m^3 holds the minimum at –1.6 m, close to the design criterion of –2.0 m.

6.5.3 By-pass arrangement for inflow to an air vessel

The rate at which water enters and leaves an air vessel may be controlled by an appropriate configuration of valves in the riser pipe connecting the pipeline to the air vessel as shown in Figure 6.11(f). Usually, the flow out of the air vessel is unrestricted, thus allowing the maximum continuation flow into the pipeline as the pump slows down. For the return flow, generated once the column between the pump and a downstream reservoir boundary has come to rest, some throttling of the inflow to the air vessel is desirable in order to both damp the pressure fluctuations and limit the maximum pressure head generated when the reversing water column is eventually brought to rest. The optimum loss coefficient for flow into the air vessel may be obtained from the results of a series of computer simulations by varying the outflow and inflow head loss coefficients and these optimim values would be utilised in equation 6.28. As shown in Figure 6.11(f) the variation in air vessel inflow and outflow loss coefficients may most easily be achieved by the introduction of a throttled by-pass to allow inflow to the chamber to be controlled. Outflow is unrestricted through a non-return valve. This configuration allows rapid emptying of the chamber and a controlled filling during the return phase of the pipeline flow oscillation. The degree of throttling necessary, once determined, may be set by providing a lockable throttle valve on the by-pass as shown.

In the majority of pumping installations an air vessel is introduced to prevent the formation of significant sub-atmospheric pressures. This is particularly the case with pipelines conveying drinking water where it is necessary for the line pressures to remain positive at all times under both steady and transient flow conditions, to avoid infiltration.

Controlling the rate of change of flow conditions at a system boundary controls the transients generated. Thus reducing the inflow rate to the air vessel during the return flow phase will reduce the positive pressures generated as the column is brought to rest.

Figure 6.11(g) illustrates the effect of introducing a throttled inflow by-pass to the air vessel incorporated in the pumped main shown in Figure 6.11(b) during the pump trip simulation. The air vessel volume has been maintained at the value of 28.0 m^3 suggested by the Graze and Horlacher (1986) design charts. It is clear that, by restricting the air vessel inflow during the return flow phase, the maximum static pressure head experienced along the pipeline has been reduced. (As the minimum static pressure head experienced along the length of the pipeline will occur as a result of the first transients generated on pump trip this lower envelope boundary remains as illustrated for this air vessel volume by Figure 6.11(d).) It is stressed that in an actual design simulation the by-pass throttle loss coefficient would be determined by successive computer runs to establish its optimum value consistent with the maximum static pressure head to be tolerated in any system.

Figures 6.11(h) to 6.11(j) illustrate the variation of pipeline pressure at the air vessel connection, air vessel air volume and inflow rate, taken positive into the air vessel, with time. It is clear from Figure 6.11(h) that the fall in pressure at the air vessel to pipeline junction following pump trip is unaffected by the introduction of the throttled by-pass as the non-return valve protected outflow connection flow dominates. However, once the flow in the pipeline reverses, in this case at about 15 seconds, the effect of the throttle becomes apparent as the inflow to the air vessel is reduced, relative to the unthrottled case, Figure 6.11(j), and the pressure experienced on the pipeline side of the connection rises due to the increased loss coefficient introduced into equation 6.28. As the inflow to the air vessel decreases, so does the rate of change of air volume, equation 6.29. The throttle thus reduces the peak positive pressure experienced on the pipeline side of the junction and also, as clearly shown in Figures 6.11(h) to 6.11(j), increases the period of the flow oscillation. Figure 6.11(h) also confirms that the minimum static pressure head experienced along the pipeline is determined by the first transient cycle following pump trip.

This simulation thus demonstrates the application of both the method of characteristics model and the underlying principles of transient propagation to a practical example of surge control, emphasising the degree of design flexibility offered by this computer based approach.

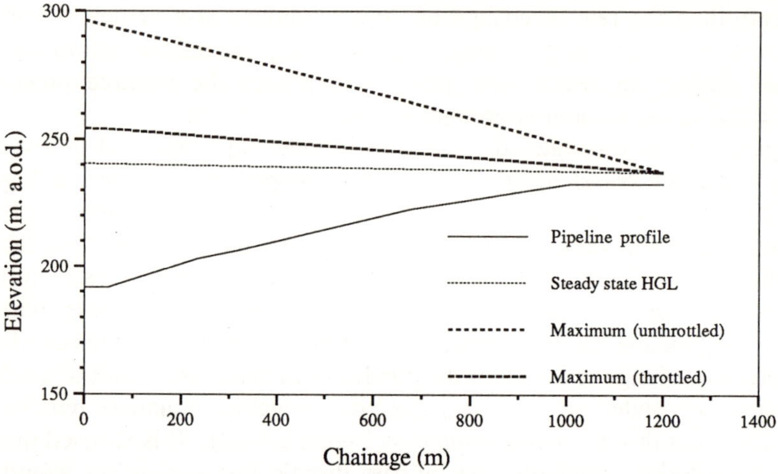

Figure 6.11(g) Comparison of maximum static pressure head envelope for throttled and unthrottled cases

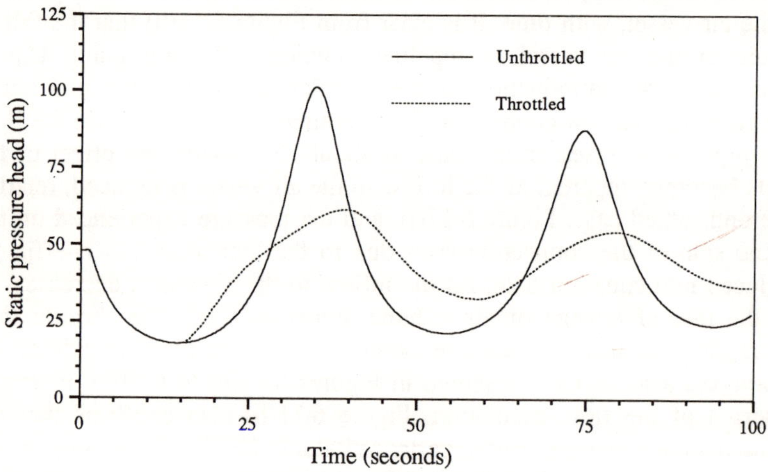

Figure 6.11(h) Comparison of static pressure head envelope at air vessel entry junction for throttled and unthrottled cases

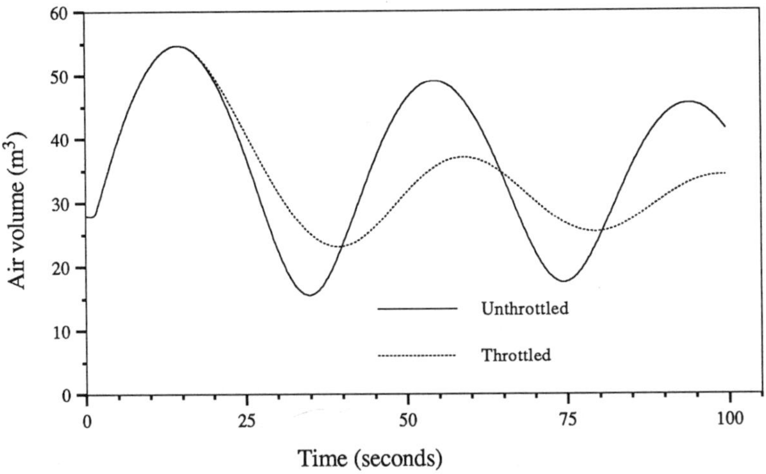

Figure 6.11(i) Comparison of air volume in the air vessel for throttled and unthrottled cases

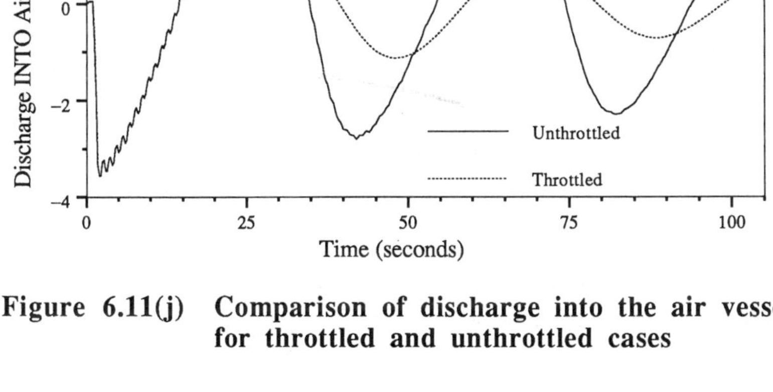

Figure 6.11(j) Comparison of discharge into the air vessel for throttled and unthrottled cases

6.6 Case study: Worsham to Burford pumping main (Courtesy of Thames Water Utilities Ltd)

Prior to this investigation water was pumped at a rate of 16 l/s from Worsham Treatment Works to Burford Reservoir via a 6″ asbestos cement pipeline incorporating a booster pumping station at Swinbrook. The proposed arrangement involved laying a new 8″ high strength PVC pipeline from Worsham to Burford to duplicate the existing 6″ pipeline. The new pumps (duty and standby) at Worsham were rated for the year 2011 peak demand of 30 l/s to be pumped via the new pipeline from Worsham as far as Swinbrook and then via the new and existing pipelines from Swinbrook to Burford. The hydraulic profile of the proposed pipeline arrangements, together with the nodes used for the pressure surge simulations are shown in Figures 6.12 and 6.13.

An existing air vessel is connected to the pipeline downstream of the pumping station. The usable volume of the air vessel was estimated to $0.74m^3$ with an initial steady state air volume of $0.37m^3$.

The pressure surge simulations were undertaken to establish if the existing air vessel is capable of preventing the formation of significant sub-atmospheric (not below −2.0m) pressures during the transient flow following a pump trip. A selection of the simulations are presented in order to highlight some of the fundamental principles of pressure surge phenomena, together with the main conclusions.

6.6.1 Simulations - no pressure surge protection

The first series of simulations of the pressure surges generated in the proposed pipeline arrangement were undertaken to demonstrate the effect of the value of the transient wave speed specified for the new PVC pipeline, whilst maintaining the same value of 930 m/s for the transient wave speed appropriate to the existing asbestos cement pipeline. The significant parameters and results for these simulations are summarised in Table 6.1, together with a listing of the relevant figures.

Following a pump trip the pump check valve closes in approximately 1.5 seconds due to the very small value of the inertia of the pump and motor. Therefore, the Joukowsky pressure head drop is generated downstream of the pump as shown in Figures 6.15, 6.17 and 6.19. Further reduction in the transient pressure drop occurs due to the dissipation of the friction in the pipeline.

The static pressure head against time relationships, Figures 6.15, 6.17 and 6.19, also demonstrate that the reflected pressure surge waves arrive at the pump at approximately the time given in Table 6.2.

Extensive discussions with the manufacturers of the PVC pipeline eventually established values for the pipe geometry and material properties which resulted in a calculated value for the transient wave speed of 310.0 m/s. Table 6.1 and Figure 6.18 shows that for a transient wave speed of 300 m/s the minimum transient pressure in the pipeline falls to vapour pressure with all the pipeline downstream of node 56 subjected to a very high sub-atmospheric transient pressure.

Table 6.1 Pump trip - no pressure surge protection

Run No	Wave speed in a PVC pipe (m/s)	Joukowsky Pressure Head cV_o/g (m)	Pressure Wave Return time 2L/c(s)	Static Head m. Envelopes Max	Min	Max/Min Static Head	Static Head vs. Time. Plot
150	200.0	19.5	64.7	68.4	−1.8	Fig 6.14	6.15
101	250.0	24.3	51.8	68.4	−5.9	Fig 6.16	6.17
102	300.0	20.2	43.1	70.5	−10.0	Fig 6.18	6.19

The Joukowsky pressure change following an instantaneous change in flow velocity is expressed as $c\Delta V/g$ in pressure head terms, m of flowing fluid.

6.6.2 Simulations - existing air vessel at node 2

Figures 6.20 and 6.21 show the maximum and minimum static pressure head envelopes and static pressure head against time plots respectively predicted for the proposed pipeline following a pump trip both with the existing air vessel connected to the pipeline at node 2 and with a value of 300 m/s for the wave speed. The minimum transient static pressure head is limited to −2.1 m at node 59 and the air vessel damps out the pressure surges, Figure 6.21.

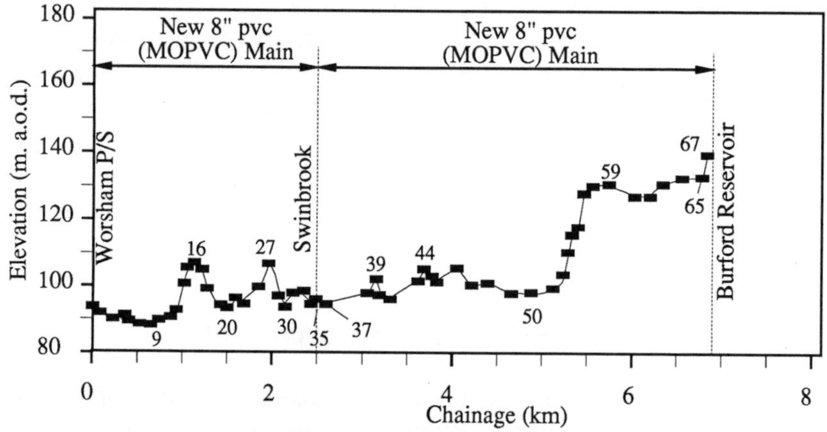

Figure 6.12 Worsham to Burford water main: new line, pipeline profile and node numbers

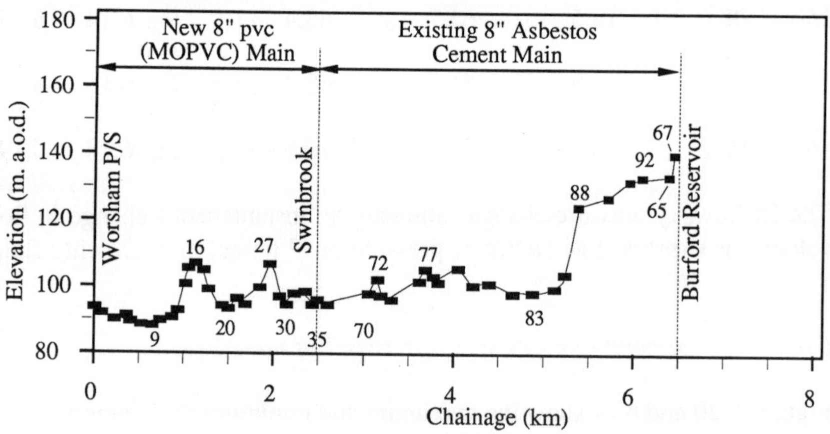

Figure 6.13 Worsham to Burford water main: new/existing line, pipeline profile and node numbers

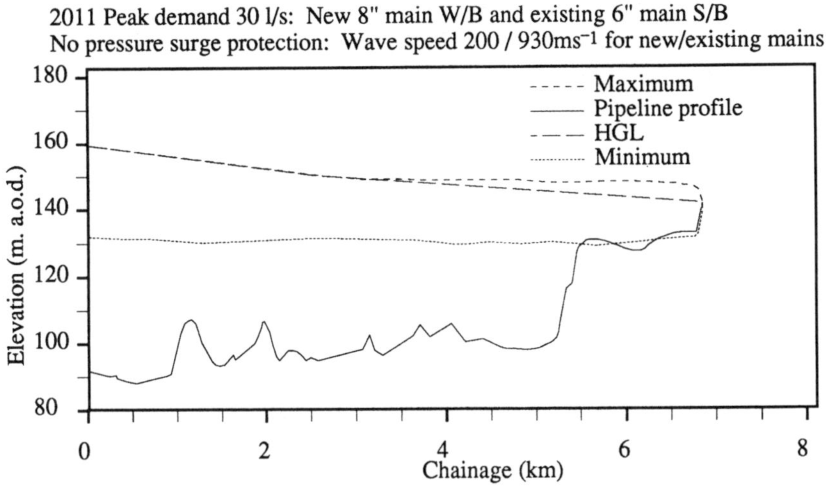

Figure 6.14 New main maximum and minimum static pressure head envelopes

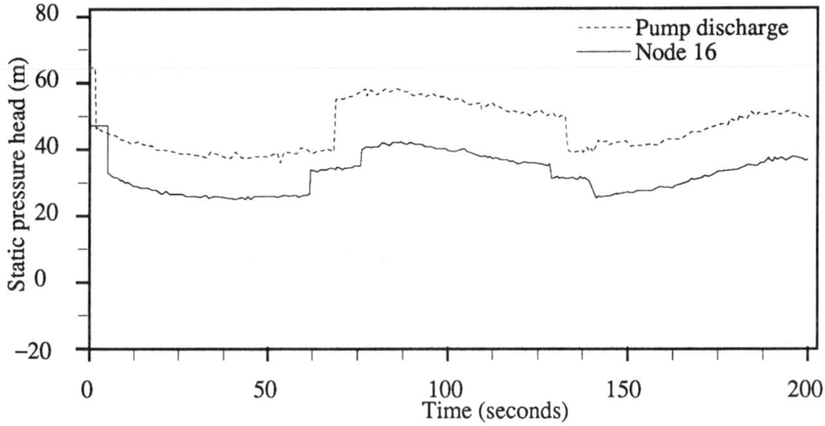

Figure 6.15 Static pressure head

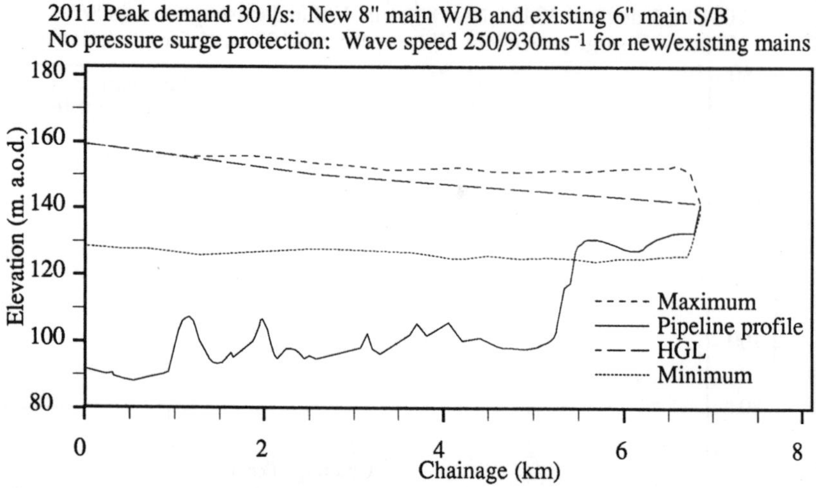

Figure 6.16 New main maximum and minimum static pressure head envelopes

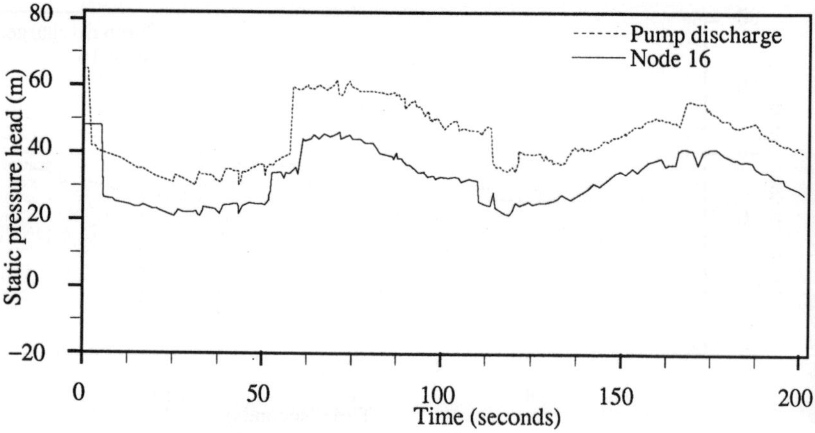

Figure 6.17 Static pressure head

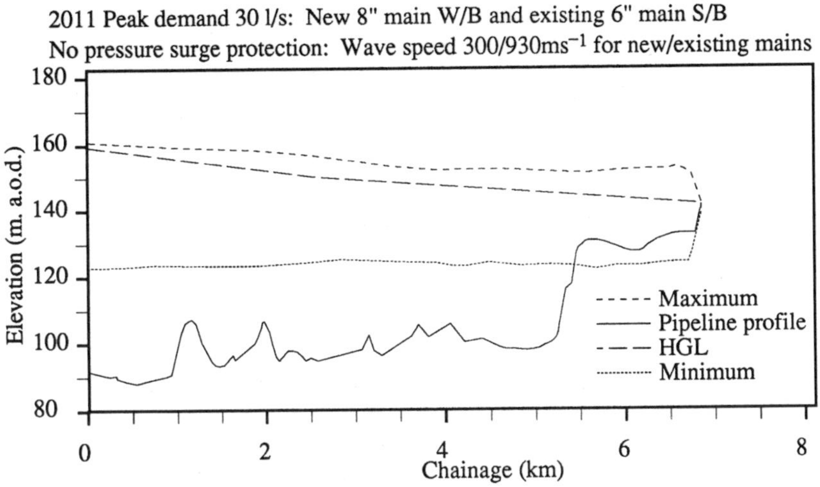

Figure 6.18 New main maximum and minimum static pressure head envelopes

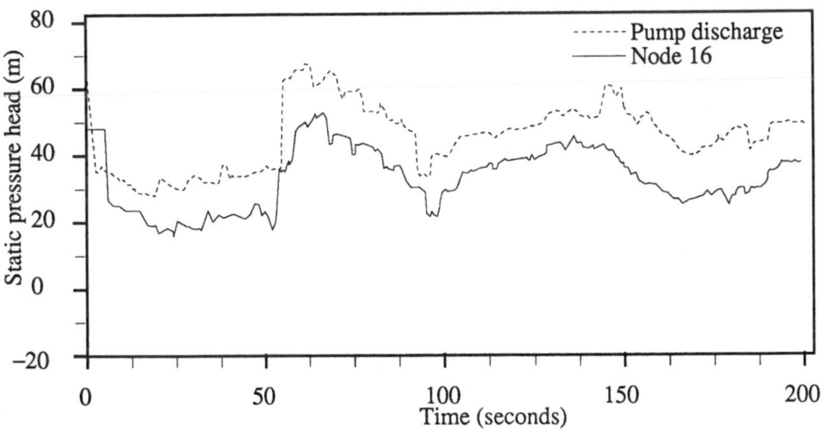

Figure 6.19 Static pressure head

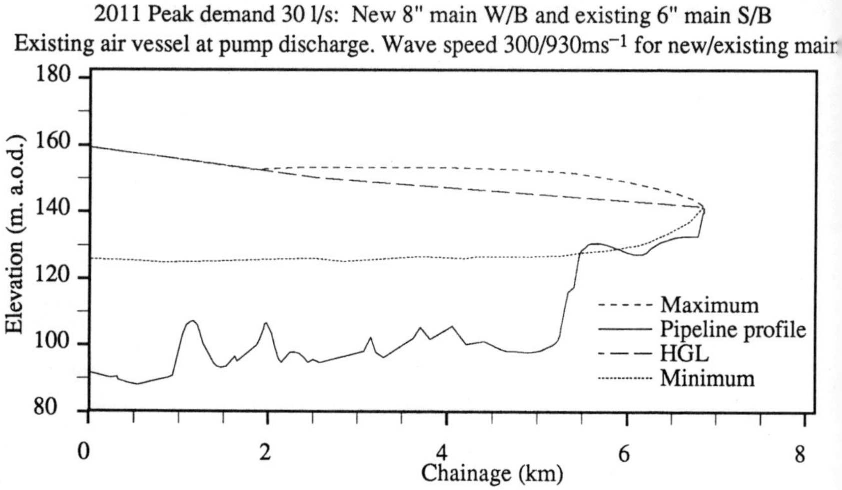

Figure 6.20 New main maximum and minimum static pressure head envelopes

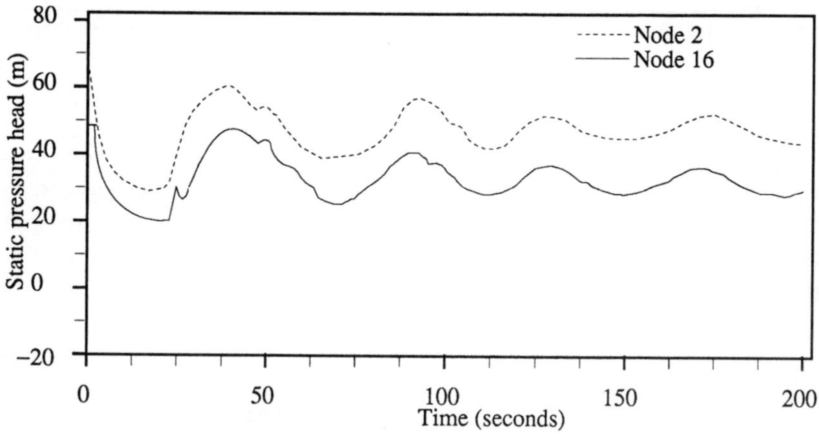

Figure 6.21 Static pressure head

6.7 Air inlet/outlet pressure surge alleviation valve

In any internal fluid flow system, incorporating valves, transient pressure fluctuations will be generated upstream and downstream of the valve during the closure of the valve. The maximum amplitude of the transient pressure fluctuations is a function of both the type of valve and the rate of closure of the valve, particularly during the last 20% of the closure.

The negative pressure surge wave which travels downstream causes the static pressure in the pipeline to fall rapidly during the last 20% of the valve closure, and depending upon the profile of the pipeline and the location of the valve, sub-atmospheric pressure may be generated over a section of the pipeline. In some situations, where the steady state static pressure head is less than the Joukowsky pressure head, the transient static pressure downstream of the valve may well fall to vapour pressure even if the valve closure time is of the order of 1500 seconds.

The most appropriate and economical means of minimising the magnitude of the transient sub-atmospheric pressure is to connect an air inlet/outlet valve to the pipeline downstream of the valve. When the pipeline pressure at the location of an air valve drops below atmospheric pressure, the air valve opens and allows air to enter the pipeline quickly thereby minimising the magnitude of the sub-atmospheric pressure at the valve location.

If the pressure in the pipeline recovers and rises above atmospheric pressure, air is vented through the valve, at a much slower rate than the inflow rate, in order to prevent significant secondary transient pressure when the air valve closes.

If the pressure in the pipeline does not recover sufficiently to cause the air valve to close, then the downstream water column will separate and eventually come to rest with a free water surface at approximately atmospheric pressure at an elevation corresponding to the downstream reservoir water surface elevation.

6.7.1 *Representation of air inlet/outlet valves*

The mass rate of flow of air through the valve depends upon the values of atmospheric absolute pressure p_o and absolute T_o outside the pipe, as well as the absolute pressure p and temperature T within the pipe. The following four equations are required to describe the zones of operation

Zone 1: Critical Inflow

$$m = C_{in} A_{in} \frac{0.686}{\sqrt{RT_o}} p_o \qquad (6.36)$$

for

$$p \leq 0.53 p_o$$

Zone 2: Subsonic Inflow

$$m = C_{in} A_{in} \left(7 p_o \rho_o \left(\frac{p}{p_o} \right)^{1.4286} \left(1 - \left(\frac{p}{p_o} \right)^{0.286} \right) \right)^{0.5} \qquad (6.37)$$

for

$$p_o \succ p \succ 0.53\, p_o$$

Zone 3: Subsonic Outflow

$$m = - C_{out} A_{out} \left(\frac{7}{\sqrt{RT}} \left(\frac{p}{p_o} \right)^{-1.4286} \left(1 - \left(\frac{p}{p_o} \right)^{-0.286} \right) \right)^{0.5} \qquad (6.38)$$

for

$$\frac{p_o}{0.53} \succ p \succ p_o$$

Zone 4: Critical Outflow

$$m = - C_{out} A_{out} \frac{0.686}{\sqrt{RT}} p \tag{6.39}$$

for

$$p \succ \frac{p_o}{0.53}$$

where

m	: mass rate of air flow (kg s^{-1})
C_{in},C_{out}	: valve discharge coefficients
A_{in}, A_{out}	: area of valve opening (m^2)
ρ_o	: mass density of air (kg m^{-3})
R	: gas constant (Nm $kg^{-1o}K^{-1}$)
T	: absolute temperature (°K)
p	: pressure in the pipe (N m^{-2})

By defining the critical mass rate of inflow of air as a constant (CON) and the pressure ratio p/p_o as p', equations 6.36 and 6.37 can be written as

$$m = CON \tag{6.40}$$

and

$$m = CON\ 3.8568 \left((p')^{1.4286} - (p')^{1.714} \right)^{0.5} \tag{6.41}$$

Similarly, by rearranging equation 6.39 for the critical mass rate outflow of air as

$$m = - C_{out} A_{out} \frac{0.686}{\sqrt{RT}} p_o p' = - COU\ p' \tag{6.42}$$

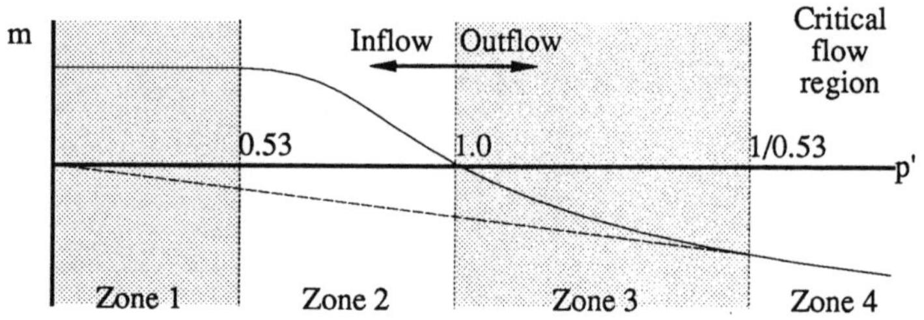

Figure 6.22 Schematic relationship of mass rate of air flow through an air inlet/outlet valve as a function of the pressure ratio

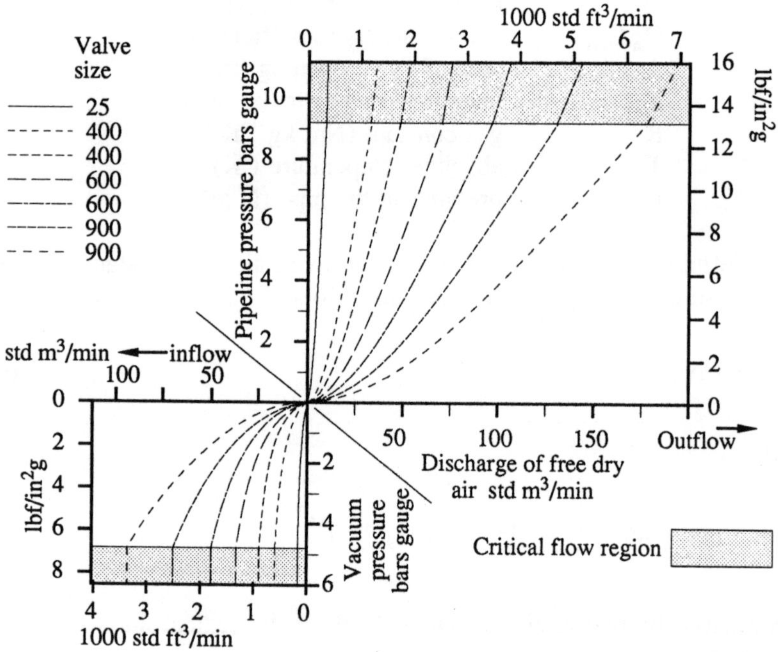

Figure 6.23 Discharge capacity of Biwater air inlet/outlet relief valves

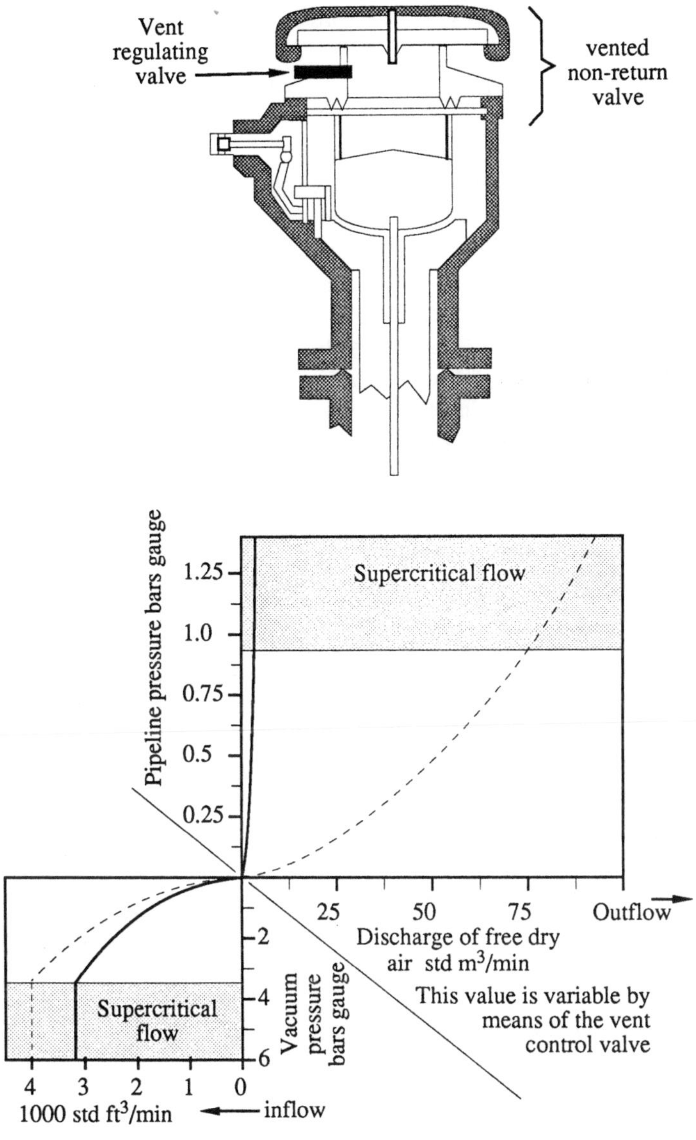

Figure 6.24 Discharge capacity of Biwater air inlet/outlet pressure surge alleviation valves

where COU is a constant multiplier (slope). Equation 6.38 for the subsonic mass rate outflow of air can be expressed as

$$m = -\ COU\ 3.8568 \left((p')^{-1.4286} - (p')^{-1.714} \right)^{0.5} \tag{6.43}$$

For specified values of CON and COU, equations 6.40 through to 6.43 represent a continuous curve of m as a function of p′, Figure 6.22.

Figures 6.23 and 6.24 are reproduced from Biwater Valves Limited Water Industries Manual and forms the basis for determining the values of CON and COU for the required cluster of air inlet/outlet valves. These values of CON and COU, together with the valve throat elevation, are specified as the input data for the simulation of the transient response of the air inlet/outlet air valve boundary condition. At any time in the simulation the appropriate section of the curve shown in Figure 6.22 may be expressed by the following linear equation by calculating values of m corresponding to values of $p_t' \pm 0.05$, where p_t' is the pressure ratio at the start of the time interval

$$m = sp' + c \tag{6.44}$$

The boundary condition may then be solved using the characteristic equations either side of the boundary junction, together with the normal equations of state.

6.7.2 *Calculation of CON and COU*

The following procedure for the calculation of the value of the parameters CON and COU was established in collaboration with Biwater Valves.

6.7.3 *Critical mass rate of inflow of air*

The number associated with any curve in Figure 6.23 indicates the recommended maximum pipeline diameter (mm) for a given air valve size.

Therefore, from the curve marked 900 (largest valve) the value of the critical rate of inflow of air for a large orifice air valve is 93 m^3/min. This value is reduced by 20% to allow for the fitting of a vented non-return valve in place of the normal valve cover and cowl and then converted to the required units to give a value of CON for a single air inlet/outlet pressure surge alleviation valve of designated size 900mm as

$$CON = \frac{93.0 \cdot 0.8 \cdot 0.293}{60.0} = 1.603 \quad Kg/s$$

If a cluster of more than one valve is connected to the pipe at any location, through a manifold arrangement, the combined value of CON is further reduced by 10% to allow for the manifold losses. Therefore, for a cluster of four air inlet/outlet valves, the combined value of CON is given by:

$$CON = 1.603 \cdot 4 \cdot 0.9 = 5.772 \quad Kg/s$$

6.7.4 *Critical mass rate multiplier for outflow of air (COU)*

The vent regulating valve on the vented non-return valve can be adjusted to obtain the required ratio of CON/COU.

6.8 Typical pumping main

Figure 6.25 shows the hydraulic profile and steady state hydraulic gradeline for a 400mm diameter main discharging a storm raw water flow of 150 l/s from the pump at junction 1 to a treatment works collecting tank at junction 75. With no pressure surge protection sub-atmospheric pressures, falling to vapour pressure, are generated in the main following a pump trip due to power failure, as shown in Figure 6.26.

The addition of an air vessel, with an initial air volume of $10m^3$, located downstream of the pump check valve, maintains positive pressure in the main after the pump trip, Figure 6.27. In order to minimise the size of the air vessel it is possible in some cases, to allow sub-atmospheric pressure at the high points, as shown by the minimum static pressure head envelope in Figure 6.27 for an air vessel with an initial air volume of $5m^3$.

The sub-atmospheric pressure at the high points may then be minimised by incorporating air inlet/outlet relief valves at these junctions. The values for the constants CON and, in particular, COU, have a significant influence on the transient flow in the main as shown by the simulations in Table 6.2.

With COU set at 0.043, Figures 6.29 and 6.30 show that secondary surges are created at junction 8 at the instant the air is vented from the pipeline.

These secondary pressure surges have a marked effect on the maximum and minimum static pressure head envelope presented in Figure 6.28.

With COU set at 0.009, to give a ratio of CON/COU of 80, the air is vented from the pipeline at a much slower rate, as shown in Figure 6.33, thereby minimising the secondary pressure surges presented in Figure 6.32 and producing acceptable envelopes of maximum and minimum static pressure head (see Figure 6.31).

Table 6.2 Simulations incorporating air inlet/outlet valves

Run No.	Air Valve CON	Air Valve COU	Max/Min Envelope Figure	Static Pressure Figure	Air Volume Figure
280	0.711	0.043	6.28	6.29	6.30
281	0.711	0.009	6.31	6.32	6.33

This simulation verifies that the performance characteristics of air inlet/outlet valve may be defined by two constants, simplifying the simulation of the valve boundary during transient flow calculations. It is necessary to optimise the outflow constant of the valve in order to avoid large secondary pressure surge when the air is vented from a pipeline.

6.9 Flow continuation as a form of surge suppression

To prevent line pressures falling too low downstream of a closing valve, or a tripped pump, it may be possible to introduce some form of flow continuation that effectively reduces the rate of change of the flow conditions at this boundary. Figure 6.34 illustrates two simple techniques that may be employed, namely a by-pass valve or a feed tank, kept full by the system initial flow condition. Both of these solutions prevent the ingress of air into the pipeline and are particularly suitable in low head systems.

Figure 6.34(a) illustrates a by-pass valve network. The pump is protected against backflow by the non-return valve in the main. A parallel, smaller bore, by-pass pipe includes a control valve that opens as the pump trips and is subsequently allowed to close slowly. (Note that 'slow' should be thought of in pipe periods.) The by-pass valve must not open too early.

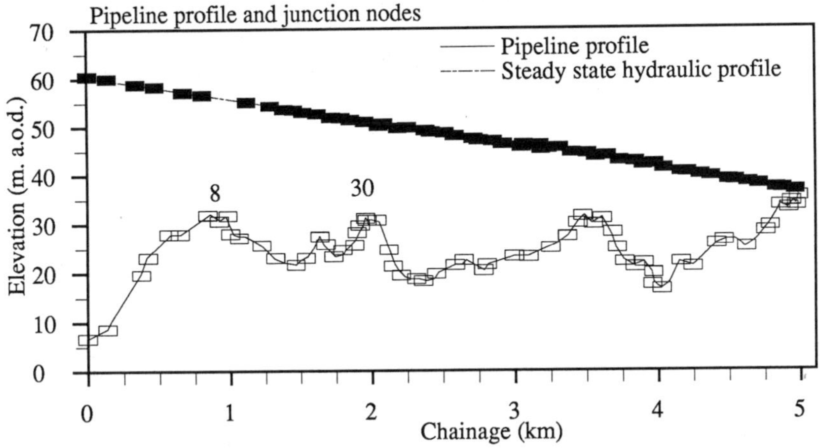

Figure 6.25 Profile and steady state hydraulic gradeline for a pumping main

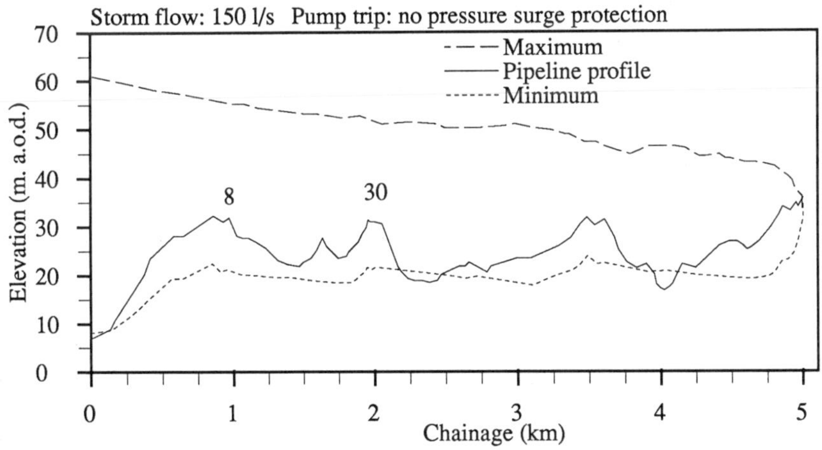

Figure 6.26 Maximum and minimum static pressure head envelopes: no pressure surge protection

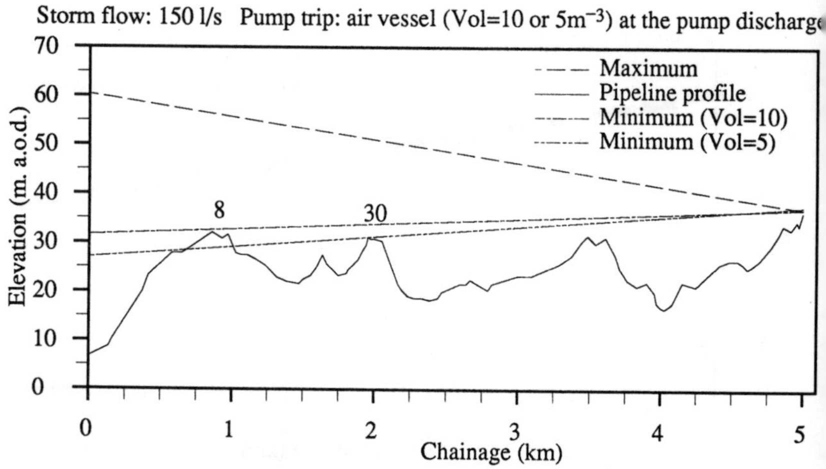

Figure 6.27 Maximum and minimum static pressure head envelopes: air vessel included

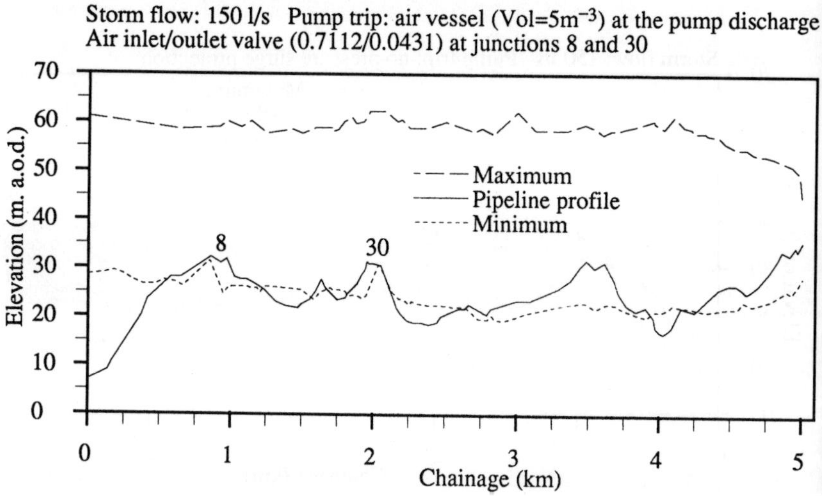

Figure 6.28 Maximum and minimum static pressure head envelopes: air valves (COU = 0.043) included

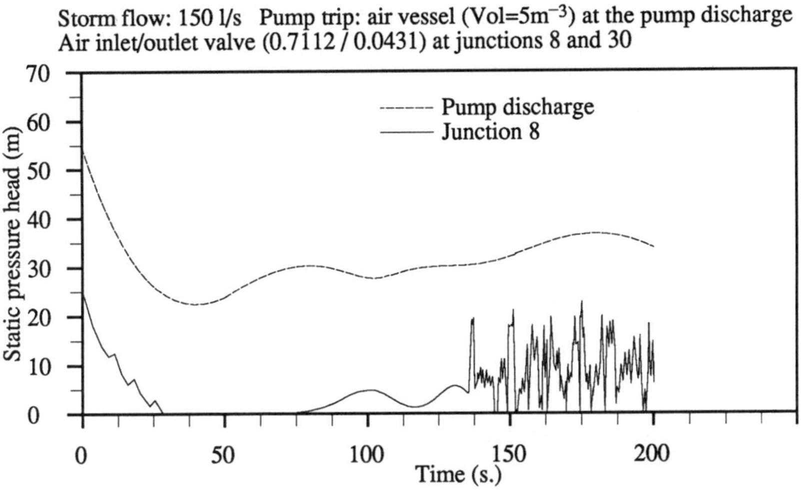

Figure 6.29 Static pressure heads: air valves (COU = 0.043) included

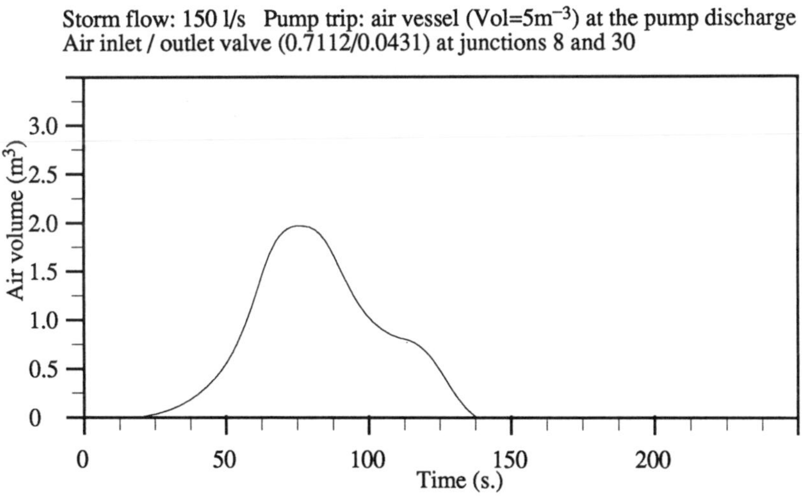

Figure 6.30 Air volume at air inlet/outlet location: air valve COU = 0.043

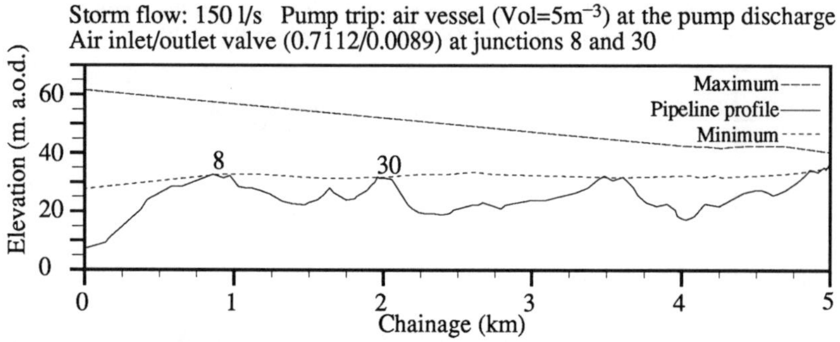

Figure 6.31 Maximum and minimum static pressure head envelope: air valves (COU = 0.009) included

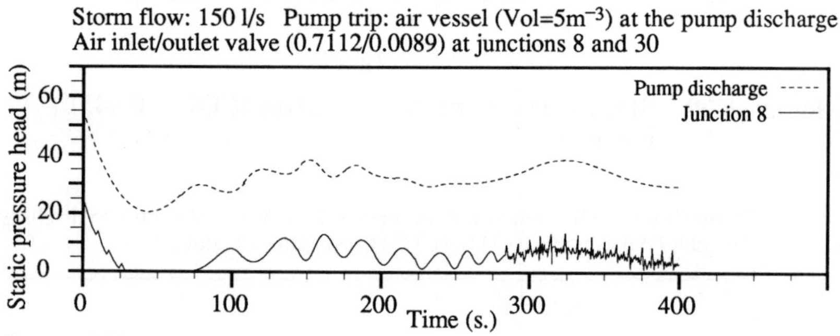

Figure 6.32 Static pressure heads: air valves (COU = 0.009) included

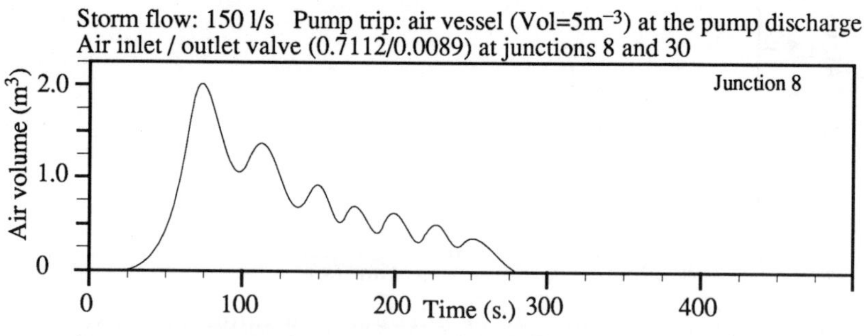

Figure 6.33 Air volume at air inlet/outlet location: air valve COU = 0.009

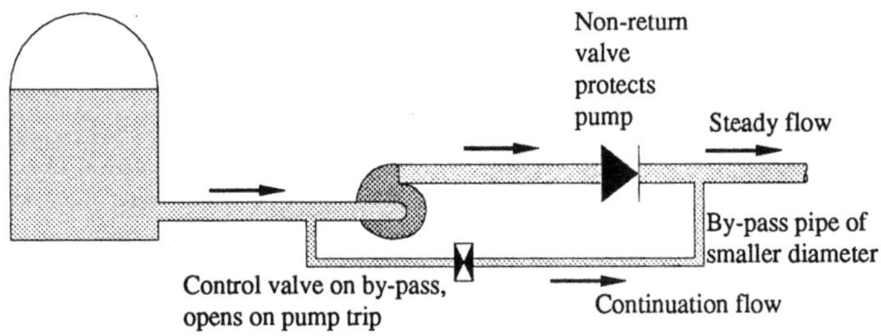

Figure 6.34(a) Pump by-pass to control downstream surge on pump trip

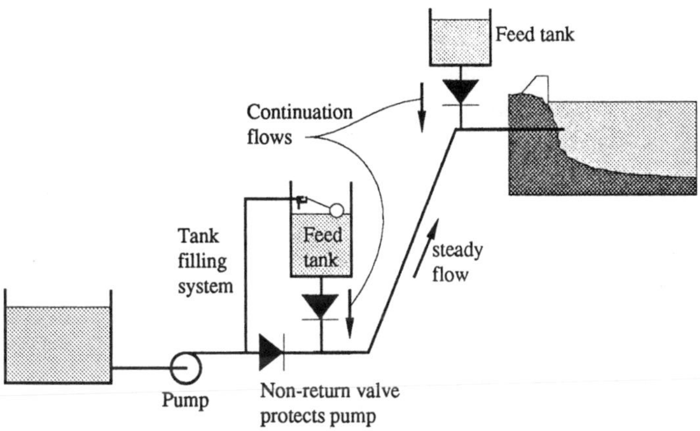

Figure 6.34(b) Feed tank locations to provide protection following pump trip

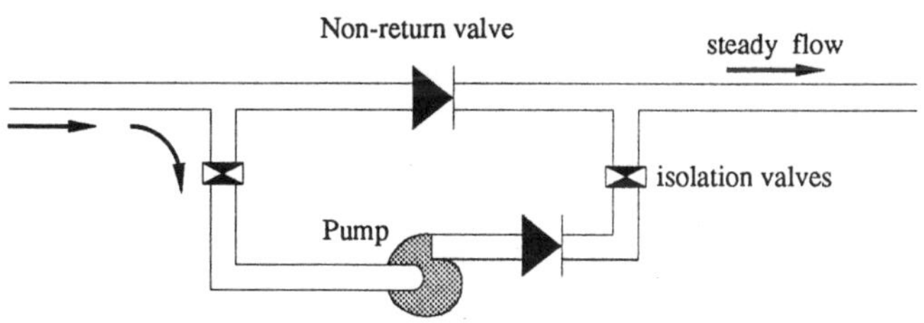

Figure 6.34(c) By-pass to cater for localised booster pump failure in long pipelines

Figure 6.34(b) illustrates a feed tank installed to provide an inflow into the main following pump trip. This is particularly useful if air ingress to the pipeline is to be avoided. Fluid flows into the pipeline as the pressure falls downstream of the pump, thus preventing cavity formation and thus providing protection against both sub-atmospheric pressures and the subsequent high pressure surge following cavity collapse. This is effectively the same as the solution proposed, Figure 6.6, in the aircraft ground refuelling simulation. Feed tanks may also be used at high points in a pipeline profile to prevent column separation, Figures 5.17 and 5.33.

The volume of fluid required in the tank may be determined by simulations similar to those already presented for the modelling of column separation on valve closure and pump trip. Piping arrangements must be made to ensure that the feed tank is kept filled prior to the next transient event.

In the case of long pipelines it may be necessary to introduce booster pumps. Local booster pump trip may be the cause of transient problems as the pump becomes a major flow obstruction. By-pass networks, as illustrated in Figure 6.34(c), allow continuation of system use at lower flow rates and minimise the surge implication of local pump failure. In this case the by-pass piping diameter should be similar to the main.

6.10 Surge pressure simulations

While this chapter has been concerned with surge control and suppression, it must be appreciated that an essential first stage in any discussion of control or alleviation device selection is an estimate of the severity of the problem. The case studies discussed in this chapter, involving pumping main protection, air vessel specification and the use of air/fluid inwards relief valves, illustrate the level of advice that can be obtained as a result of a method of characteristics based computer simulation. Similarly, in cases where it is not possible to introduce surge alleviation devices, for example the aircraft fuel system analysis, presented in Chapter 5, a simulation can indicate the optimum valve closure mode to minimise the surge event.

The method of characteristics has been established as the most effective methodology upon which to base such simulations. In particular the method's ability to incorporate specialised system boundary conditions, while retaining the general pipe and more common boundary equations unaltered, allows a wide range of system simulations to be undertaken with minimal re-writing of the basic model. The development of computing power now makes this approach to simulation readily accessible.

Chapter 7:

Free surface wave and air pressure transient propagation

The method of characteristics is presented in this text as a general solution technique for a pair of hyperbolic quasi linear partial differential equations and as such can be applied to a wide range of unsteady flow conditions, certainly not limited to the traditional 'waterhammer' problem. This Chapter will apply the developed methodology to open channel, or partially filled pipe flows, and to small differential air pressure transient propagation, concentrating upon the development of suitable boundary conditions for both these cases.

The applications of the method of characteristics developed so far in this text have concentrated upon conditions where the wave speed has remained sensibly constant along a particular length of pipe. Variations in the C^+ and C^- characteristic slopes have to be considered when the local flow velocity becomes large enough to affect the slope term.

There are two common applications of the method of characteristics where this simplification cannot be sustained, namely in the study of transient propagation through gas flows and the propagation of surge waves through free surface flows. These two cases will be dealt with below and the necessary boundary conditions will also be presented later in parallel with the more traditional treatment of pressure transients.

Despite the apparent differences between the applications presented in this chapter and the more traditional application of the method of characteristics it will be shown that the fundamental approach is identical, being again concerned with boundary conditions and stability.

7.1 Open channel unsteady flow

The derivation of the momentum and continuity equations in Chapter 2 were general, no assumptions being made as to the nature of the enclosing conduit. Figure 7.1 illustrates the free surface flow element and the development of both the momentum and continuity equations. Thus the momentum equation for an element of unsteady open channel flow may be written as

$$-\rho g \frac{\partial h}{\partial x} A\Delta x - \tau_o P\Delta x + \rho g\Delta x \sin\alpha$$
$$= \frac{\partial}{\partial x}(\rho A V^2)\Delta x + \frac{\partial}{\partial t}(\rho A V)\Delta x \qquad (7.1)$$

Introducing the hydraulic mean depth, m = A/P, expanding terms and dividing by $\rho A\Delta x$, yields

$$g\frac{\partial h}{\partial x} + \frac{\tau_o}{\rho m} - g\sin\alpha + 2V\frac{\partial V}{\partial x} + \frac{V^2}{A}\frac{\partial A}{\partial x} + \frac{V}{A}\frac{\partial A}{\partial t} + \frac{\partial V}{\partial t} = 0 \qquad (7.2)$$

The shear stress, τ_o may be expressed as

$$\tau_o = f\,\tfrac{1}{2}\rho V^2 \qquad (7.3)$$

where f is the friction factor.

The friction slope, S, may be derived from the Chezy relationship where the local mean velocity, V, and the flow depth are utilised,

$$S = V^2 f / (2gm) \qquad (7.4)$$

Thus the shear stress may be expressed in terms of the friction slope S

$$\tau_o = \rho g m S \qquad (7.5)$$

Equation 7.2 becomes

$$g\frac{\partial h}{\partial x} + g(S - S_o) + 2V\frac{\partial V}{\partial x} + \frac{V^2}{A}\frac{\partial A}{\partial x} + \frac{V}{A}\frac{\partial A}{\partial t} + \frac{\partial V}{\partial t} = 0 \qquad (7.6)$$

where $S_o = \sin \alpha$, the slope of the pipe invert.

The equation of continuity may be developed from Figure 7.1, where it will be seen that

Inflow – Outflow = Rate of increase of control volume storage

$$AV - [AV + \frac{\partial}{\partial x}(AV)\,\Delta x] = \frac{\partial A}{\partial t}\,\Delta x \tag{7.7}$$

Simplifying

$$A\frac{\partial V}{\partial x} + V\frac{\partial A}{\partial x} + \frac{\partial A}{\partial t} = 0 \tag{7.8}$$

From Figure 7.1, as T is the flow surface width, it follows that

$$\partial A = T\partial h$$

and

$$A\frac{\partial V}{\partial x} + VT\frac{\partial h}{\partial x} + T\frac{\partial h}{\partial t} = 0 \tag{7.9}$$

The momentum and continuity equations are together known as the St Venant unsteady flow equations, expressed here in their partially filled pipeflow form.

Equation 7.6 may be further simplified by introducing $\delta A = T\delta h$, and subtracting equation 7.9, multiplied through by V/A, to yield.

$$g\frac{\partial h}{\partial x} + g\,(S - S_o) + V\frac{\partial V}{\partial x} + \frac{\partial V}{\partial t} = 0 \tag{7.10}$$

Equations 7.9 and 7.10 are a pair of quasi linear hyperbolic partial differential equations and as such are not directly amenable to solution by finite difference techniques.

However, if a suitable combination can be formed to replace the partial derivative terms by total derivatives, then such a solution becomes feasible.

Equations 7.9 and 7.10 may be combined as follows

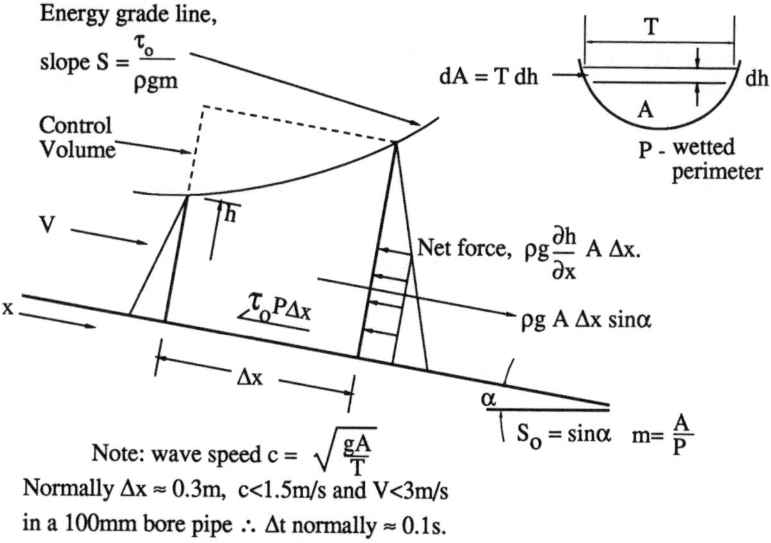

Figure 7.1 Development of the equations of continuity and motion for an element of unsteady, partially filled pipe flow

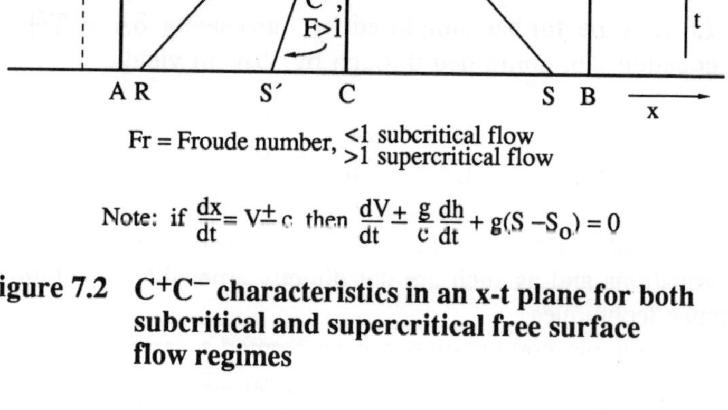

Figure 7.2 C^+C^- characteristics in an x-t plane for both subcritical and supercritical free surface flow regimes

$$L_1 = A\frac{\partial V}{\partial x} + VT\frac{\partial h}{\partial x} + T\frac{\partial h}{\partial t} = 0$$

$$L_2 = V\frac{\partial V}{\partial x} + \frac{\partial V}{\partial t} + g\frac{\partial h}{\partial x} + g\,(S - S_o) = 0$$

$$\lambda L_1 + L_2 = 0$$

where λ is an appropriate factor to render the partial derivatives terms into total derivatives.

Collecting terms yields

$$\lambda T\,[\frac{\partial h}{\partial t} + (V+\frac{g}{\lambda T})\frac{\partial h}{\partial x}] + [\frac{\partial V}{\partial t} + (\lambda A+V)\frac{\partial V}{\partial x}] + g(S-S_o) = 0 \qquad (7.11)$$

Now the total derivatives with respect to time are

$$\frac{dV}{dt} = \frac{\partial V}{\partial t} + \frac{\partial V}{\partial x}\frac{dx}{dt} \qquad (7.12)$$

$$\frac{dh}{dt} = \frac{\partial h}{\partial t} + \frac{\partial h}{\partial x}\frac{dx}{dt} \qquad (7.13)$$

Thus for the partial derivative terms in equation 7.11 to be replaced by the total derivations of V and h with respect to time, it follows by inspection of equations 7.11, 7.12 and 7.13 that

$$\frac{dx}{dt} = \lambda A + V \qquad (7.14)$$

and

$$\frac{dx}{dt} = g/\lambda T + V \qquad (7.15)$$

may be introduced in order to allow the replacement of the partial differential terms by total derivatives.

From equations 7.14 and 7.15 the value of λ may be taken as

$$V + g/\lambda T = \lambda A + V$$

$$\lambda = \pm\sqrt{g/AT}$$

and hence

$$\frac{dx}{dt} = V \pm \sqrt{gA/T} \tag{7.16}$$

The term $(gA/T)^{1/2}$ is the wave propagation velocity, reducing to $(gh)^{1/2}$ for the special case of the rectangular channel.

Provided that equation 7.16 is true,

$$\frac{dx}{dt} = V \pm c \tag{7.17}$$

then the unsteady partially filled pipeflow equations, 7.9 and 7.10, may be reduced to a total differential equation

$$\frac{dV}{dt} \pm \frac{g}{c}\frac{dh}{dt} + g\,(S - S_o) = 0 \tag{7.18}$$

where

$$c = \sqrt{gA/T}$$

Equation 7.17 represents two lines drawn in an x-t grid, as shown in Figure 7.2, the slopes being dependent upon the Froude Number, V/c. Equation 7.18 represents two simultaneous equations in V and h that apply only along the characteristics lines defined by equation 7.17.

Thus equations 7.17 and 7.18 provide a basis for a finite difference solution. Referring to Figure 7.2, if the variables V and h are known at R and S at time t_o then four equations may be written in terms of the unknowns at P.

$$V_P - V_R + g \int_{h_R}^{h_P} \frac{1}{c}\, dh + \int_{t_R}^{t_P} g\,(S - S_o)\; dt = 0 \tag{7.19}$$

$$x_P - x_R = \int_{t_R}^{t_P} (V + c)\; dt \tag{7.20}$$

$$V_P - V_S - g \int_{h_S}^{h_P} \frac{1}{c}\, dh + \int_{t_S}^{t_P} g(S - S_o)\; dt = 0 \tag{7.21}$$

$$x_P - x_S = \int_{t_S}^{t_P} (V - c)\; dt \tag{7.22}$$

These equations, forming two pairs of relationships are known as the C^+ and C^- characteristics, respectively linking conditions at P to known earlier conditions at R and S, Figure 7.2.

In the special case where V << c, ie pressure transients in full bore pipes, similar characteristics through P would pass through the adjacent grid nodes A and B, Figure 7.2. If V and c are comparable, as is the case with all open channel or partially filled pipeflow applications, then it will be necessary to interpolate between known conditions to evaluate the base values at R and S at each time step. This presents particular difficulties as it immediately implies the need for interpolation with the consequent danger of rounding errors.

The presence of the local flow velocity as an unknown in the characteristic slope calculation also adds complexity to such interpolation improvements as the time line technique. However in most applications a linear interpolation technique has been found satisfactory.

Referring to Figure 7.2, in subcritical flow the local wave speed, expressed as $c = (gA/T)^{1/2}$, exceeds the local mean flow velocity V, therefore the C^+ and C^- characteristics intersecting at P originate in both adjacent grid sections. This implies that information can flow to P from both directions, as already discussed a prerequisite of subcritical flows. However in supercritical flow, then c < V and both the C^+ and C^- characteristics originate in the grid section upstream of P, as shown, implying that information can only reach P from upstream.

For a rectangular grid, the length Δx may be fixed, thus removing x_P as an unknown from equations 7.20 and 7.22. However, as both R and S, or S′ must lie within adjacent grid sections for a stable solution, the size of the time step Δt cannot be greater than

$$\Delta t = \frac{\Delta x}{(V + c)_{max}} \tag{7.23}$$

where the maximum values of V and c at time t, Figure 7.2, are used. Equation 7.23 is known as the Courant Criterion.

The introduction of a fixed grid length and time step, via equation 7.23 removes t_P as an unknown in equations 7.19 to 7.22 and defines the technique to be used as the 'specified grid' version of the method of characteristics. While this technique introduces interpolations for R and S, it also imposes an order on the solution of equations 7.19 to 7.22.

Referring to Figure 7.2, expressions may be derived for base conditions at R and S in subcritical flow or at S′ in supercritical flow. Note the expression for R remains the same in both flow regimes and is not stated twice.

For a single node it follows that

$$x_P = x_C$$

and by geometry that

$$x_P - x_R = (V_R + c_R) \Delta t$$

hence

$$\frac{c_C - c_R}{c_C - c_A} = \frac{V_C - V_R}{V_C - V_A} = \frac{x_C - x_R}{x_C - x_A} = (V_R + c_R) \frac{\Delta t}{\Delta x} = \frac{h_C - h_R}{h_C - h_A}$$

Elimination results in the following interpolation relationships

$$V_R = \frac{V_C + \theta (V_A c_C - c_A V_C)}{1 + \theta (V_C - V_A + c_C - c_A)} \tag{7.24}$$

$$c_R = \frac{c_C - V_R \theta (c_C - c_A)}{1 + \theta (c_C - c_A)} \tag{7.25}$$

$$h_R = h_C - \theta (V_R + c_R) (h_C - h_A) \tag{7.26}$$

where
$\theta = \Delta t/\Delta x$

Similarly

$$V_S = \frac{V_C - \theta (V_C c_B - c_C V_B)}{1 - \theta (V_C - V_B - c_C + c_B)} \tag{7.27}$$

$$c_S = \frac{c_C + V_S \theta (c_C - c_B)}{1 + \theta (c_C - c_B)} \tag{7.28}$$

$$h_S = h_C + \theta (V_S - c_S) (h_C - h_B) \tag{7.29}$$

For the supercritical flow regime the equations relating to R remain unchanged, those for S′ may be shown to be

$$V_{S'} = \frac{V_C - \theta (V_C c_A + V_A c_C)}{1 + \theta (V_C - V_A + c_A - c_C)} \tag{7.30}$$

$$c_{S'} = \frac{c_C + V_{S'} \theta (c_A - c_C)}{1 + \theta (c_A - c_C)} \tag{7.31}$$

In representing equations 7.19-7.22 by a first order finite difference approximation a number of points need to be expanded.

Frictional representation is via the slope of the energy line S, determined from the Chezy equation. For large open channels it is acceptable to use the Manning equation where C is given by $m^{1/6}/n$, Swaffield and Bridge (1983).

However for partially filled pipe flows, certainly in cases with a pipe diameter < 1.0m, the Manning 'n' does not adequately represent the frictional losses and it is preferable to use a modified version of the Colebrook White equation where D is replaced by D = 4m. The value of friction factor f may then be utilised in $C = (2g/f)^{1/2}$.

In equations 7.19 and 7.21 the integral (dh/c) may be approximated by

(1/c)(integral dh) if the change in c, the local wave speed, is small over the depth change likely in the calculation. If this is not the case, and particularly for non-rectangular section channels, such as circular or elliptic cross section sewers, Henderson (1966) suggested replacing the depth by a 'stage' variable linked to depth by the equivalence

$$d\omega = \sqrt{g \frac{T}{A}}\, dh = \frac{1}{c}\, dh \tag{7.32}$$

For any cross section where T and A are known as functions of h values of ω corresponding to depth can be found by evaluating the integral

$$\omega = \int_o^h \sqrt{g \frac{T}{A}}\, dh \tag{7.33}$$

Equation 7.33 may be used to evaluate ω for any h value at each node at each time step within the solution. Equation 7.18 thus becomes

$$\frac{dV}{dt} \pm \frac{d\omega}{dt} + g (S - S_o) = 0 \tag{7.34}$$

with a consequent simplification in equations 7.19 and 7.21. It is of course necessary to 'translate' ω back into a depth, h, prior to the output of results. The effect of the ω correction on depth prediction is not significant in the 0.075 to 0.3 m diameter pipes.

7.2 Gas flow low amplitude pressure transient propagation

The propagation of pressure transients in gas or air ducts or pipelines is of interest in a range of applications, from distribution networks to a variety of venting systems. Much of the available literature refers to the problems of tunnel venting as a result of high speed train operation.

Figure 7.3(a) illustrates an elemental volume of compressible gas flow for which the continuity equation may be written as:

Inflow – Outflow = rate of change of mass in the control volume

$$\rho A u - (u + \delta u)(\rho + \delta\rho)A = \frac{\partial}{\partial t}(\rho A\, \delta x)$$

reducing, when second order terms are neglected, to

$$\rho\ \delta u + u\ \delta\rho + \frac{\partial}{\partial t}\rho\ \delta x = 0$$

The final form of the continuity equation becomes

$$\rho\frac{\partial u}{\partial x} + u\frac{\partial \rho}{\partial x} + \frac{\partial \rho}{\partial t} = 0 \tag{7.35}$$

The momentum equation for the same elemental volume may be expressed as

Force = rate of change of momentum

thus

$$pA - (p + \delta p)A - \tau P\ \delta x = \frac{d}{dt}(\rho A\ \delta x\ u)$$

$$-\ \delta pA - \tau P\delta x = \frac{\partial}{\partial t}(\rho A\ \delta x\ u) + \frac{\partial}{\partial x}(\rho A\ \delta x\ u^2)$$

Dividing by A δx yields

$$-\frac{\partial p}{\partial x} - \tau\frac{P}{A} - \frac{\partial}{\partial t}(\rho u) - \frac{\partial}{\partial x}(\rho u^2) = 0$$

$$\frac{\partial p}{\partial x} - \tau\frac{P}{A} - \rho\frac{\partial u}{\partial t} - u\frac{\partial \rho}{\partial t} - u^2\frac{\partial \rho}{\partial x} - 2\rho u\frac{\partial u}{\partial x} = 0$$

$$-\frac{\partial p}{\partial x} - \tau\frac{P}{A} - u\left[\frac{\partial \rho}{\partial t} + u\frac{\partial \rho}{\partial x} + \rho\frac{\partial u}{\partial x}\right] - \rho\frac{\partial u}{\partial t} - \rho u\frac{\partial u}{\partial x} = 0 \tag{7.36}$$

From equation 7.35 it follows that

$$\frac{\partial p}{\partial x} - \tau\frac{P}{A} - \rho\frac{\partial u}{\partial t} - \rho u\frac{\partial u}{\partial x} = 0 \qquad (7.37)$$

where τ is the wall shear stress acting around the wetted perimeter P of the drain or vent pipe. Substituting for shear stress, $\tau = \frac{1}{2}f\rho u^2$, where f is the appropriate friction factor for the induced air flow, yields

$$\frac{\partial p}{\partial x} + \rho u\frac{\partial u}{\partial x} + \rho\frac{\partial u}{\partial t} + 4\rho f\, u\, |u| \,/\, 2D \;=\; 0 \qquad (7.38)$$

Equations 7.35 and 7.38 will be recognised as a pair of quasi linear hyperbolic differential equations clearly belonging to the same family of relationships as the equations already derived for the free surface flow conditions.

Air pressure and density may be linked if it is assumed that the fluid is a perfect gas and that the flow is isentropic. This is acceptable in a drainage and vent system application as the changes in temperature and pressure are extremely small during transient propagation, so that the process may be considered reversible. As the temperature gradients are small, and the process is rapid, no heat is transferred, hence the process is isentropic and air pressure and density are related throughout the flow by

$$p/\rho^{\gamma} \;=\; K \qquad (7.39)$$

where K is a constant.

The acoustic velocity, c, of a plane, weak, pressure wave in a gas at low absolute pressures may be determined from the equations of continuity and momentum applied across a wave front travelling in a duct, Figure 7.3(b). If shear forces are neglected then, from the momentum equation

$$(p+dp)A \;-\; pA \;=\; cA\rho(-(c-du)-(-c))$$

so that

$$dp \;=\; \rho c du$$

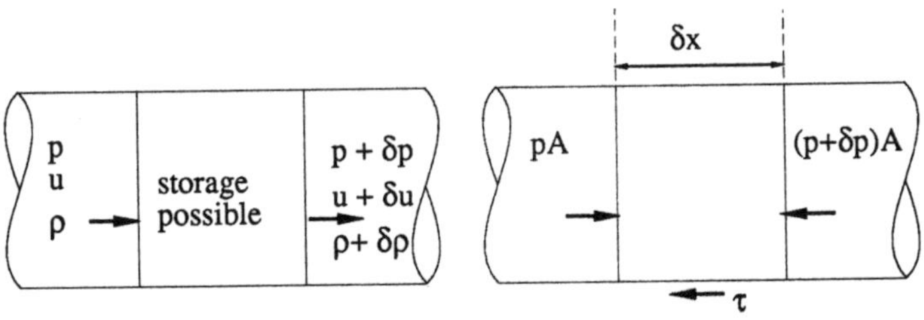

Figure 7.3(a) Development of continuity and momentum equations, full bore induced air flow

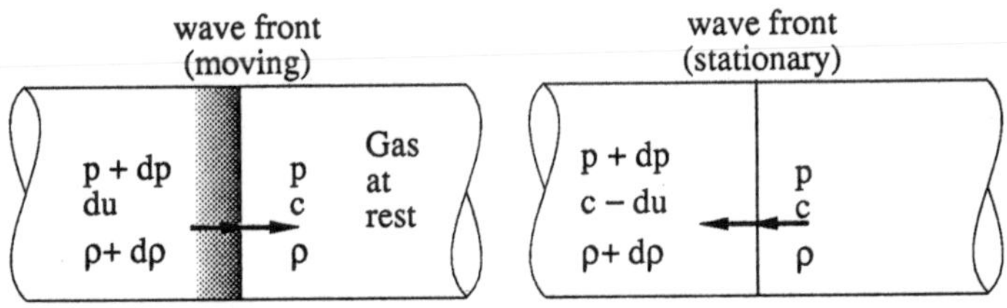

Figure 7.3(b) Weak pressure wave in a drain or vent of area A brought to rest by superposition of a –c air velocity

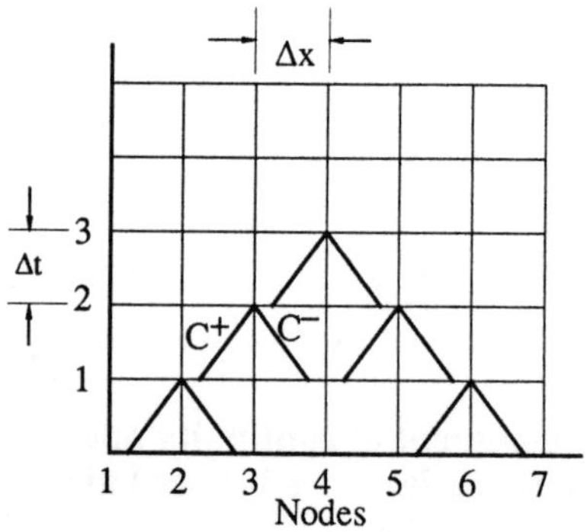

Figure 7.4(a) Limits of characteristic solution in subcritical flow

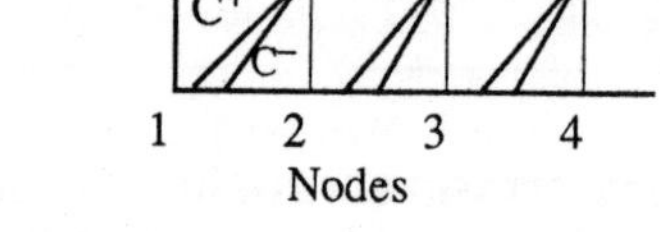

Figure 7.4(b) Limits of characteristic solution in supercritical flow

From continuity, with the wave brought to rest by the superposition of an equal and opposite wave propagation velocity

$$cA\rho = (c-du)A(\rho+d\rho)$$

When second order terms are neglected, $d\rho/\rho = du/c$ and, from equation 7.39

$$c = \sqrt{dp/d\rho} = \sqrt{K\rho^{\gamma-1}} = \sqrt{\gamma\, p/\rho} \qquad (7.40)$$

These wave speed and pressure/density equations are strictly only applicable to inviscid flows. However previous work has indicated close agreement between experimental measurements of air pressure under transient conditions and predictions utilising this major specification, Henson and Fox (1972) and Woodhead, Fox and Vardy (1976).

This approach is particularly suitable where the ambient pressure is low and the transient pressure excursions small. The application discussed here clearly meets these criteria as the ambient pressure is atmospheric and the fluctuations may be measured in mm of water gauge.

In cases where this is not a viable approach it is necessary to introduce the energy equation leading to an iterative solution of the characteristic equations defining transient flow, Vardy (1976). As will be demonstrated this simplification allows the direct solution of the characteristic equations derived from the equations of continuity and momentum and its use is supported by experimental validation.

From equations 7.39 and 7.40 the following substitutions

$$\partial\rho = \left(\frac{2}{\gamma-1}\right)\frac{\rho}{c}\,\partial c$$

and

$$\partial p = \left(\frac{2}{\gamma-1}\right)\rho\, c\, \partial c$$

may be developed, thus allowing substitutions into the momentum and continuity equations as follows

$$L_1 = \left(\frac{2}{\gamma-1}\right) \rho c\frac{\partial c}{\partial x} + \rho u\frac{\partial u}{\partial x} + \rho\frac{\partial u}{\partial t} + 4\rho f u \mid u \mid / 2D = 0$$

$$L_2 = \rho\frac{\partial u}{\partial x} + \left(\frac{2}{\gamma-1}\right) \frac{\rho}{c} \left[u\frac{\partial c}{\partial x} + \frac{\partial c}{\partial t}\right] = 0$$

Combining as before

$$L = L_1 + \lambda L_2 = 0$$

yields

$$L = \frac{\partial u}{\partial x}(\lambda+u) + \frac{\partial u}{\partial t} + \left(\frac{2}{\gamma-1}\right)\left[\frac{\partial c}{\partial x} \left(\lambda\frac{u}{c}+c\right)+\frac{\lambda}{c} \frac{\partial c}{\partial t}\right] + 4fu \ |u| \ / \ 2D = 0$$

where

$$\frac{dc}{dt} = \frac{\partial c}{\partial x}\frac{\partial x}{\partial t} + \frac{\partial c}{\partial t}, \qquad \frac{du}{dt} = \frac{\partial u}{\partial x}\frac{dx}{dt} + \frac{\partial u}{\partial t}$$

thus reducing to

$$\frac{du}{dt} \pm \frac{2}{\gamma-1}\frac{dc}{dt} + \left(4f\, u \mid u \mid / \ 2D \right) \Delta t = 0 \tag{7.41}$$

provided that

$$\frac{dx}{dt} = u \pm c \tag{7.42}$$

These expressions will be recognised as in every way the equivalent of the corresponding expressions in free surface surge propagation.

Thus the C^+ characteristic may be expressed as

$$u_P - u_R + \frac{2}{\gamma-1} (c_P - c_R) + 4 f_R u_R \mid u_R \mid \frac{\Delta t}{2D} = 0 \qquad (7.43)$$

when

$$\frac{dx}{dt} = u + c \qquad (7.44)$$

and the corresponding C^- characteristic

$$u_P - u_S - \frac{2}{\gamma-1} (c_P - c_S) + 4 f_S u_S \mid u_S \mid \frac{\Delta t}{2D} = 0 \qquad (7.45)$$

when

$$\frac{dx}{dt} = u - c \qquad (7.46)$$

These equations may be reduced to a pair of first order equations to be solved simultaneously for the unknowns, wave speed and flow velocity

$$u_P = K1 - K2 \; c_P$$

for the C^+ characteristic, and

$$u_P = K3 + K4 \; c_P$$

for the C^- characteristic, where

$$K1 = u_R + (\frac{2}{\gamma-1}) c_R - 4 f u_R \mid u_R \mid \Delta t/2D \qquad (7.47)$$

$$K2 = K4 = 2/(\gamma-1) \qquad (7.48)$$

$$K3 = u_S - \left(\frac{2}{\gamma-1}\right)c_S - 4\,f\,u_S\,|\,u_S\,|\,\Delta t/2D \tag{7.49}$$

These equations link conditions at node P to conditions one time step earlier at points R and S, located within one distance increment upstream and downstream of P respectively.

The local pressure at P is then calculated as

$$p_P = \left[\,(p_o\,/\,\rho_o^{\gamma})\,(\gamma\,/\,c^2\,)^{\gamma}\,\right]^{1/(1-\gamma)} \tag{7.50}$$

The characteristic lines, C^+, C^- again fall in the adjacent grids to the node P, Figure 7.2, which can be seen to describe both free surface wave and air pressure transient propagation; interpolation only being required if the value of the local mean flow velocity is appreciable when compared to the local acoustic velocity. However, due to the pressure dependence of the acoustic velocity the necessary interpolation expressions are as follows.

For a single node it follows that

$$x_P = x_C$$

and by geometry that

$$x_P - x_R = (\,u_R + c_R\,)\,\Delta t$$

hence

$$\frac{c_C - c_R}{c_C - c_A} = \frac{u_C - u_R}{u_C - u_A} = \frac{x_C - x_R}{x_C - x_A} = (\,u_R + c_R\,)\,\frac{\Delta t}{\Delta x} = \frac{p_C - p_R}{p_C - p_A}$$

Elimination results in the following interpolation relationships

$$u_R = \frac{u_C + \theta\,(-\,u_C\,c_A + c_C\,u_A\,)}{1 + \theta\,(u_C - u_A + c_C - c_A\,)} \tag{7.51}$$

$$c_R = \frac{c_C\,(1 - u_R\,\theta) + c_A\,u_R\,\theta}{1 + c_C\,\theta - c_A\,\theta} \tag{7.52}$$

$$p_R = p_C - \theta\,(u_R + c_R)\,(p_C - p_A) \tag{7.53}$$

where

$\theta = \Delta t/\Delta x$

Similarly

$$u_S = \frac{u_C - \theta\,(u_C\,c_B - c_C\,u_B)}{1 - \theta\,(u_C - u_B - c_C + c_B)} \tag{7.54}$$

$$c_S = \frac{c_C\,(1 + u_S\,\theta) - c_B\,u_S\,\theta}{1 + \theta\,c_C - \theta\,c_B} \tag{7.55}$$

$$p_S = p_C + \theta\,(u_S - c_S)\,(p_C - p_B) \tag{7.56}$$

It will be noted that again the Courant Criterion must be satisfied, expressed as

$\Delta t \leq \Delta x \;/\; (V{+}c)_{\max}$

where the maximum values of wave speed and flow velocity in the duct are utilised to calculate the size of the time step at each successive time step, leading to the possibility of interpolation to determine the base conditions at R and S in Figure 7.2.

For both the free surface and gas unsteady flow cases the same general points apply concerning the need to avoid excessive interpolation and the need to introduce suitable boundary conditions. Figures 7.4(a) and 7.4(b) illustrate the limits to the method of characteristics based finite difference scheme if no boundary condition information is available. The overall solution process will be identical for all the cases set out in this chapter and individual boundary condition equations will be presented.

7.3 Open channel and partially filled pipe unsteady flow analysis

Unsteady flows in free surface liquid flows may be characterised by modifications to the inflow vs. time profile as the surge propagates through the network. In general these modifications may be observed as an attenuation in the peak flow rate as the wave moves downstream and a lengthening of the overall flow duration downstream. These effects are also described by a steepening of the wave leading edge and a flattening of the wave trailing edge.

Both these effects are illustrated in Figure 7.5. If the surge wave is imagined as a series of incremental waves each having an individual depth and hence an identifiable wave speed dependent upon that depth, increasing as depth increases, then the modifications to the surge may be predicted as the peak depth flow propagates faster than either the leading or trailing edges, resulting in the overall wave spread and peak attenuation observed.

The actual phenomenon is complex as it depends upon channel size, roughness, slope and cross sectional shape as well as the base flow rate over which the surge propagates and the shape of the surge itself, defined in terms of an initial flow hydrograph. Accurate prediction of these attenuation effects would be advantageous in a range of applications, including flood routing, urban storm water system design and building drainage schemes. Analytical solutions are again limited and steady flow models are inappropriate as they lead to gross oversizing, however a numerical solution approach is practical and the method of characteristics is among the easier of the available techniques to apply due primarily to the ease with which complex boundary conditions may be handled.

Figure 7.2 and equations 7.19 through 7.22 defined the C^+, C^- characteristics that may be utilised to describe the propagation of surge waves through free surface flows. The characteristic equations may be expressed for ease of manipulation as, for the C^+ characteristic, RP,

$$V_P = V_R - \frac{g}{c_R}(h_P - h_R) - g(S_R - S_o)\Delta t = 0 \tag{7.57}$$

and for the C^- characteristic SP or S′P,

$$V_P = V_S + \frac{g}{c_S}(h_P - h_S) - g(S_s - S_o)\Delta t = 0 \tag{7.58}$$

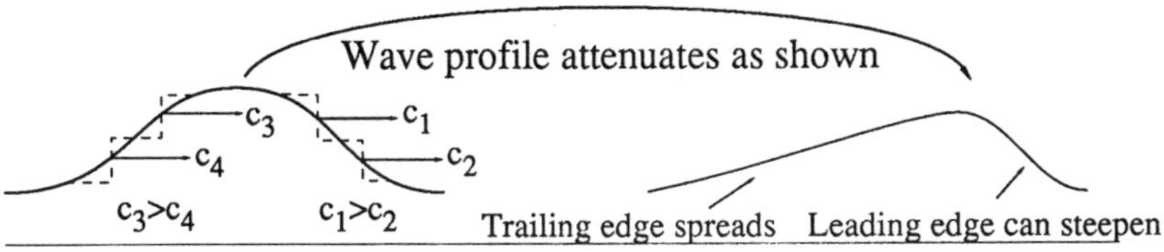

Figure 7.5 Unsteady flow wave attenuation in open channels or partially filled pipeflows

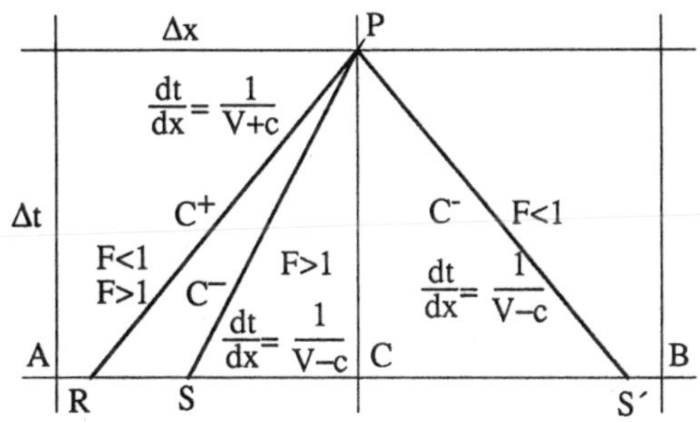

Figure 7.6 C^+ and C^- characteristics in both subcritical, F<1, and supercritical, F>1, flows

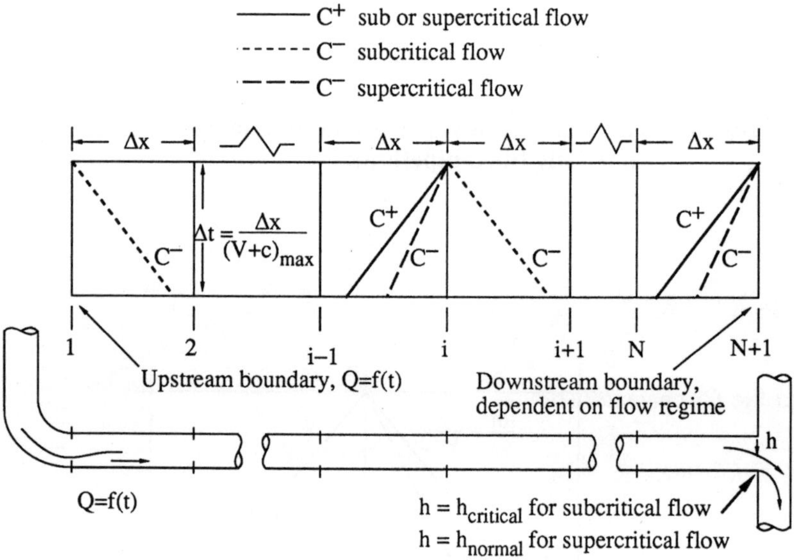

Figure 7.7 Application of specified time interval grid to partially filled pipe flow with known entry and exit boundary equations

where dx/dt = V + c for the C^+ case and dx/dt = V − c for the C^- case.

$$K1 = V_R + \frac{g}{c_R} h_R - g (S - S_o) \Delta t$$

$$K3 = V_S - \frac{g}{c_S} h_S - g (S - S_o) \Delta t$$

$$K2 = g / c_R \qquad K4 = g / c_S$$

Note the suffix S represents both points S and S′ for subcritical and supercritical flows.

Thus a first order finite difference approximation yields two equations in V_P and h_p based upon conditions known at a time step earlier, namely for the C^+ characteristic

$$V_P = K1 - K2\, h_p \tag{7.59}$$

and the C^- characteristic

$$V_P = K3 + K4\, h_p \tag{7.60}$$

yielding solutions for the mean velocity and depth at P

$$h_P = \frac{K1 - K3}{K2 + K4} \tag{7.61}$$

$$V_P = \frac{K1.K4 + K2.K3}{K2 + K4} \tag{7.62}$$

The values of K1 to K4 depend upon the interpolated values of V, h and c at R, S or S′.

Equations 7.59 and 7.60 will be used as the standard notation for C^+ and C^- characteristics in the remainder of the text.

The slope of the C^- characteristic may be positive or negative in the flow direction, dependent upon the relative value of flow velocity and wave speed, ie whether the flow is supercritical or subcritical. It will be seen that as the flow varies along the channel the slope of the characteristic lines will vary so that R, S or S′ may lie anywhere within one Δx either side of P. Thus the values of flow depth and velocity utilised in equations 7.57 and 7.58 to define the local time values of K1 to K4 depend upon values interpolated between A and C and B and C, Figure 7.6. In common with other applications of the method of characteristics the degree of interpolation should be kept to a minimum to avoid undue rounding errors.

In common with the application of the method of characteristics to pressure transient propagation the analysis requires that values for flow depth, velocity and wave speed along the network at time zero are assigned. In the case of steady uniform flow existing prior to the arrival of a surge this presents no difficulty or inaccuracy, however in regions of steady non-uniform flow, such as the backwater profile to a junction or the drawdown profile at a free discharge, the interpolations necessary to assign initial values at point R, S or S′, may result in rounding errors leading to a progressive collapse of the backwater profile over several time steps, Swaffield and Maxwell-Standing (1986). The initial flow backwater profile will be curved, thus a distance based linear interpolation will either under or over estimate the flow depth dependent upon whether the backwater profile is convex or concave, leading to errors in the flow velocity based on a steady continuous flow. Swaffield and Galowin (1992) reported a correction to the characteristic equations that prevented this collapse by introducing a multiplier to the $(S - S_0)$ term in equations 7.57 and 7.58.

These correction factors calculated for all points within a steady non uniform flow regime at time zero were defined as

$$V_P = V_R - \frac{g}{c_R}(h_P - h_R) - C1\ g\ (S_R - S_o)\Delta t = 0$$

$$V_P = V_S + \frac{g}{c_S}(h_P - h_S) - C2\ g\ (S_S - S_o)\Delta t = 0$$

Obviously both C1 and C2 are set to unity for regions of initial steady uniform flow. Suffix S again applies to both sub and supercritical flow conditions.

Thus a systematic application and solution of the characteristic equations, commencing with a known initial flow condition will yield flow depth and velocity results throughout a network over any time range, provided that the discharges to the system are known and that the various boundary conditions throughout the system can be modelled within the method of characteristics solution. The inclusion of boundary conditions is central to any method of characteristics model and the method is particularly suited to the inclusion and development of the boundaries, both fixed and moving, that exist within free surface flow network applications.

7.4 Boundary condition modelling within free surface flow networks

The boundary conditions met within a free surface flow network include entry and exit conditions, junction conditions and most importantly interface conditions between supercritical and subcritical flows that take the form of moving hydraulic jumps whose position relative to junction boundaries depend upon the approach flows. The flow regime in building drainage systems, due to the relatively steep pipe slopes, commonly 1 in 50 to 1 in 100, is supercritical, while free surface flows in canals is normally subcritical. Thus the presence of junctions generates a local subcritical regime that must be terminated upstream by a hydraulic jump in each of the channels terminating at the junction.

7.4.1 Entry conditions

Boundary conditions as developed in Chapter 5 depend upon the solution of the available C^+ or C^- characteristic with some other equation that defines the flow conditions at the boundary, most commonly in free surface flow applications a relationship between flow rate and time or between flow rate and local depth. At the upstream entry to a channel in supercritical flow no characteristic is available as the C^- characteristic slopes downstream. In this case it is necessary to relate flow depth at entry to the inflow rate at that particular instant. Two inflow regimes should be considered

1. Tranquil flow that enters the channel at the normal flow depth appropriate to that flow rate. The Chezy equation for fully developed steady uniform flow, Q, in a channel of slope S and hydraulic mean depth, m, defined as the flow cross sectional area, A, divided by the wetted perimeter of the channel, may be stated as

$$Q = A\,(2g\,/\,f\,)^{0.5}\,(mS)^{0.5} \tag{7.63}$$

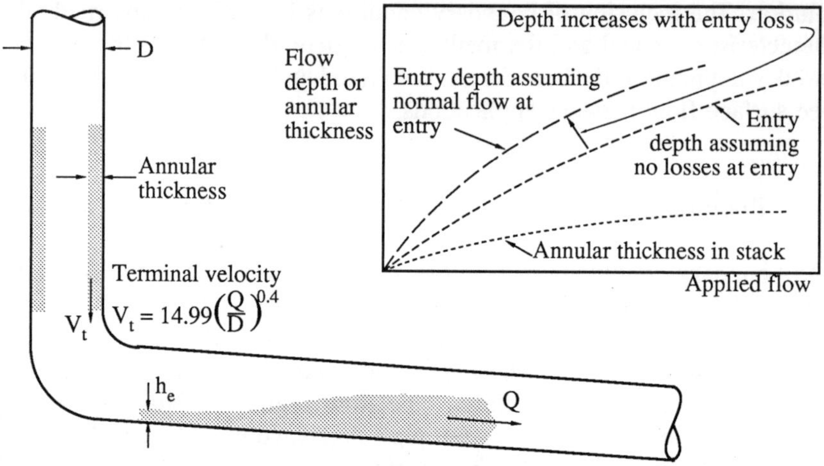

Figure 7.8 Relationship between entry depth and annular thickness. Note that a normal flow entry condition is equivalent to a total loss of kinetic energy at entry

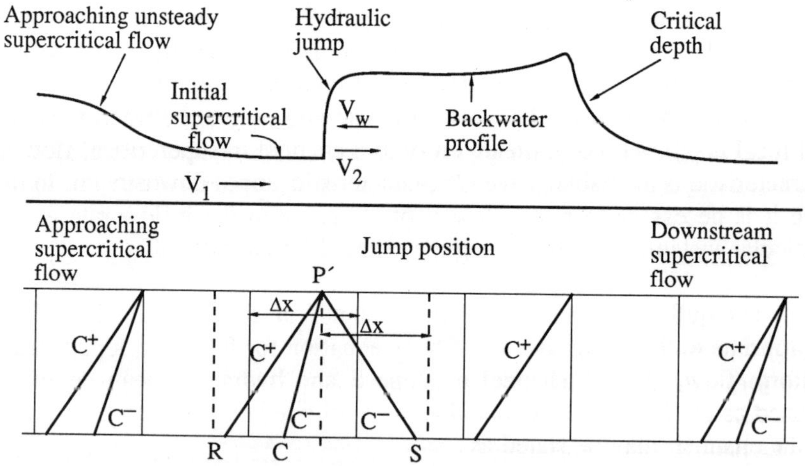

Figure 7.9 Summary of available characteristic equations in the junction region

where f is the flow friction factor, best calculated from the Colebrook-White relationship modified to account for the channel shape. It will be seen that equation 7.63 is in reality a relationship in terms of functions of depth and thus a depth at channel entry may be developed by the solution of

$$1 - Q / (A (2g / f)^{0.5} (mS)^{0.5} = 0 \tag{7.64}$$

The bisection technique will yield the required solution when the depth functions replacing A and m are introduced and Q is defined in terms of inflow at each time step

$$Q = \phi(t) \tag{7.65}$$

2. For discharges possessing a high kinetic energy such a solution is inappropriate as the flow enters as a shallow high speed layer, examples would be at the base of vertical stacks in drainage systems or following a sudden change in channel slope. In these cases entry depth may be determined by equating the entry flow energy, E, to the downstream entry specific energy, again provided that the inflow profile is known, equation 7.65, and the channel shape is defined as a function of depth, thus

$$E - (h_1 + (Q_t/A_{h_1})^2 / 2g) = 0 \tag{7.66}$$

In subcritical flow conditions a C^- characteristic does exist and the inflow profile of equation 7.65 may be solved with this characteristic, thus

$$Q_t - A_1 (K3 + K4\, h_1) = 0 \tag{7.67}$$

where 1 is the channel entry node and A is the flow cross sectional area, again a function of depth. The bisection technique yields a solution.

Entry to a downstream channel from a junction of two or more channels may be determined by reference to the change in flow regime forced by the junction. As mentioned a junction in normally supercritical flow will generate a local upstream subcritical zone in each of the approach channels. Downstream of the junction the combined flow will revert to supercritical. Thus the entry to the downstream channel may be approximated by the introduction of a critical flow depth section at the channel entry. At a critical depth section the wave speed and flow velocity are equal, thus it may be shown that

$$V_1 = (gA_1 / T_1)^{0.5} \tag{7.68}$$

where T is the surface width of the free surface flow, or, incorporating the known discharge to the channel from the junction, Q_t, as a result of the solution of the appropriate characteristics at the termination of the upstream joining channels

$$1 - Q_t^2 T_1 / (gA_1^3) = 0 \tag{7.69}$$

It will be seen that no characteristic is available at an entry node critical section as the C^- characteristic from the next node downstream slopes downstream.

Figures 7.7 and 7.8 illustrate these entry conditions.

7.4.2 Exit boundary conditions

Free discharge from a channel under supercritical conditions requires no special boundary condition as in this flow regime both characteristics slope downstream and the flow is 'unaware' of the channel termination. Equations 7.59 and 7.60 thus yield flow depth and velocity at discharge.

Free discharge from a channel under subcritical flow conditions does require the introduction of a boundary equation as only the C^+ characteristic is available. Flow observations of the drawdown curve for subcritical flows indicates that the flow passes through a critical depth section just prior to channel termination, Figure 7.7, Henderson (1966), and the C^+ characteristic may then be solved with this relationship, equations 7.59 and 7.68

$$\sqrt{(gA/T)_{n+1}} - (K1 - K2h)_{n+1} = 0 \tag{7.70}$$

where n+1 is the channel exit node and it is assumed that the area and surface width terms, A and T, are defined as geometric functions of depth.

7.4.3 *Junction boundary conditions*

Pipes or channels terminating at junctions of two or more normally supercritical flows feature a region of subcritical flow as illustrated in Figures 7.9 and 7.10. For level invert junctions empirical measurements have indicated that the common junction depth, h, is governed by the summation of the junction inflows

$$h_j = k \sum Q_j^n \tag{7.71}$$

where j is the exit node for each of the pipes discharging to the junction, Q is the discharge from each pipe and k and n are empirical coefficients that depend upon the geometry of the junction.

Thus for a junction of three pipes, where the junction depth is assumed to be common for all the joining pipes, and pipes 1 and 2 carry the approach flows while pipe 3 carries away the combined flow, Q_j, the boundary equation to be solved becomes an amalgamation of equation 7.70 with the C^+ characteristics for the exit node of pipes 1 and 2

$$h_j - k((K1 - K2h_j)_1 A_j + (K1 - K2h_j)_2 A_j)^n = 0 \tag{7.72}$$

an expression again solvable by the bisection technique as the A terms are known as functions of h. The exit depth for both approach pipes are equal. Junctions featuring an inflow from above are more complex in that the flow from above may be regarded as terminating in a supercritical free discharge. The flow depth at entry to the combined downstream flow may again be taken as critical, while the upstream horizontal channel termination depth has been shown empirically to be modelled under steady conditions by an expression of the form

$$h_j = h_{mc} (1 + ((h_{dc} - h_{mc})/h_{mc})^n) \tag{7.73}$$

where

mc denotes upstream flow rate critical depth,
dc denotes combined downstream flow rate critical depth,

and

j is the junction depth.

Solve at jump:		Solve at junction:	
Upstream	C^+, C^- (supercritical)	Upstream	C^+ (both inflow pipes)
Downstream	C^- (subcritical)		Depth vs. junction flow eqn
Jump	Flow continuity Momentum eqn		
Yields	$V_u, h_u, V_d, h_d, V_{wave}$	Yields	V_{uj}, h_{uj} (both pipes)

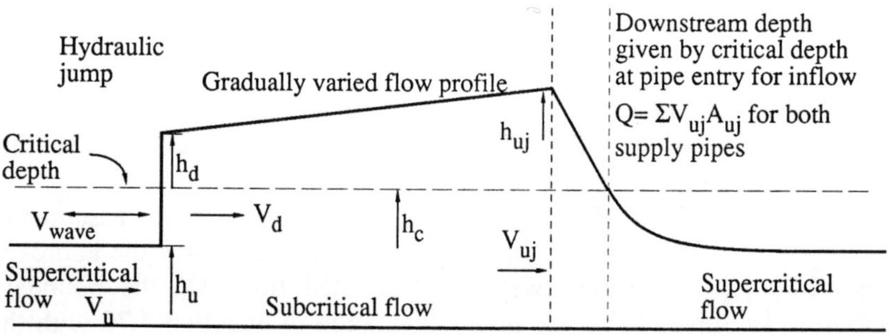

Figure 7.10 Junction model

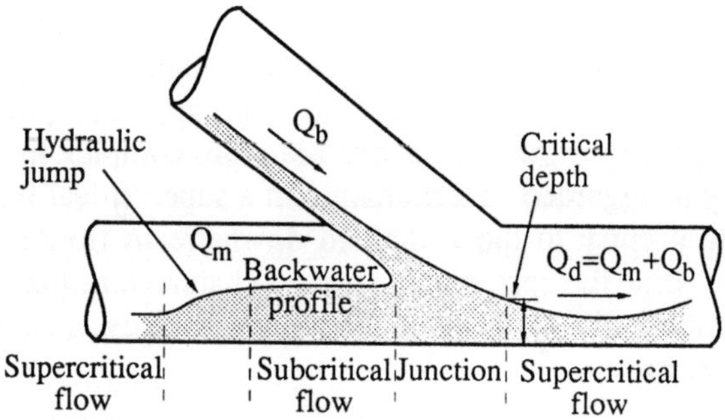

Figure 7.11 Schematic diagram of flow regime for top entry junctions

The solution of equation 7.73 is somewhat cumbersome and must be achieved by an iterative process. Referring to Figure 7.11 the discharge flow from pipe 2, ie from above, may be regarded as a supercritical discharge so that a solution of the C^+ and C^- characteristics at the exit to pipe 2 yields a flow Q_2.

The combined flow downstream is given by

$$Q_3 = Q_1 + Q_2$$

where

$$Q_1 = A_j\,(K1 - K2h_j)$$

Thus expressions may be generated for the critical flow depths associated with both the downstream combined flow and the approach flow along pipe 1. There is only one real value of junction depth that will satisfy these equations and this may be determined iteratively.

In both cases the upstream channels feature a hydraulic jump positioned to transform the supercritical approach flow into the forced subcritical flow through the junction. The position of these upstream jumps, as there is one in each approach channel, is determined by the junction throughflow. Initially the jump positions are determined for the initial steady base flow by standard backwater profile gradually varied flow calculations. The interpolations necessary to provide a zero time condition throughout the network have already been mentioned. As the calculation proceeds the position of the hydraulic jumps relative to the junction changes as the approach flows vary, moving upstream as the approach flow increases. This jump movement effectively provides a storage zone upstream of the junction that empties as the junction depth falls in response to a fall in the combined throughflow. In this way the model accurately predicts the smoothing effect of junctions on surges arriving along any one of the joining channels. The form of the junction equations, ie the solution of the appropriate C^+ equation for each joining channel simultaneously with the junction depth expression, equation 7.71, allows for the possibility of reverse flow to be established back up any one of the approach channels with a lower flow rate than any other. This is an important feature of the model.

Figure 7.10 also illustrates the equations to be solved to trace the movement of the hydraulic jump. It is necessary to solve for the flow depth and velocity on either side of the jump, together with the jump velocity. Characteristic equations do not apply across the jump, which is similar to a shock wave in supersonic/subsonic gas flows. Thus five equations are necessary, namely the C^+ and C^- equations on the upstream side of the jump, a C^- equation immediately downstream of the jump, ie in the forced subcritical regime, and the equations of continuity of flow and momentum across the jump.

1. the continuity equation applied across the wave

$$(V_1 - V_w)A_1 = (V_2 - V_w)A_2 \tag{7.74}$$

2. the force equation applied across the wave

$$gA_1\,\overline{h}_1 - gA_2\,\overline{h}_2 = \rho A_2(V_2 - V_w)^2 - \rho A_1(V_1 - V_w)^2 \tag{7.75}$$

3. C^+ characteristic P′R associated with the supercritical flow

$$V_{P'} - V_R + \frac{g}{c_R}(h_{P'} - h_R) + g(S_R - S_o)\Delta t = 0 \tag{7.76}$$

4. C^- characteristic P′C, also associated with supercritical flow

$$V_{P'} - V_C - \frac{g}{c_C}(h_{P'} - h_C) + g(S_C - S_o)\Delta t = 0 \tag{7.77}$$

5. C^- characteristic P′S associated with downstream subcritical flow

$$V_{P'} - V_S - \frac{g}{c_S}(h_{P'} - h_S) + g(S_S - S_o)\Delta t = 0 \tag{7.78}$$

The first two expressions apply to rapidly varied flow and are valid across the wave, the remaining expressions apply to the flow either side of the wave. Solution of these boundary equations at each time step allows the motion of the jump to be predicted and the flow depths and velocities on either side to be calculated. The position of the jump at the next time step may thus be determined. The path of the jump through the x–t plane may be appended to the 'normal' characteristics grid by the introduction of a pseudo-characteristic having a slope of $1/V_w$.

The flow at the junction is subcritical in both the joining upstream pipes. Thus two C^+ characteristics are present, terminating pipes identified as i and k below. Thus the available equations are

1. the C^+ characteristics for pipe i

$$V_i = K1_i - K2_i \, h_{junction}$$

2. the C^+ characteristics for pipe k

$$V_k = K1_k - K2_k \, h_{junction}$$

3. the junction common depth relationship

$$(V_i A_i) + (V_k A_k) = a \, h^b_{\;junction}$$

where A_i, A_k are the flow cross sectional areas in these two pipes at the common junction depth.

Substitution for V_i and V_k from the C^+ characteristics yields an equation in h_j that may be solved by the bisection technique. This formulation allows for the establishment of reverse flow in either of the joining supply pipes.

The flow conditions at entry to the pipe downstream of the junction may be calculated based on the assumption of critical depth at this point, the critical depth value being that appropriate to the combined throughflow from both supply pipes,

$$Q_t = \sum (AV)_{\text{local values at pipe/junction boundary}}$$

and critical depth is given by

$$1 - Q_t^2 \, T/gA^3 = 0$$

Figure 7.10 illustrates the solution technique described above.

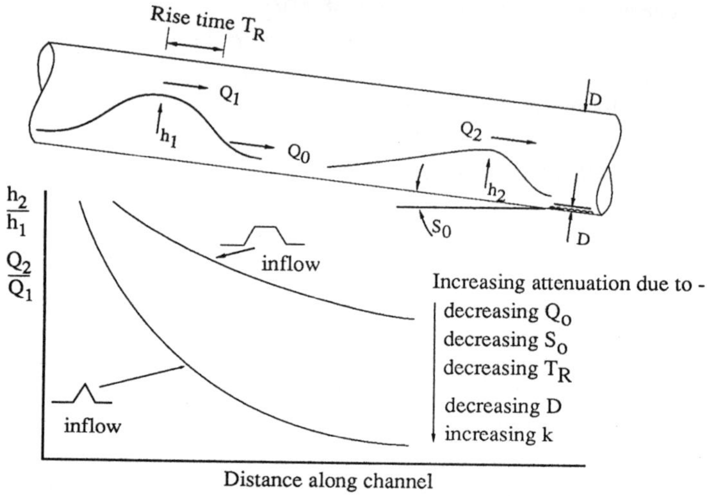

Figure 7.12 Dependence of wave attenuation on pipe and flow parameters

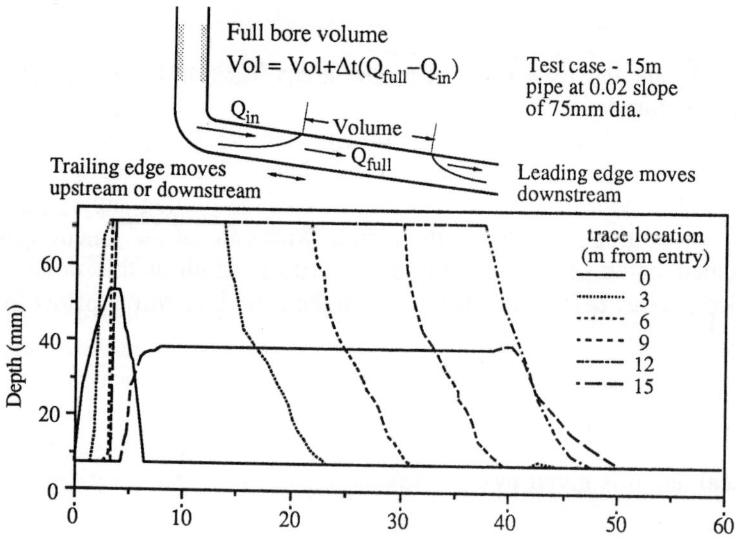

Figure 7.13 Flow depth along a surcharged collection drain. Note full bore flow regime moves downstream

The introduction of moving boundary conditions, such as the oscillating hydraulic jump discussed above, may be extended to include the onset of full bore flow as a result of pipe or junction surcharging, the necessary boundary condition at the junction being supplied by a knowledge of the flow necessary to cause a full bore flow event and the subsequent discharge of the junction under pressure flow conditions.

7.5 Applications of the method of characteristics free surface unsteady flow model

Applying the boundary conditions discussed above to a simple single channel with a free discharge and a tranquil entry flow allows the effect upon wave propagation of the various channel parameters to be determined, together with the effect of dynamic variables such as the initial steady flow over which the surge propagates and the shape of the inflow surge itself.

Figure 7.12 illustrates the resulting dependencies, the degree of attenuation rises as the pipe or channel slope decreases, the attenuation increases as the channel roughness increases, attenuation increases if the inflow profile is triangular w.r.t. time rather than trapezoidal, etc. The effect of ignoring attenuation upon channel sizing is clear.

Figure 7.13 illustrates the inclusion of a full bore flow condition to replicate the possibility of a surcharged flow. The limiting effect on flow rate is clearly shown by this example. Figure 7.14 illustrates the use of the model, incorporating the junction boundary equations to predict backflow conditions. Note that the depth increases in pipe 4 ahead of those in pipe 2 can only be explained in terms of a backflow from the junction, as confirmed by the predicted negative flow rates for pipe 4.

In common with its application to pressure transient analysis the method of characteristics has much to offer the free surface unsteady flow modeller. In particular it offers considerable flexibility in the representation of boundary conditions and is particularly suitable for the modelling of moving boundary conditions. It retains the shortcomings of its application to pressure transients, namely an inherently small time step determined by the local wave speed and the possibility of considerable rounding errors if interpolation techniques are used to determine the base conditions for each characteristic at each time step. Nevertheless the method's advantages outweigh its difficulties in this application as in pressure transient modelling and make its use as a design tool a realistic proposition in many free surface unsteady flow applications as the availability of computing power increases.

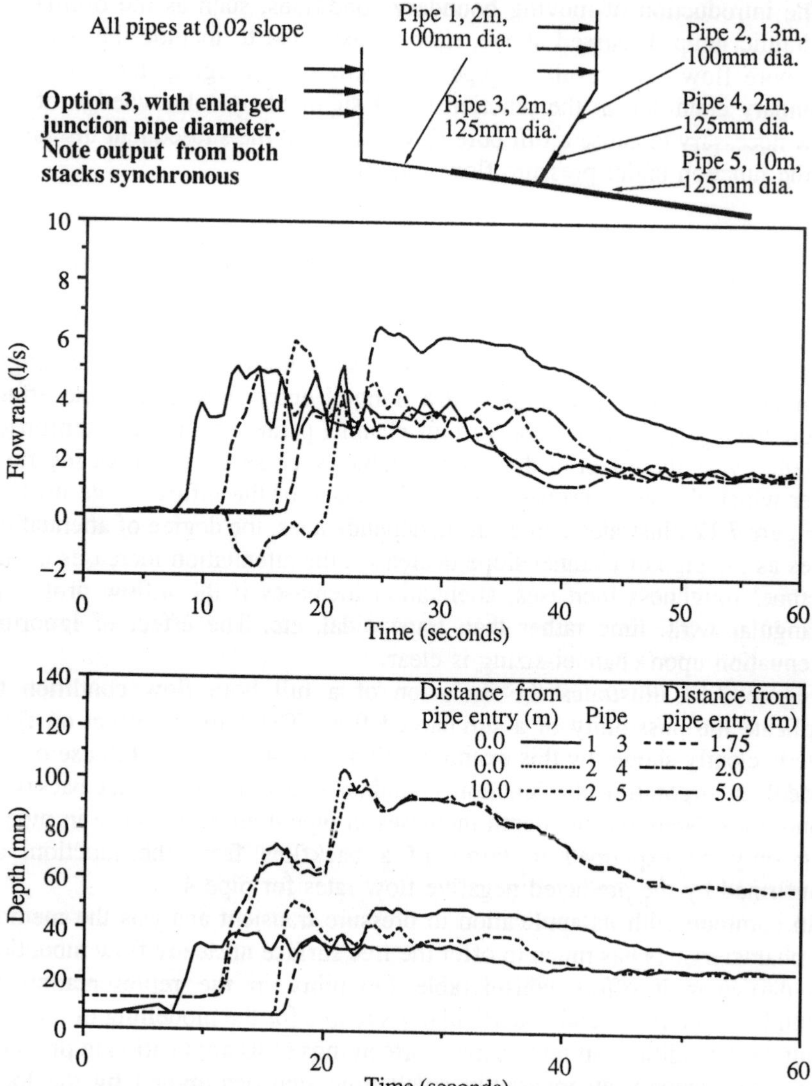

Figure 7.14 Application of the model to guide building drainage system design. The individual floor to stack discharges were also modelled, based on an assumed appliance usage pattern and a knowledge of individual appliance discharge profiles

7.6 Application of the method of characteristics to the modelling of small amplitude air pressure transients

Section 7.2 developed the general equations for the propagation of air pressure transients. This section will illustrate the application of those equations to the prediction of air pressure transient fluctuations in vent systems, an application that features low amplitude pressure variations.

Equations 7.43 through 7.46 represent the C^+ and C^- characteristics determining air pressure transient propagation. As shown in Section 7.2 these equations may be reduced to

$$u_P = K1 - K2\ c_P \tag{7.79}$$

for the C^+ characteristic and

$$u_P = K3 + K4\ c_P \tag{7.80}$$

for the C^- characteristic, where u is the gas mean velocity at a section, P, Figure 7.3, and c is the gas wave speed.

The local wave speed and gas absolute pressure are linked by equation 7.50, defining wave speed in terms of local pressure and some datum gas pressure and density, normally taken as atmospheric.

The values of the terms K1 to K4 are defined by equations 7.47 to 7.49 and depend upon the local conditions in the duct within one Δx of P at a time step earlier. The time step magnitude is again determined by the Courant Criterion introduced in Chapter 4 and the precautions necessary against undue interpolation discussed earlier continue to apply.

As with the other examples of method of characteristics applications to unsteady flow problems the development of suitable boundary conditions is the key to successful modelling. In this case the boundaries most commonly met involve open pipe ends to atmosphere, with and without local losses, junctions of two or more ducts, with or without an annular water film present, and terminations featuring a moving boundary, for example a water trap seal.

Figure 7.15 illustrates these potential boundaries and their respective reflection and transmission coefficients, developed in the same way as those described for liquid systems in Chapters 1 and 3. Despite the fact that such air pressure transient have low amplitude, the mechanisms defining transient propagation and the techniques necessary to predict pressure excursions within the network remain analogous to those already discussed.

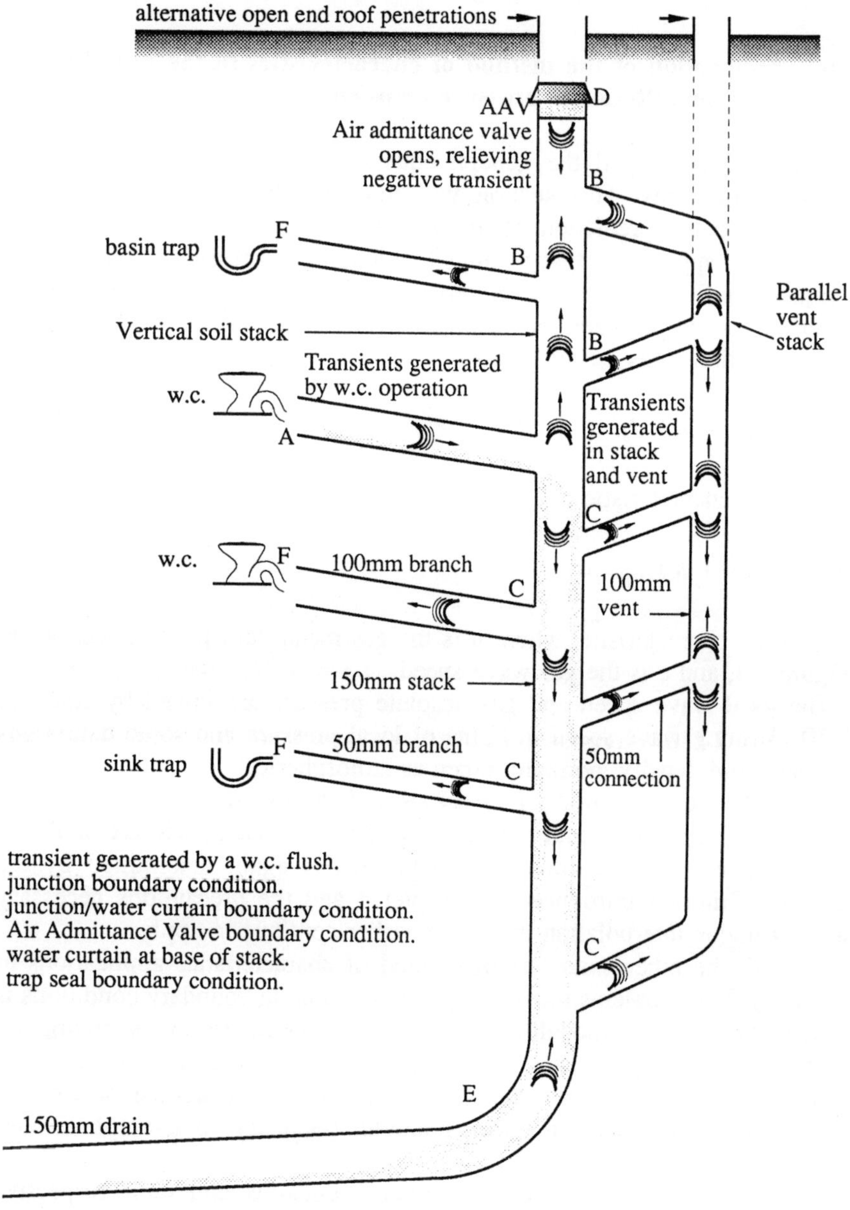

A - transient generated by a w.c. flush.
B - junction boundary condition.
C - junction/water curtain boundary condition.
D - Air Admittance Valve boundary condition.
E - water curtain at base of stack.
F - trap seal boundary condition.

Figure 7.15(a) Synthesis of the transient air pressure boundary conditions into a drain and vent model, applied to a modified one pipe system

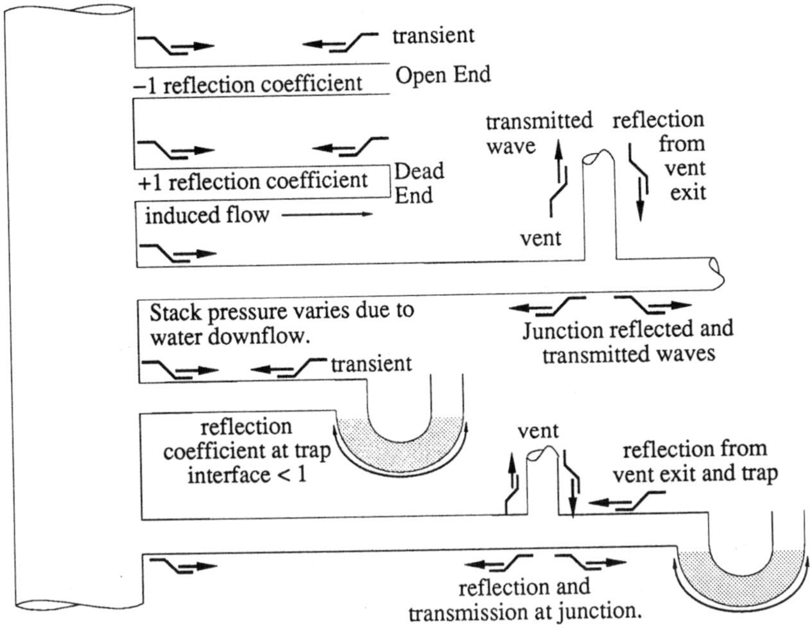

Figure 7.15(b) Transient propagation and trap seal response within a horizontal drain or vent system

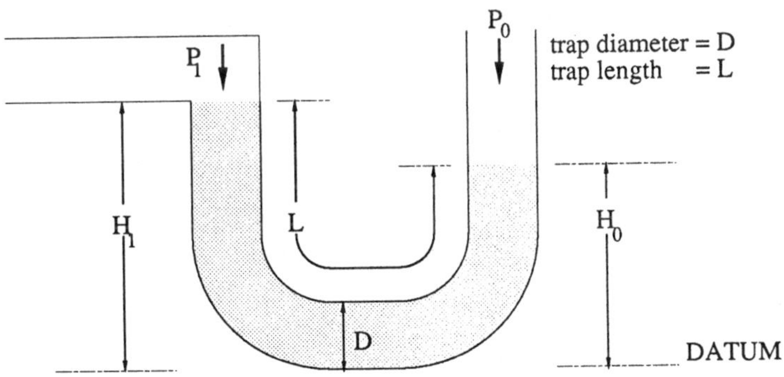

Figure 7.16 Definition of trap parameters. Note that if $P_1 < P_0$ then H_1 is +ve

7.6.1 Entry boundary conditions

The upstream entry is analogous to the constant pressure reservoir discussed in Chapter 5. The available C^- characteristic may be solved with the imposed entry pressure condition. In its simplest form this involves the solution of

$$u_1 = K3 + K4\ c_1 \tag{7.81}$$

with

$$p_1 = p_{Res} - K\ 0.5\rho\ u\ |\ u\ | \tag{7.82}$$

and the pressure-wave speed equation, 7.50. Note that the absolute value of flow velocity ensures that the entry separation loss is correctly represented for a reverse flow condition.

Note that it may be necessary to employ a separate entry loss coefficient, k, to represent inflow or reverse flow situations. In many cases the local separation loss may be ignored.

7.6.2 Exit and junction boundary conditions

Two main boundary condition types need to be represented, namely discharge to atmosphere and termination in a junction of two or more ducts. The additional boundary formed by a liquid trap seal will be discussed separately.

The available equations at a free discharge exit from a duct to a zone of known pressure, either constant or varying with time in a known manner, are the appropriate C^+ equation together with the local separation loss equation and the general wave speed/pressure expression

$$u_{n+1} = K1 - K2\ c_{n+1} \tag{7.83}$$

and

$$p_{n+1} = p_{Res} - K\ 0.5\rho u_{n+1}\ |\ u\ |_{n+1} \tag{7.84}$$

In the case of a dead-ended branch, zero velocity at the dead end provides the boundary condition, hence

$$c_{n+1} = K1_{n+1} / K2_{n+1} \tag{7.85}$$

In the case of a junction of two or more ducts the boundary equation at the junction consist of

1. the C^+ characteristics for all the ducts supplying initial flow to the junction,

$$u(j,n(j)+1) = K1(j,n(j)+1) - K2(j,n(j)+1)\ c(j,n(j)+1) \tag{7.86}$$

2. the C^- characteristics for all the ducts initially carrying flow away from the junction,

$$u(i,1) = K3(i,1) + K4(i,1)\ c(i,1) \tag{7.87}$$

3. continuity of flow through the junction, expressed as

$$\sum_{j=1}^{M} u(j,n(j)+1)\ A(j,n(j)+1) + \sum_{i=1}^{N} u(i,1)\ A(i,1) = 0 \tag{7.88}$$

where M is the number of inflow ducts and N is the number of outflow ducts in terms of the initial set up of the flow network: note that the chosen sign convention determines these definitions of duct type,

4. the pressure at the junction must be common for all the ducts in the absence of a junction separation loss,

$$p(j,n(j)+1) = p(i,1) \tag{7.89}$$

5. the general wave speed/pressure equation, 7.50, is again utilised for all nodes calculated.

The presence of a local separation loss at the junction may be included in equation 7.89 or the likely increased loss may be included in an enhanced friction loss for each pipe length. The presence of a water film may also be included in equation 7.89 as a local pressure loss.

7.6.3 *Moving boundary conditions - trap seal response*

Figures 7.16 and 7.17 illustrate a common duct termination in a fluid trap seal capable of displacement in response to pressure fluctuation at the air to water interface. The response of the trap may be analysed by application of Newton's Second Law of Motion to the fluid column

$$A\ (p(j,n(j)+1)-p(ref)) + \rho_w(H_1-H_0) - \tau_o LP = \rho_w LA\ dU_w/dt \tag{7.90}$$

where A is the trap cross sectional area, P is its wetted perimeter, L is the liquid column length, H_1 and H_0 are the water surface elevations, p(j,n(j)+1) and p(ref), which may be taken as atmospheric, are the gas pressures exerted upon the trap liquid, of density ρ_w, and τ_o is the wall shear stress.

A suitable replacement for the wall shear stress in terms of a friction factor for the liquid flow in the oscillating trap reduces equation 7.90 to

$$p(j,n(j)+1)-p(ref)+\rho_w[H_1-H_0-L(4f\ U_w|U_w|\ /2D-AdU_w/dt)] = 0 \tag{7.91}$$

where U_w is the liquid column velocity and f is a friction factor approximated from the Colebrook White expression for a circular tube of diameter D. Note that it is necessary to reduce the length of the water column, L, if the trap water level overflows the entry to the drain, Figure 7.16.

The p(j,n(j)+1) term in equation 7.91 is dependent upon the transient conditions at the pipe to trap interface and may be replaced via the C^+ equation at the duct exit

$$u(j,n(j)+1) = K1(j,n(j)+1) - K2(j,n(j)+1)\ c(j,n(j)+1)$$

where

$$u(j,n(j)+1) = U_w$$

The use of the absolute value of liquid velocity in equation 7.91 again ensures that the friction opposes the trap motion. The use of a steady state friction factor again naturally underestimates the damping of any trap oscillation caused by a transient pressure wave arriving at the liquid-gas interface.

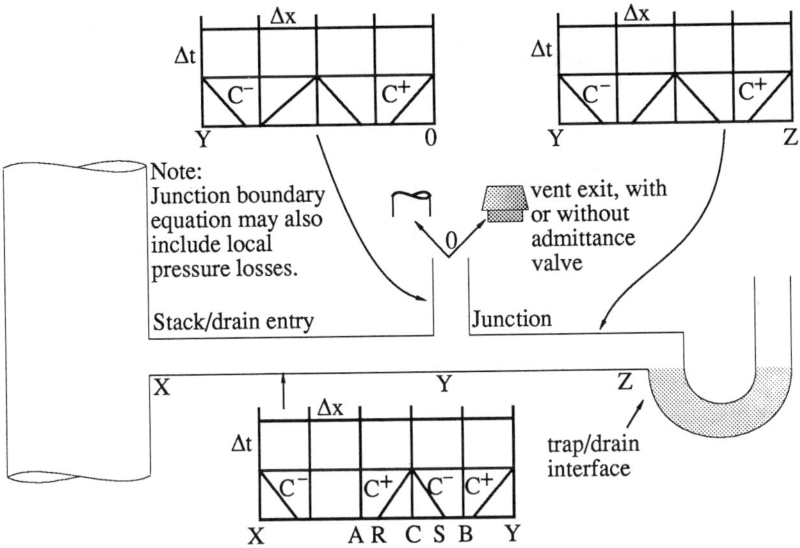

Figure 7.17 Method of characteristics treatment of stack to appliance drain with attached vent. Note Δt constant for all pipes and, as c constant, Δx also constant

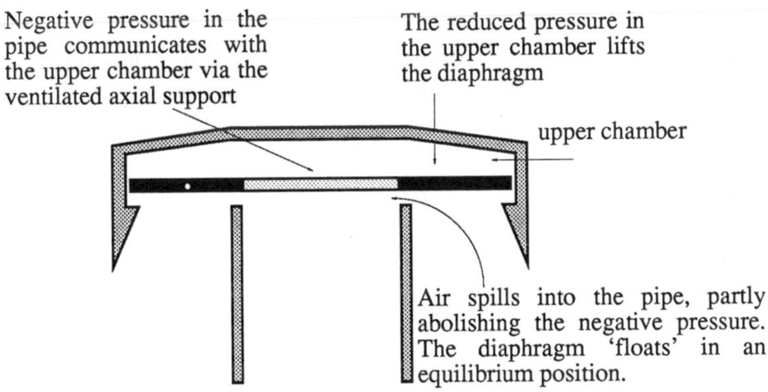

Figure 7.18 A generalised cross-section drawing of an air admittance valve. The diaphragm usually has a rigid plastic support to prevent flexing. This valve is shown in the open position

7.6.4 *Sign conventions*

For the purposes of the developed model it was assumed that open stack terminations, trap seals and air admittance valves were situated at pipe exits. This also implies a sign convention such that distance is measured positive away from the system assumed entry connection and hence any airflow into the network via the stack termination, air admittance valve or depleted trap is assumed to have a negative velocity. This is merely a convenient sign convention and does not affect the model, however it will be shown that it has considerable computational advantages in the later treatment of actual drainage and vent networks as it allows a wide range of systems to be dealt with by identifying the base of the vertical stack as a fixed reference and equivalent single entry point.

7.6.5 *Air admittance valves*

As a boundary the valve may be described as a variable entry/exit loss, the value of the loss coefficient being dependent upon the degree of opening of the valve, which in turn is determined by the pressure differential across the valve. Figure 7.18 illustrates the mechanism commonly found.

The valve remains closed as long as the pressure in the vent stack is insufficient to overcome the mass force of the diaphragm. Once the diaphragm lifts, it is free to rise until the valve is fully open. Equally the diaphragm may 'hover' at some intermediate setting, dependent upon the pressure-time history in the stack. Thus the boundary condition representing the value may be expressed by a series of three simple equations

Closed valve:- $$u_{exit} = 0.0 \tag{7.92}$$

Partially open:- $$p_{atm} - p_{exit} = K\, 0.5\, u_{exit}^2 \tag{7.93}$$

Fully open:- $$p_{atm} - p_{exit} = K_{min}\, 0.5\, u_{exit}^2 \tag{7.94}$$

K_{min} is the appropriate loss coefficient for the fully open valve.

Equations 7.92 to 7.94 may be solved with the appropriate C^+ or C^- equation, in this case of the chosen sign convention the C^+ equation.

It is clear that a relationship linking K to valve position and/or pressure differential is a necessary prerequisite for the application of the method of characteristics model.

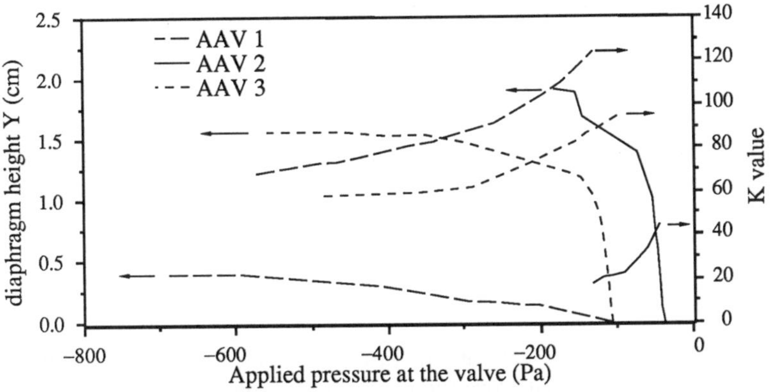

Figure 7.19 Change in diaphragm height Y and loss factor K with applied pressure for three admittance valves

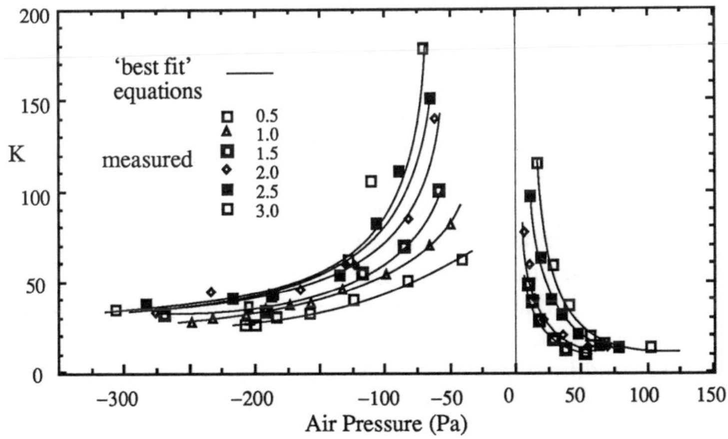

Figure 7.20 Variation in water curtain K value over a range of water flow rates, for both +ve and −ve air pressures

Figure 7.19 illustrates the variation of K values with differential pressure, DP, and valve position, Y, for three valve types, Swaffield and Campbell (1992). In each case a third order polynomial allows information describing the valve to be integrated directly into each of the valve boundary models

$$K = C1 + C2\ DP + C3\ DP^2 + C4\ DP^3 \qquad (7.95)$$

It will be noted that when fully open the valve K value reaches a minimum. A closed valve obviously has an infinite loss coefficient K.

7.6.6 Water curtain effects at a branch to stack junction

The annular flow within a vertical stack, Figure 7.15, provides a water film that covers the entry or exit from the pipes discharging to the stack on each floor. The presence of this water film acts as a boundary condition at the drain to stack junction that has to be modelled within the system. The effect of the film is to provide a variable reflection/transmission coefficient that depends upon film thickness.

The pressure differential necessary to force an air flow through a pipe entry or exit covered by a water film, whose thickness is dependent upon the stack annular downflow rate, may thus be described by a pressure differential equation of the form

$$dP = function\ (q_{stackflow}) \qquad (7.96)$$

where dP is the differential existing between the branch side and stack side pressures across the water film. In every way this is conceptually identical to the pressure differential necessary to force air through an air admittance valve when the system pressure falls below atmosphere. The main difference however is that air may pass from the appliance drain, or vent, to the stack or from the stack to the appliance connection dependent upon the local pressure conditions in either pipe. The effect of the water film upon the transmission and reflection of transients at this boundary thus becomes extremely important.

In common with the other complex boundary conditions already covered it will be sufficient to develop boundary equations applicable at the water film based upon steady state measurements of the pressure differentials across the film for a range of flow conditions, including a range of annular film thicknesses, determined by annular downflow rate and air flow rates in the appliance connection, either positive or negative. This formulation will allow the boundary to model incoming positive and negative transients generated by appliance discharge, within the network. Figure 7.20 illustrates the variation of film loss coefficient for both positive and negative branch air pressures, relative to the stack air pressure. The form of the defining equation is

$$K = C_1 + C_2 \Delta p + C_3 \Delta p^2 + C_4 \Delta p^3 \tag{7.97}$$

where $C_{1\text{-}4}$ depends on the stack annular water rate. It should be noted that at $\Delta p = 0$ both curves are asymptotic to infinity, which is supportable as with no pressure differential there will be no forced air flow through the water film. The film does not however act as a dead end, once a Δp is established by the arrival of a transient the K value will fall and an air flow will be set up.

The water film acts as a duct termination in exactly the same manner as the air admittance discussed above or the other alternative vent/drain terminations. The degree of reflection, and hence transmission, of an incoming transient at the water curtain will depend upon the thickness of the curtain, and thus on the annular flow being carried by the vertical stack. Observations suggest that the appropriate loss coefficients for the water film, upon which the mechanism of transient response depends, is also dependent upon the direction of the induced air flow. The water film is more resistive to branch applied suction than to branch applied positive pressures, Figure 7.20 illustrates this. While the use of steady state measured loss coefficients is open to discussion in a transient response situation, experience in the wider area of pressure transient simulation suggests that this is an acceptable technique provided that the overall transient wavelength, or the time taken for the transient input to be completed, may be shown to be an order greater than the system pipe period. In many vent system applications the lengths of the vent will be less than 50 m, a length for which the pipe period would be approximately 0.3 seconds, thus it may be seen that the use of the steady state loss coefficients for annular films or air admittance valves is justified.

7.6.7 *Positive pressure transient propagation on air path closure*

Positive air pressure transients may be generated within a drainage and vent network whenever there is a full or partial closure of the air path through the network. Such closures can occur as a result of surcharging the collection drain at the base of the vertical stack or as a result of an overloading of a branch to stack junction following appliance discharge.

If the air flow is reduced to zero, then subsequent events are identical to the mechanisms accepted in classical waterhammer with the simplifying condition that the magnitude of the resulting transients is insufficient to distend the pipe walls. The Joukowsky relationship defining pressure change as a result of an instantaneous flow stoppage applies

$$\Delta p = \rho_{air} c u_{air} \tag{7.98}$$

Thus a 1 m/s air flow velocity with density 1.2 kg/m^3 and a wave speed of 325 m/s will generate on stoppage a transient at the closure location of 390 N/m^2 or 40 mm water gauge. This transient will propagate throughout the network and will be reflected or transmitted at each boundary. An open termination will have a −1 reflection coefficient, thus generating a relieving transient. A closed end, for example an air admittance valve which remains closed in response to the arrival of a positive transient, will have a +1 reflection coefficient that generates an enhanced positive pressure in the stack. It should however be noted that these Joukowsky values will only be attained if the flow closure is achieved in less than one pipe period, defined as the time taken for a transient to travel from the closure location to the first major reflecting boundary and back, in this case to the top of the stack. Pipe period for a stack height L may be thus defined as

$$T_p = 2L/c$$

The second air path closure condition is somewhat more complex as transients are propagated both up and down the stack as a result of an air path closure at a branch to stack junction. The upper section of the stack is subjected to a positive pressure transient while the lower portion of the stack is subjected to an equal but negative transient. The simulation can deal with this condition by redefining the air velocities on either side of the closure as zero for the duration of the blockage, essentially identical to the treatment of valve closure between two pipes, Chapter 5. It will be appreciated that this implies the separation of the network into sections above and below the closure, however if the closure is assumed to occur at an already modelled junction this presents no increase in model complexity. Once the closure is considered to have ended, indicated by the duration of peak water flows in the branch or the collection drain at the base of the stack, the local boundary conditions revert to those in place prior to the closure.

It is worth noting that for the lower stack subjected to a negative transient it is possible that trap seal depletion will occur immediately. This effectively transforms the trap boundary to an open pipe and as such this becomes a relieving vent. As the method of characteristics deals with each boundary separately the model is automatically capable of dealing with this eventuality. This point will be demonstrated in later simulations. Similarly once the positive pressure transient has displaced the trap water up into the appliance this boundary also becomes a possible vent, however a vent with a high loss coefficient and as such not explicitly modelled.

7.7 Network modelling

The identification, development and validation of the boundary conditions discussed above confirmed that the method of characteristics solution for the unsteady flow equations would allow an air pressure transient model for building drainage and vent systems to be generated, capable of representing all the boundary conditions illustrated in Figure 7.15. A model may be based on the supposition that variations in annular film flow rate generates a variable suction in the stack and that these variations are communicated to the system as transients within the air core in the vent and horizontal branch networks, Figure 7.15.

The suction available to entrain an airflow in the stack may be considered to be dependent on the following parameters

$$\Delta P = \Phi(\rho_a, \rho_w, Q_a, V_w, D, t, \mu_a, \mu_w) \tag{7.99}$$

where

ρ_a	=	air density
ρ_w	=	water density
Q_a	=	entrained air core volumetric flow rate
V_w	=	annular water film terminal velocity
D	=	stack diameter
t	=	annular water flow terminal thickness
μ_a	=	air viscosity
μ_w	=	water viscosity

A traditional dimensional analysis utilising Buckingham's π Theorem reduces equation 7.99 to six dimensionless groupings

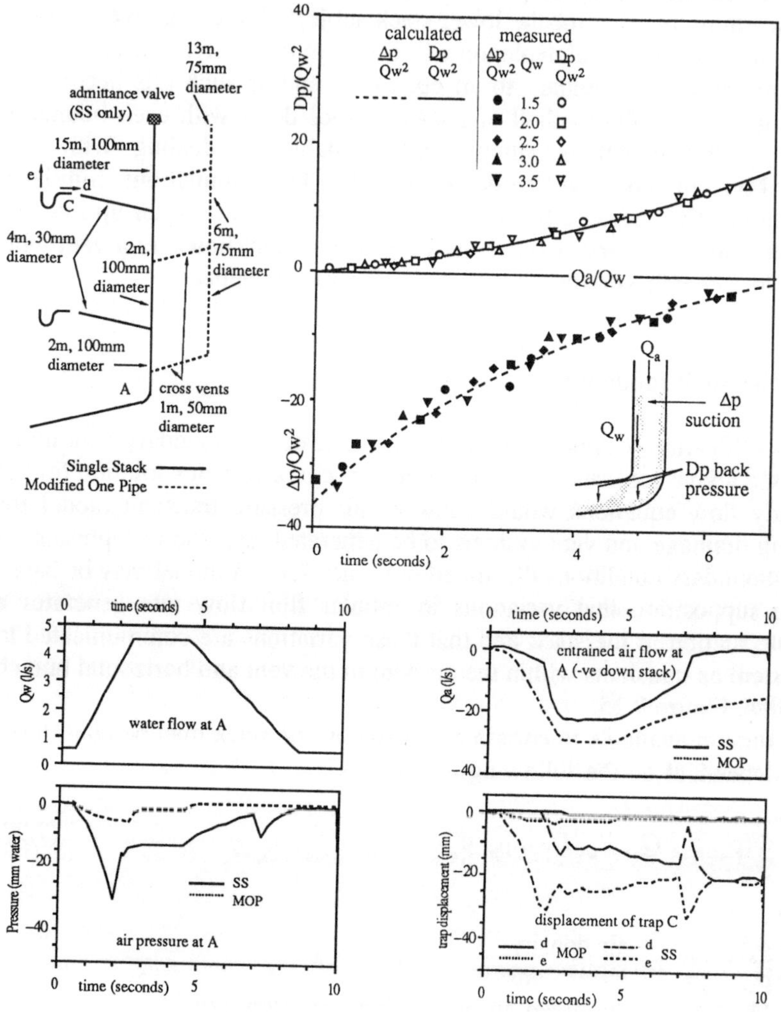

Figure 7.21 **Comparison of the air pressure transient propagation within a Single Stack (SS) fitted with an Air Admittance Valve and a Modified One Pipe (MOP) drainage and vent system as a result of representative stack loading.**

$$\frac{\Delta P D^2}{\rho_w Q_w^2} = \Phi_1\left(\frac{\rho_a}{\rho_w}, \frac{t}{D}, \frac{Q_a}{Q_w}, \frac{\mu_w}{\mu_a}, Re_a\right) \tag{7.100}$$

where Re_a is the entrained airflow Reynolds Number. Equation 7.99 may be simplified considerably as a single stack diameter of 100 mm would cover a wide range of applications and the density and viscosity ratios may be assumed constant. The entrained airflow Reynolds Numbers, normally above 10^5, lies in the fully turbulent region, and with a constant pipe roughness Reynolds Number effects may be ignored. For the range of typical water downflows, from 2 to 5 litres/sec, the variation of terminal film thickness is small.

Thus equation 7.100 may be reduced to

$$\Delta P = Q_w^2 \, \Phi_2\left(\frac{Q_a}{Q_w}\right) \tag{7.101}$$

Figure 7.21 illustrates the degree of agreement between laboratory measurements of stack suction pressure and the model dimensional groupings suggested by equation 7.101. It is clear from these results that the suggested dependencies may be represented by a polynomial to be solved with the available C^- characteristic at the base of the stack. Transient generation may then be modelled provided that the annular water downflow, Q_w, is known as a function of time.

$$\Delta P = Q_w^2 \left(b_1 + b_2\left(\frac{Q_a}{Q_w}\right) + b_3\left(\frac{Q_a}{Q_w}\right)^2 + b_3\left(\frac{Q_a}{Q_w}\right)^3\right) \tag{7.102}$$

Figure 7.21 illustrates the application of the modelling techniques developed to a simple single stack network with and without air admittance valve stack termination. The model provides predictions for air flow, air pressure and trap displacement. A number of conclusions may be drawn:

1. The water flow rate at the base of the stack controls the entrained air flow via the boundary condition previously developed. Information concerning changes in water flow rate is propagated up the stack leading to changes in the line air pressure as shown. If the stack has an open termination then the upper boundary remains atmospheric

pressure, explaining the reduced suction in the upper most stack section.

2. Propagation of these transients into the horizontal branches lead to trap oscillation and depletion.

3. In each case the air flow is seen to continue to increase once the water flow has reached its maximum and become as constant. In parallel it will be seen that the maximum suction pressures of the base of the stack fall back as this increase in air flow occurs. As the Q_w value rises the solution to the boundary condition is carried out between the C^- characteristic and a succession of stack suction pressure versus air flow characteristics that feature greater suction pressures as Q_w rises, Figure 7.21. While Q_w increases the value of stack pressure is likely to continue to decrease as the solution points moves onto the next 'pseudo-fan' curve. Air flow rate also increases as shown.
However once Q_w stabilises the solution is constrained to that particular Δp versus Q_a curve, and as the stack base pressure is lower than that further up the stack, the tendency will be for the air flow to increase, leading to a reduction in stack base suction pressure as the air flow rate rises. This is effectively a transfer of energy from static suction pressure to kinetic pressure in the air flow.
As the water flow rate abates, stack base pressures move back towards atmospheric, however for the open termination single stack the air flow continues after the water flow ceases. This is due to the inertia of the system. The only force to reduce this residual air flow is stack friction and this acts to reduce the air flow as shown in a manner clearly dependent upon the reciprocal of Q_a^2.

4. The effect of the air admittance valve is clearly shown in Figure 7.21 and may be summarised as follows

    a. The presence of the air admittance valve restricts air flow into the top of the stack at any given water flow rate. At pressures above the air admittance valve opening pressure, in the case illustrated -40 N/m^2, no air is allowed into the stack. These two effects reduce the static pressure and the air flow eventually achieved. Figure 7.21 clearly illustrates this effect, the minimum stack pressure being approximately double with an air admittance valve in place, whilst the maximum air flow decreases by some 25%.

b. As the water flow rate decreases the stack static pressure differential to atmosphere falls below that necessary to open the air admittance valve. The air flow is cut off leading to a small positive air pressure transient being generated in the network, an effect clearly shown in the trap seal oscillation prediction. As the air flow is very low as the air admittance valve closes this effect is small but predictable.

c. The mechanism of air admittance valve opening does affect the air pressure levels simulated in the stack. As the air admittance valve moves to fully open, at pressure levels between 4 and 10mm of water suction, an air flow is generated which increases gradually, then rises sharply once the air admittance valve is fully open. This effectively reduces the time taken for the air flow to increase from 20% to 90% of its final value from 0.05 seconds to 0.02 seconds in Figure 7.21. This reduction is, in terms of the transient round trip time, or stack pipe period, a major reduction from approximately 4 to 2 pipe periods and moves the events into the 'fast' boundary change category. Thus the pressure simulations achieved with the air admittance valve stack termination, Figure 7.21, illustrate discrete and identifiable pressure waves both at the stack base and in the upper reaches of the stack. For the open stack termination case this effect is attenuated and the pressure changes appear as a 'smooth' reduction in pressure.

7.7.1 *Air flow stoppages - the generation of positive transients*

If the collection drain at the base of the stack becomes surcharged, or if any stack offset or branch to stack junction becomes closed due to appliance discharge, then the entrained air flow velocity in the network is reduced to zero. This may be considered to occur instantaneously, or over a short time period.

Both cases may be handled at either type of location by the model. The boundary air velocity is set to zero and this condition is solved with the appropriate available C^+ or C^- characteristic.

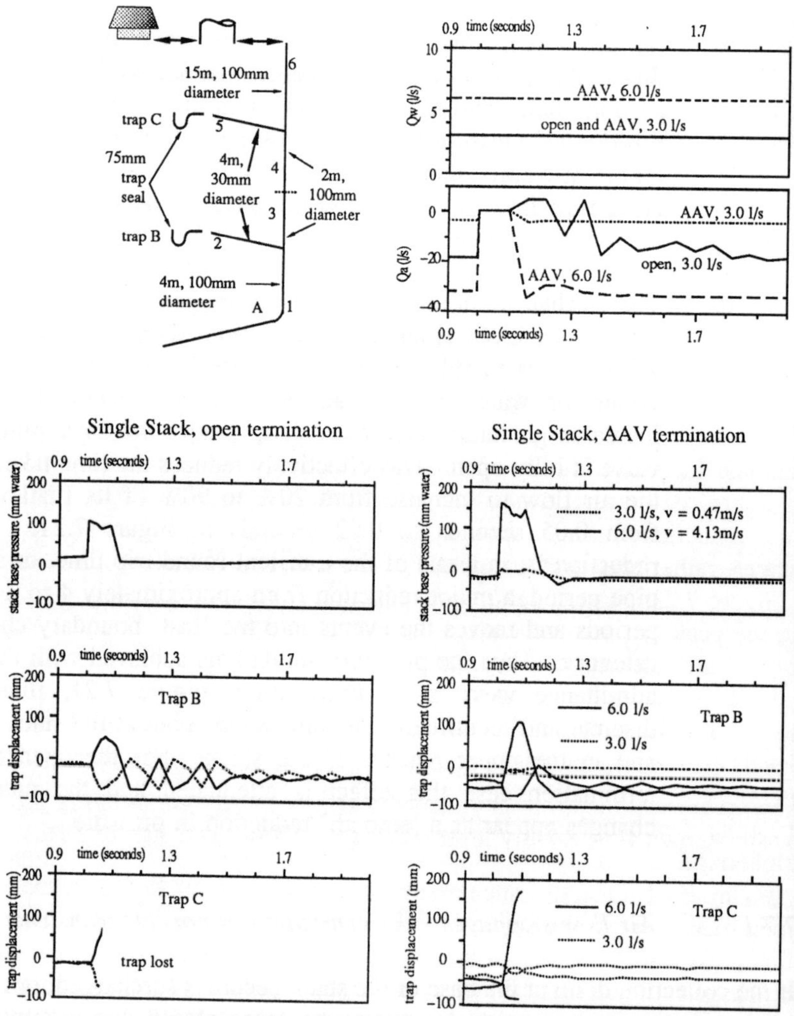

Figure 7.22 Comparison of air pressure, airflow and trap seal displacement following an air path closure at the base of the stack, with and without an Air Admittance Valve termination. Note that an open upper termination allows positive stack airflows.

Figure 7.22 illustrates the effect of an instantaneous flow stoppage at the base of the stack on a single stack system with and without an air admittance valve upper termination.

It is worth noting that as the air admittance valve effectively reduces entrained air flow rate at any water flow, the air admittance valve system will, for any water flow rate, display lower air pressure positive surge pressures following a flow stoppage. This mitigates the obvious problem with air admittance valve application, namely that it cannot relieve positive pressure transient propagation.

The effect of rate of reduction of air flow during a drain blockage is a variable that must also be considered. Classical pressure transient theory states that, provided the flow stoppage is completed in less than one pipe period, in this case approximately 0.14 seconds, then the full Joukowsky pressure change will be observed.

In the simulation the effect of the 'slow' flow stoppage was covered by defining the reduction in air flow velocity over each time step. This is a suitable but not fully accurate model. As the air flow velocity drops, the pressure rises thus forcing more air through the blockage. The actual flow rate reduction during a stoppage will therefore be greater over its final stages, rather than uniform over the whole time involved.

Figure 7.23 illustrates the slow closure case, demonstrating the reduction in the peak pressure generated. These reductions clearly depend upon the rate of air velocity reduction assumed per time step.

If the air path closure occurs in the stack, due to an overload at a branch to stack junction, or the presence of a stack offset, then the lower stack levels also become affected. Downstream of a closure a negative Joukousky pressure wave is generated by flow stoppage, ie a –40mm per m/s of air velocity destroyed wave that subjects the lower level trap seals to induced siphonage.

Figure 7.24 illustrates such an event following an air flow stoppage at an identified point in the stack, ie at the junction of pipes 3 and 4. It is clear that pressures at the base of the stack are reduced, while pressures above the blockage rise, and that the lower level trap, on pipe 2, is completely depleted by the passage of the negative wave. Following negative wave depletion of this trap, pipe 2 acts as a relief valve and this case is modelled by the simulation, explaining the lower limit imposed upon the negative surge in the lower stack levels.

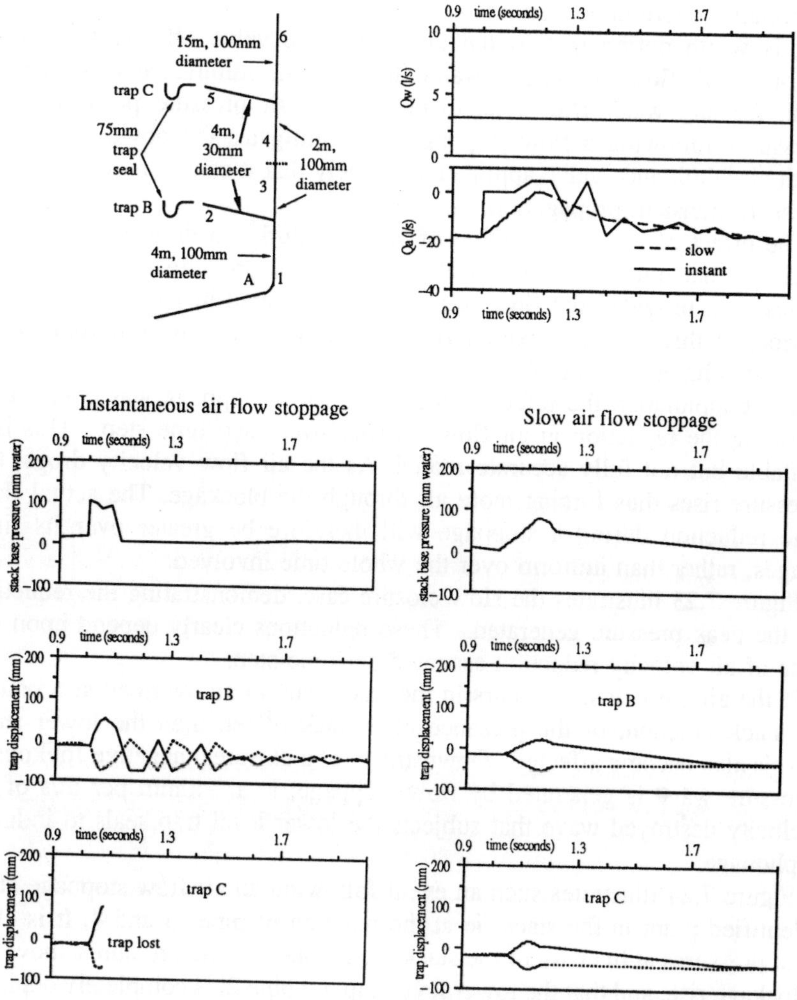

Figure 7.23 Comparison of air pressure, airflow and trap seal displacement for a rapid or gradual air path blockage at the base of the stack.

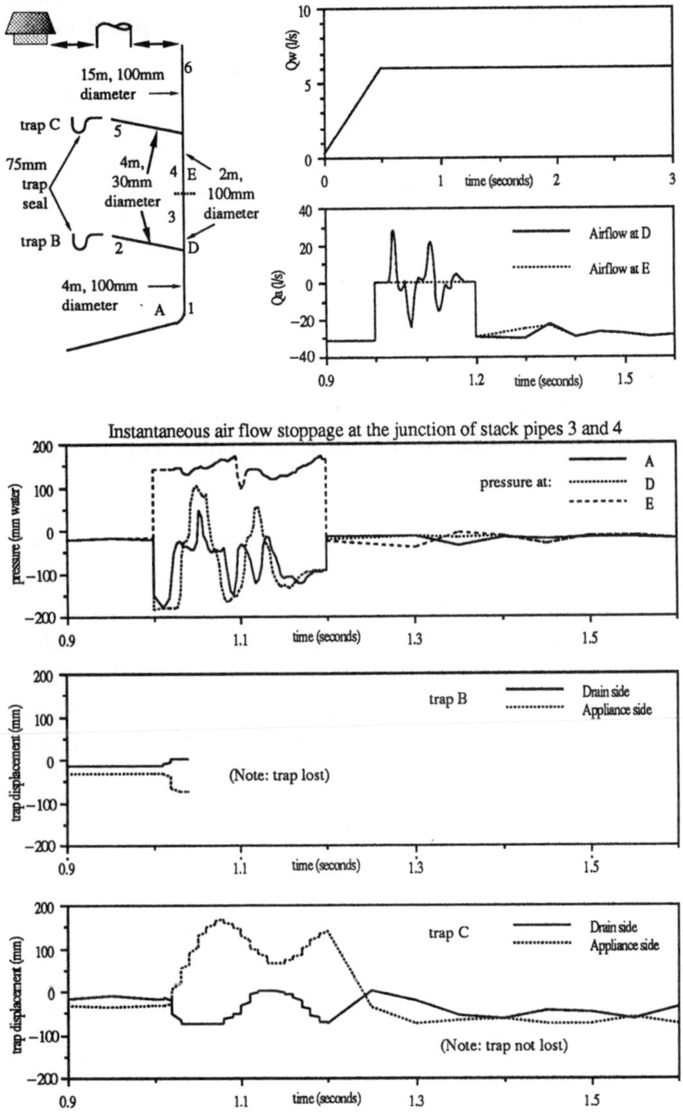

Figure 7.24 Pressure, airflow and trap seal displacements as a result of air path closure at a stack junction

7.8 Application of the method of characteristics to free surface wave and low amplitude air pressure transient propagation

This Chapter has concentrated on the introduction of the method of characteristics to areas not generally considered to be within the main stream of applications. However the treatment and the model simulations provided have demonstrated that these areas are within the broad unsteady flow spectrum within which the method of characteristics is probably the most suitable technique. Conditions necessary for the efficient use of these techniques include complex piping and channel networks where the duration of the transient event is relatively short in terms of the calculated pipe periods. Thus while the method is probably not suited to the prediction of long duration surge events in rivers it is ideal for the consideration of wave propagation and attenuation within building drainage systems. Similarly the application to low amplitude air pressure transients meets the requirements of time step and event duration.

The development of computing to support models of the type proposed in this and accompanying Chapters is however probably the greatest influence on the acceptability of these techniques. The simulations presented in this text were undertaken on PC type computers, ending previous reliance on main frame computing. The method of characteristics has already achieved industry standard status in the traditional pressure surge applications and there can be little doubt that its application to the areas covered by this Chapter will also be accepted by design and consulting organisations.

Afterword:

The computer based simulation of transient propagation

This text set out its objectives as providing an understanding of the mechanisms of transient propagation as an essential pre-requisite for the development of modelling techniques based upon the method of characteristics. It is essential that the interrelationships between the system and fluid properties, determining wave speed, and the mode of operation of the system components, effectively responsible for the generation or alleviation of pressure surge, be understood. In this context the text developed the wave speed relationships and the reflection/transmission coefficients applicable at the system's entry, exit and internal boundaries. The frictionless model, featuring reflection and transmission coefficients operating upon transients generated by near instantaneous valve closure, has much to offer in providing both an understanding of the mechanisms by which transients propagate within any network and a means of 'translating' the sometimes complex pressure traces that record the sequence of events at any location within a network.

The concept of transient analysis as an example of a communications network problem is useful. This approach allows the central significance of wave speed and pipe period, ie the time taken for a reflection to return from the closest boundary, to be appreciated. This model is also extremely helpful when introducing the method of characteristics as the characteristic equations linking flow conditions across both a time and distance increment may be conceptualised as communication links that allow information, in the form of pressure transients, to propagate throughout a network. This approach is particularly appropriate in the treatment of free surface wave propagation where the concept of supercritical flow, represented by both characteristics sloping downstream, precludes the transmission of waves, ie information, upstream.

However it is the introduction of computing methods that has transformed the study of pressure surge into a practical and usable technology to be incorporated into the design of fluid engineering systems. In the acceptance of any computing model by designers it is essential that confidence is established through a thorough validation of the mathematical model to be utilised. The pressure transient research effort in the 1960-1980 period successfully provided this confidence level and has led to the acceptance of such models, and the method of characteristics in particular, as the industry standard.

The method of characteristics is of course only one of a range of methods that could have been used. However, it has a range of particular advantages. It is particularly suitable for applications where the event to be simulated is of relatively short duration, due to the constraints imposed by the Courant Criterion. The method deals with networks in terms of internal nodes and boundaries as separate types of calculation and, when linked to a first order finite difference scheme, only involves the simultaneous solution of two equations at each internal pipe node and, at boundaries, the solution of one characteristic per pipe with some equation defining the boundary condition. This is a particular strength of the method as it allows the development and solution of complex boundary conditions in 'isolation' from the calculations for all other boundaries and internal nodes. Similarly this compartmentalised approach allows the introduction of dormant subroutines, dealing for example with the boundary condition representing cavity formation or surge control and suppression devices, into generalised network models, these boundary solutions only being activated when the simulated pressure conditions in the network reach pre-set levels. Thus the method of characteristics model allows for the treatment of the three types of boundary identified in the text, namely active boundaries, such as valves or pumps; passive boundaries, such as pipe junctions and reservoirs; and system initiated boundaries, including vapour and released gas during and following column separation. Thus the model can identify for the designer surge problems not apparent from the physical description of the system alone.

While these modelling considerations are important it is still true that the mathematical methods underlying a simulation will have to remain transparent to the general user. What is essential is an understanding of the mechanisms that generate the transient response simulated and a confidence level based upon proven validation. This text has attempted to demonstrate the capability of the method through a series of case studies drawn from a range of applications. The pumping main and air relief valve simulations demonstrate a major application of the methodology presented. The applications to aircraft fuel systems, whether in the ground or in flight refuelling mode, is rarer but introduces a different perspective on the problems of surge control; component weight and relatively low maximum pressure levels become important, dictating the careful use of valve closure modes as a first line of defence against surge. The applications of the method of characteristics to drainage free surface flows and to low amplitude air pressure transients open a new design area to the approach already established in the other areas treated in this text. Again it is the availability of computing power, coupled to the inherent attributes of the method of characteristics, that has made these applications possible; the need to provide the designer with confidence through validation remains a priority, however the simulations included in the text demonstrate the advantages such an approach will bring.

By presenting both a historically based fundamental treatment of transient propagation and an exposition of the numerical analysis underlying the computer simulations, based upon the method of characteristics, the text has sought to provide the reader with an understanding of the mechanisms of transient propagation and the prediction of their effects upon a range of fluid engineering systems.

Appendix I:

Wave speed calculations

Based on the derivation of wave speed presented in Chapter 1, the FORTRAN program below calculates wave speed for a range of fluid and pipe wall variables, including the influence of free gas.

```
C     WAVE SPEED CALCULATION PROGRAM.
C
C     THIS PROGRAM CALCULATES TRANSIENT WAVE SPEED,
C     INCLUDING THE EFFECTS OF PIPE ELASTICITY, FLUID
C     PROPERTIES AND THE PRESENCE OF FREE GAS. THE WAVE
C     SPEED CALCULATIONS PRESENTED IN CHAPTER 1 ARE
C     UTILISED IN THIS PROGRAM. THE SPECIALISED TUNNEL
C     AND NON-CIRCULAR WAVE SPEED VALUES MAY BE
C     DETERMINED BY SIMPLE EXTENSIONS TO THIS PROGRAM,
C     UTILISING THE EQUATIONS DEVELOPED IN CHAPTER 1.
C
C     THE FOLLOWING READ / WRITE CHANNELS ARE USED ....
C
C     1       WRITE TO FILE FOR001.DAT
C     5       WRITE TO THE SCREEN
C     6       INPUT FROM THE KEYBOARD
C
C
```

```
C     PIPE MATERIAL PROPERTIES AND CROSS SECTION
C     DIMENSIONS....
C
      WRITE(5,*)'IS PIPE ELASTICITY TO BE INCLUDED ? INPUT'
      WRITE(5,*) '1 =  RIGID PIPE.'
      WRITE(5,*) '2 =  ELASTIC PIPE'
      READ(6,*) IE
      E=1.0
      D=0.0
      T=1.0
      C1=0.0
      IF(IE.EQ.1) GOTO 1
C
C     PIPE ELASTICITY IMPORTANT....
C
      WRITE(1,*)'PIPE ELASTICITY CONSIDERED...'
      WRITE(5,*)'INPUT PIPE DIA. AND WALL THICKNESS, M.'
      READ(6,*) D,T
      WRITE(5,*)'INPUT PIPE YOUNGS MODULUS, E*10^XN, N/M^2'
      READ(6,*) E1,XN
      E=E1*10.0**XN
      WRITE(5,*)'INDICATE PIPE RESTRAINT MECHANISM.....'
      WRITE(5,*)          '1  C1=1.0-0.5*PR'
      WRITE(5,*)          '2  C1=1.0-PR^ 2'
      WRITE(5,*)          '3  C1=1.0'
      READ(6,*) IP
      WRITE(5,*)'INPUT MATERIAL POISSONS RATIO...'
      READ(6,*) PR
      C1=1.0
      IF(IP.EQ.2) C1=1.0-0.5*PR
      IF(IP.EQ.3) C1=1.0-PR**2
      WRITE(1,*)'PIPE PROPERTIES AND DIMENSIONS.'
      WRITE(1,21) E1,XN1,PR,C1,D,T
21    FORMAT(/,'YOUNGS MODULUS = ',F5.2,' *10^ ',F4.1,
     1 /,'POISSONS RATIO = ',F5.2,
     2 /,'C1 = ',F5.2,
     3 /,'PIPE DIA. = ',F5.2,' M'
     4 /,' WALL THICKNESS = ',F6.3,' M')
C
1     CONTINUE
```

```
C     FLUID PROPERTIES....
C
      WRITE(5,*)'INPUT FLUID DENSITY, KG/M^3'
      READ(6,*) RHO
      WRITE(5,*)'INPUT FLUID BULK MODULUS, K*10^XN, N/M^2'
      READ(6,*) XK1,XN
      XK=XK1*10.0**XN
      WRITE(1,*)'FLUID PROPERTIES....'
      WRITE(1,22) XK1,XN1,RHO
2     FORMAT(/,'FLUID BULK MODULUS=',F5.2,'*10.0^',F3.0,'N/M^2'
    1 /,'FLUID DENSITY = ',F7.2,' KG/M^ 3')
C
C
C     FREE GAS CONTENT
C
      WRITE(5,*)'INDICATE PRESENCE OF FREE GAS....'
      WRITE(5,*)'0    =   NO FREE GAS'
      WRITE(5,*)'1    =   FREE GAS UNIFORMLY DISTRIBUTED'
      READ(6,*) IG
      GAS=0.0
      GASP=1.0
      RHOG=0.0
      IF(IG.EQ.0) GOTO 2
      WRITE(5,*)'INPUT FREE GAS VOL RATIO AT LINE
    1 PRESSURE...'
      READ(6,*) GAS
      WRITE(5,*)'INPUT GAS DENSITY, KG/M^3'
      WRITE(5,*)'GAS ABSOLUTE PRESSURE, KN/M^2'
      READ(6,*) RHOG,GASP1
      GASP=GASP1*1000.0
      WRITE(1,*)'FREE GAS RATIO AND PROPERTIES...'
      WRITE(1,23) GAS,RHOG,GASP1
23    FORMAT(/,' FREE GAS RATIO BY VOLUME = ',F7.4,
    1 /,' GAS DENSITY AT LINE PRESSURE = ',F7.3,' KG/M^ 3'
    2 /,' GAS ABSOLUTE PRESSURE = ',F8.3,' KN/M^ 2')
C
C     CALCULATION OF WAVE SPEED FOR RIGID OR ELASTIC
C     PIPES, WITH OR WITHOUT FREE GAS....
C
C
```

```
2      CONTINUE
       RHOEFF=GAS*RHOG+(1.0-GAS)*RHO
       BULKEFF=1.0/((1.0-GAS)/XK+GAS/GASP+D*C1/(E*T))
       C=SQRT(BULKEFF/RHOEFF)
       WRITE(1,20) BULKEFF1,RHOEFF,C
20     FORMAT(/,' EFFECTIVE BULK MODULUS = ',F10.4,' KN/M^ 2',
     1 /,' EFFECTIVE DENSITY = ',F10.4,' KG/M^ 3',
     2 //,' WAVE SPEED = ',F10.4,' M/S')
       END
```

Appendix II:

Three pipe system computer model

A simple three pipe system method of characteristics model is presented, written in FORTRAN, in order to allow the reader to both investigate a range of boundary conditions and add further boundary conditions to reflect individual interests.

The system presented consists of up to three pipes, one pipe being provided with an upstream supply while the other two feature downstream boundaries, including valves, dead ends, trapped air and constant pressure reservoirs. Both outwards and inwards relief valves are provided as optional boundary conditions, and a junction is provided between two or three pipes.

Column separation both upstream and downstream of an operated valve are included, the possibility of gas release to the downstream cavity case being included as an option. Variable modes of valve operation are automatically included by defining the valve closure in terms of the appropriate valve discharge coefficient at any time during closure. This option will allow the reader to experiment with two or more stage closures. Valve opening transients may also be considered, as may valves that are cycled through an opening/closing operation.

Table AII.1 and Figure AII.1 illustrate the range of cases that may be investigated. Care should be taken in the simulation of pressure levels within a trapped air pocket following the opening of an upstream valve as the program will expect a two pipe series configuration.

Output from this program has been used within the text to illustrate various boundary condition effects, the program being run on a 386 PC.

Table AII.1 Options possible with the three pipe system model

PIPE 1		PIPE 2		PIPE 3	
ENTRY	**EXIT**	**ENTRY**	**EXIT**	**ENTRY**	**EXIT**
Reservoir	Closing Valve	-	-	-	-
Entry valve closing, with/ without gas release, with/ without IRV.	Reservoir	-	-	-	-
Entry valve closing, with/ without gas release, with/ without IRV.	Junction 1-2	Junction 2-1	Reservoir	-	-
Reservoir	Junction 1-2	Junction 2-1	Exit Valve, close/open ORV option	-	-
Entry valve opening	Junction 1-2	Junction 2-1	Dead end with/without ORV or trapped gas.	-	-
All the above combinations may then be repeated with three pipes					
Entry valve closing, with/ without gas release, with/ without IRV.	Junction 1-2-3	Junction 2-1-3	Reservoir	Junction 3-2-1	Dead End, Trapped air, with/ without ORV.

(Note ORV - Outwards relief valve, IRV - Inwards relief valve, Chapter 6)

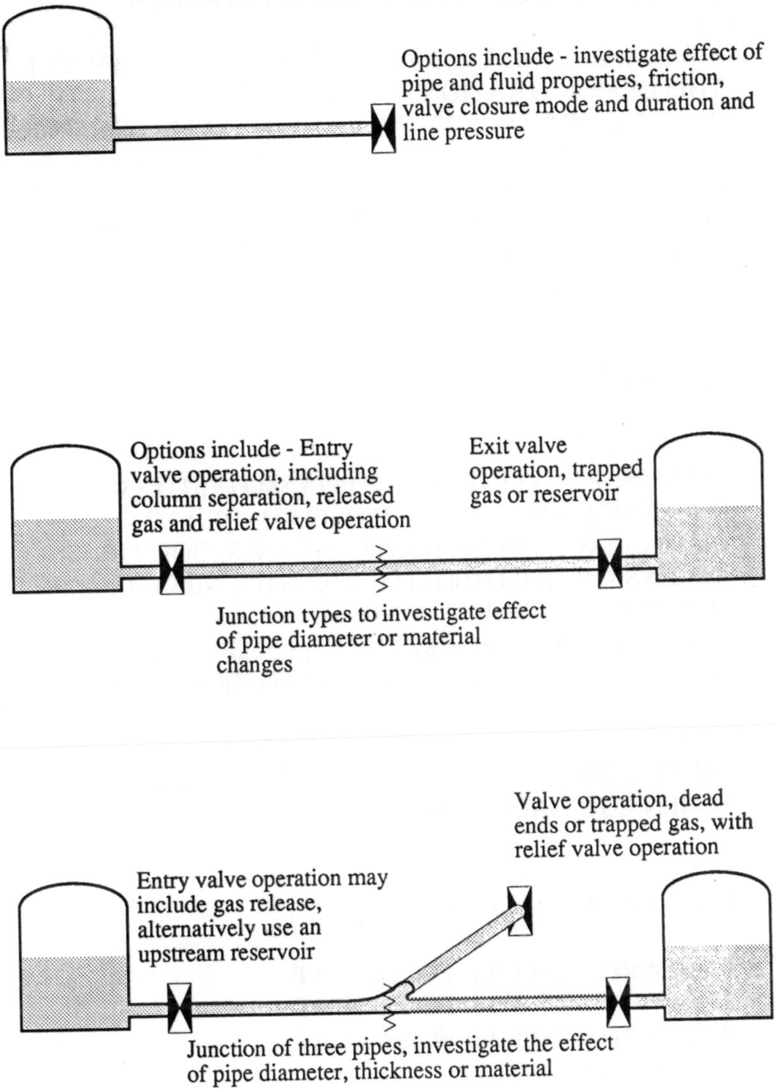

Figure AII.1 Options available within the three pipe system computer model

```
C     SURGE PRESSURE SIMULATION PROGRAM.
C
C     A NETWORK OF UP TO THREE PIPES IS CATERED FOR IN
C     THIS DEMONSTRATION PROGRAM, TOGETHER WITH A WIDE
C     RANGE OF BOUNDARY CONDITIONS, REPRESENTING BOTH
C     PASSIVE AND ACTIVE BOUNDARIES. OPENING AND
C     CLOSING VALVES, WITH AND WITHOUT COLUMN
C     SEPARATION AND GAS RELEASE, RELIEF VALVES,
C     JUNCTIONS AND DEAD ENDS, WITH AND WITHOUT
C     TRAPPED GAS, ARE PRESENTED.
C
C     EXAMPLES OF PROGRAM OUTPUT MAY BE FOUND
C     THROUGHOUT THE TEXT.
C
C
      DIMENSION VV(3,51),V(3,51),PP(3,51),P(3,51),C(3)
      DIMENSION PIRV(3), IRV(3),DPIRV(3),QIRV(3)
      DIMENSION PORV(3), ORV(3),DPORV(3),QORV(3)
      DIMENSION N(3),DT0(3),DX(3),NENT(3),NEXT(3)
      DIMENSION Q0(3),F(3),QEN(3),QEX(3),DPEN(3),DPEX(3)
      DIMENSION VR(3,51),PR(3,51),VS(3,51),PS(3,51)
      DIMENSION D(3),XK1(3,51),XK2(3,51),XK3(3,51),XK4(3,51)
      DIMENSION NPTS(3),PRES(3), TAUD(3,51),TIMD(3,51)
      DIMENSION X(3,51),VOL(3),A(3),PL(3)
      DIMENSION PX(3,51), AIR0(3),AIRVOL(3),FLVOL(3)
      DIMENSION AX(3),VX(3),FX(3)
C
C     FOR READ AND WRITE STATEMENTS THE FOLLOWING
C     CONVENTION HAS BEEN USED ....
C
C     1     WRITE TO FILE FOR001.DAT
C     2     WRITE TO FILE FOR002.DAT
C     3     WRITE TO FILE FOR003.DAT
C     5     WRITE TO THE SCREEN
C     6     INPUT DATA VIA KEYBOARD
      OPEN(UNIT=1,FILE='FOR001.PXX',STATUS='UNKNOWN')
      OPEN(UNIT=2,FILE='FOR002.PXX',STATUS='UNKNOWN')
      OPEN(UNIT=3,FILE='FOR003.PXX',STATUS='UNKNOWN')
C
C
```

```
C     SYSTEM DEFINITION SECTION.
C
C
      WRITE(5,*) 'INPUT NUMBER OF PIPES IN NETWORK, 1-3.'
      READ(6,*) NPIPE
      ATM=101.0*10**3
      PMIN=ATM
      DO 90 K=1,3
      VOL(K)=0.0
      AIRVOL(K)=0.0
      AIR0(K)=0.0
      FLVOL(K)=0.0
90    CONTINUE
C
C     PIPE AND FLOW DATA.......
C
      DO 1 I=1,NPIPE
      WRITE(5,*) 'FOR EACH PIPE SUPPLY FOLLOWING DATA..'
      WRITE(5,*)'NOTE ALL DATA IN SI UNITS, ALL PRESSURES IN
     1 KN/SQR M ABSOLUTE, ALL FLOWS IN CU. M/S.'
      WRITE(5,*) 'PIPE NUMBER = ', I
      WRITE(5,*) 'PIPE LENGTH, DIAMETER, WAVESPEED..'
      READ(6,*) PL(I),D(I),C(I)
      WRITE(5,*) 'INITIAL FLOWRATE, FRICTION FACTOR..'
      READ(6,*) Q0(I),F(I)
      WRITE(5,*) 'NUMBER OF DX SECTIONS IN PIPELENGTH..'
      READ(6,*) N(I)
C
C     ENTRY AND EXIT CONDITIONS FOR EACH PIPE.....
C
      WRITE(5,*) 'INDICATE PIPE ENTRY CONDITION...'
      WRITE(5,*)'             0 - CON. PRES. RESERVOIR'
      WRITE(5,*)'             1 - ENTRY VALVE'
      WRITE(5,*)'             2 - JUNCTION'
      READ(6,*) NENT(I)
C
C     GAS RELEASE OPTIONS......
C
      IF(NENT(I).EQ.1)WRITE(5,*)'INDICATE GAS RELEASE MODEL'
      IF(NENT(I).EQ.1)WRITE(5,*)'     0 - NO RELEASE TO CAVITY'
```

```
      IF(NENT(I).EQ.1)WRITE(5,*)'  1 - GAS RELEASE TO CAVITY'
      IF(NENT(I).EQ.1)READ(6,*) IGAS
      IF(IGAS.EQ.1) WRITE(5,*)'GAS RELEASE TO CAVITY WILL
    1 OCCUR IF THE FLUID PRESSURE DOWNSTREAM CLOSING
    2 VALVE FALLS BELOW ATM.'
C
C     INWARDS RELIEF VALVE OPTIONS AND DATA.....
C
      IF(NENT(I).EQ.1)WRITE(5,*)'INDICATE PRESENCE OF
    1 INWARDS RELIEF VALVE'
      IF(NENT(I).EQ.1)WRITE(5,*)'  0 - NO IRV PRESENT'
      IF(NENT(I).EQ.1)WRITE(5,*)'  1 - IRV MOUNTED'
      IF(NENT(I).EQ.1) READ(6,*) IRV(I)
      IF(IRV(I).EQ.1) WRITE(5,*)' INPUT FULLY OPEN
    1 THROUGHFLOW AND PRESSURE DROP, CU.M/S, KN/SQR.M'
      IF(IRV(I).EQ.1) READ(6,*) QIRV(I),DPIRV(I)
      IF(IRV(I).EQ.1) DPIRV(I)=DPIRV(I)*1000.0
      IF(IRV(I).EQ.1)WRITE(5,*)'INPUT IRV OPENING LINE
    1 PRESSURE, KN/SQR.M'
      IF(IRV(I).EQ.1) READ(6,*) PIRV(I)
      IF(IRV(1).EQ.1) PIRV(I)=PIRV(I)*1000.0
C
C     PIPE EXIT CONDITIONS.......
C
      WRITE(5,*)'INDICATE PIPE EXIT CONDITION ...'
      WRITE(5,*)'  0 - CON. PRES. RESERVOIR'
      WRITE(5,*)'  1 - EXIT VALVE'
      WRITE(5,*)'  2 - JUNCTION'
      WRITE(5,*)'  3 - DEAD END'
      WRITE(5,*)'  4 - TRAPPED AIR'
      READ(6,*) NEXT(I)
C
C     RELIEF VALVE OPTIONS AND DATA.....
C
      IF(NEXT(I).EQ.1.OR.NEXT(I).EQ.3)WRITE(5,*)'INDICATE
    1 PRESENCE OF RELIEF VALVE'
      IF(NEXT(I).EQ.1.OR.NEXT(I).EQ.3)WRITE(5,*)'   0 - NO BLOW
    1  OFF VALVE'
      IF(NEXT(I).EQ. 1.OR.NEXT(I).EQ.3)WRITE(5,*)'   1 -
    1 OUTWARDS RELIEF VALVE PRESENT'
```

```
      IF(NEXT(I).EQ.1.OR.NEXT(I).EQ.3) READ(6,*) ORV(I)
      IF(ORV(I).EQ.1)WRITE(5,*)'INPUT RELIEF VALVE OPEN
    1 THROUGHFLOW AND PRESSURE DROP, CU.M/S AND
    1 KN/SQR.M'
      IF(ORV(I).EQ.1) READ(6,*) QORV(I), DPORV(I)
      IF(ORV(I).EQ.1) DPORV(I)=DPORV(I)*1000.0
      IF(ORV(I).EQ.1)WRITE(5,*)'RELIEF VALVE OPENING
    1 PRESSURE, KN/S.M'
      IF(ORV(I).EQ.1) READ(6,*) PORV(I)
      IF(ORV(I).EQ.1) PORV(I)=PORV(I)*1000.0
C
C     TRAPPED GAS OPTIONS FOR DEAD ENDED PIPES...
C
      IF(NEXT(I).EQ.4) WRITE(5,*)'INPUT TRAPPED AIR VOLUME,
    1 AT LINE PRESSURE ASSUMED AT END OF PIPE ',I,'
    1 EXPRESSED AS NUMBER OF PIPE DIAMETERS FILLED WITH
    1 GAS...'
      IF(NEXT(I).EQ.4) READ(6,*) GASD
      IF(NEXT(I).EQ.4)VOL(I)=GASD*D(I)*3.142*D(I)**2/4.0
C
C     UPSTREAM SUPPLY RESERVOIR PRESSURE.....
C
      IF(I.EQ.1) WRITE(5,*)'UPSTREAM RESERVOIR PRESSURE ?'
      IF(I.EQ.1)READ(6,*) UPPRES
      IF(I.EQ.1)UPPRES=UPPRES*1000.0
C
C     CALCULATION OF TIME STEP FOR EACH PIPE IN SYSTEM
C
      DT0(I)=(PL(I)/FLOAT(N(I)))/C(I)
      DX(I)=PL(I)/FLOAT(N(I))
1     CONTINUE
      IF(NENT(1).EQ.1) GOTO 3
      GOTO 2
C
C     VALVE CHARACTERISTIC DATA FOR EACH VALVE,
C     IDENTIFIED BY PIPE NUMBER. NOTE ONLY ONE VALVE
C     PER PIPE...
C
3     WRITE(5,*) 'INPUT DATA FOR INFLOW VALVE, PIPE 1'
      WRITE(5,*) 'INPUT VALVE FULLY OPEN PRESSURE LOSS..'
```

```
      READ(6,*) DPEN(1)
      DPEN(1)=DPEN(1)*1000.0
      WRITE(5,*)'INPUT VALVE FULLY OPEN THROUGHFLOW,
     1 CU.M/S'
      READ(6,*) QEN(1)
2     CONTINUE
      IF(NPIPE.EQ.1.AND.NEXT(1).NE.1) GOTO 5
      DO 4 I=1,NPIPE
      IF(NEXT(I).EQ.1) GOTO 6
      GOTO 4
6     WRITE(5,*) 'INPUT FULLY OPEN DATA FOR EXIT VALVE
     1 PIPE ',I
      WRITE(5,*) 'FULLY OPEN PRESSURE LOSS, KN/SQR.M..'
      READ(6,*) DPEX(I)
      DPEX(I)=DPEX(I)*1000.0
      WRITE(5,*) ' FULLY OPEN THROUGHFLOW, CU.M/S..'
      READ(6,*) QEX(I)
4     CONTINUE
5     CONTINUE
C
C     FLUID DATA......
C
      WRITE(5,*) 'INPUT FLUID DENSITY, KG/CU.M'
      READ(6,*) RHO
      WRITE(5,*) ' FLUID VAPOUR PRESSURE'
      READ(6,*) PVAP
      PVAP=PVAP*1000.0
      IF(IGAS.EQ.1)WRITE(5,*) 'FLUID BUNSEN GAS SOLUBILITY
     1 COEFF.'
      IF(IGAS.EQ.1)READ(6,*) BUNK
C
C     VALVE MOTION DATA, TAU VS. TIME DATE INPUT.
C
      DO 10 I=1,NPIPE
      IF (NENT(I).EQ.1.OR.NEXT(I).EQ.1) GOTO 11
      GOTO 10
11    IF(NENT(I).EQ.1) WRITE(5,*) 'INPUT INLET VALVE DATA,
     1 PIPE',I
      IF(NEXT(I).EQ.1)WRITE(5,*)'INPUT EXIT VALVE DATA, PIPE',I
      WRITE(5,*) 'INPUT NUMBER OF TIME - TAU PAIRS....'
```

```
      READ(6,*) NPTS(I)
      DO 12 K=1,NPTS(I)
      WRITE(5,*) 'INPUT TIME'
      READ(6,*) TIMD(I,K)
      WRITE(5,*) 'INPUT TAU'
      READ(6,*) TAUD(I,K)
12    CONTINUE
10    CONTINUE
C
C     CALCULATION INITIAL STEADY STATE CONDITIONS...
C
      DO 7 I=1,NPIPE
      DO 7 J=1,N(I)+1
      IF(J.EQ.1) X(I,J)=0.0
      IF(J.GT.1) X(I,J)=X(I,J)+DX(I)
7     V(I,J)=Q0(I)/(3.142*D(I)**2/4.0)
      IF(NENT(1).EQ.0) P(1,1)=UPPRES
      IF(TAUD(1,1).EQ.0.0) GOTO 92
      TX=TAUD(1,1)
      IF(NENT(1).EQ.1) P(1,1) = UPPRES - DPEN(1) * (Q0(1) / (TX *
     1 QEN(1)))**2
      GOTO 93
92    IF(NENT(1).NE.1) GOTO 93
C
C     FOR AN INITIALLY CLOSED UPSTREAM ENTRY VALVE IT
C     IS NECESSARY TO DEFINE THE LINE PRESSURE.....
C
      WRITE(5,*) 'INPUT LINE PRESSURE DOWNSTREAM CLOSED
     1 VALVE AT ENTRY TO PIPE 1, KN/SQR M.....'
      READ(6,*)PLINE
      P(1,1)=PLINE*1000.0
93    CONTINUE
      DO 9 I=1,NPIPE
      IF(I.GT.1) P(I,1)=P(I-1,N(I-1)+1)
      DO 8 J=2,N(I)+1
      P(I,J)=P(I,J-1)-2.0*DX(I)*F(I)*RHO*V(I,J)*ABS(V(I,J))/(D(I))
8     CONTINUE
      IF(NEXT(I).EQ.0) PRES(I)=P(I,N(I)+1)
      IF(NEXT(I).EQ.1) TX=TAUD(I,1)
      IF(NEXT(I).EQ.1)PZ=P(I,N(I)+1)-DPEX(I)*(Q0(I)/(TX*QEX(I)))**2
```

```
      IF(NEXT(I).EQ.1) PRES(I)=PZ
      IF(NEXT(I).EQ.1.AND.TAUD(I,1).EQ.0.0) GOTO 77
      GOTO 78
C
C     FOR AN INITIALLY CLOSED EXIT VALVE IT IS
C     NECESSARY TO DEFINE THE RESERVOIR PRESSURE TO
C     WHICH THE VALVE WILL DISCHARGE AS IT OPENS.
C     NOTE THAT THIS CAN BE ZERO GAUGE.
C
77    WRITE(5,*) 'INPUT RESERVOIR PRESSURE DOWNSTREAM
     1 CLOSED VALVE'
      READ(6,*) PRES(I)
      PRES(I)=PRES(I)*1000.0
78    CONTINUE
      IF(NEXT(I).EQ.4) AIR0(I)=VOL(I)*P(I,N(I)+1)/ATM
9     CONTINUE
C
C     THE ANALYSIS TIME STEP MUST BE CHOOSEN TO BE
C     COMMON FOR ALL PIPES IN THE SYSTEM. IF IN DOUBT
C     CHOOSE SMALLEST DT LISTED...
C
      WRITE(5,*) ' TIME STEPS CALC. FOR NETWORK ARE ...'
      WRITE(5,*) (DT0(K),K=1,NPIPE)
      WRITE(5,*)'INPUT CHOOSEN TIME STEP, <= MIN.
     1 CALCULATED'
      READ(6,*) DT
C
C     TOTAL SIMULATION TIME MAY BE CHOSEN.....
C
      WRITE(5,*) 'INPUT TOTAL SIMULATION TIME'
      READ(6,*) TMAX
C
C     ANALYSIS COMMENCES AT TIME=0.0
C
      DO 95 I=1,NPIPE
      AX(I)=AIRVOL(I)*10.0**9
      VX(I)=VOL(I)*10.0**9
      FX(I)=FLVOL(I)*10.0**9
      DO 95 K=1,N(I)+1
95    PX(I,K)=P(I,K)/1000.0
```

```
C     CALCULATION OF BASE CONDITIONS AT EACH TIME
C     STEP
C
100   CONTINUE
C
C     SAVE DATA TO OUTPUT FILE FOR EACH PIPE.
C     PRESSURE AND DATA AT ENTRY, MIDPOINT AND EXIT FOR
C     EACH PIPE, PLUS THE VALUES OF CAVITY VOL., AIRVOL
C     AND FLVOL WILL BE SAVED.
C
      IF(TIME.EQ.0.0) WRITE(1,59)
      IF(TIME.EQ.0.0) WRITE(2,59)
      IF(TIME.EQ.0.0) WRITE(3,59)
59    FORMAT(//,' TIME ',' I',3(' J PRESSURE  VELOCITY '),
     1     ' CAV. VOL. ','  AIRVOL. ',' FLUID VOL.')
      NX=N(1)/2
      I=1
      TIM1=TIME/(2.0*PL(I)/C(I))
      WRITE(1,60) TIME,I,(J,PX(1,J),V(1,J),J=1,N(1)+1,NX),
     1      VOL(1),AIRVOL(1),FLVOL(1)
      IF(NPIPE.EQ.1) GOTO 97
      NX=N(2)/2
      I=2
      WRITE(2,60) TIME,I,(J,PX(2,J),V(2,J),J=1,N(2)+1,NX),
     2      VOL(2),AIRVOL(2),FLVOL(2)
      IF(NPIPE.EQ.2) GOTO 97
      NX=N(3)/2
      I=3
      WRITE(3,60) TIME,I,(J,PX(3,J),V(3,J),J=1,N(3)+1,NX),
     3      VOL(3),AIRVOL(3),FLVOL(3)
60    FORMAT(F7.4,I3,3(I3,2F10.4),3F10.7)
97    CONTINUE
C
C     UPDATE TIME..
C
      TIME=TIME+DT
      WRITE(5,*) 'TIME = ', TIME
      IF(TIME.GE.TMAX) GOTO 500
C
C
```

```
C     BASE CONDITIONS FOR CHARACTERISTICS.
C
C     IF THE CHOSEN TIME STEP IS LESS THAN DX/(V+C) THEN
C     A DEGREE OF LINEAR INTERPOLATION IS CARRIED OUT
C     HERE
C
      DO 14 I=1,NPIPE
      DO 15 J=2,N(I)+1
      VR(I,J)=V(I,J)+(DT/DT0(I))*(V(I,J-1)-V(I,J))
      PR(I,J)=P(I,J)+(DT/DT0(I))*(P(I,J-1)-P(I,J))
      XK1(I,J)=VR(I,J)*(1.0-2.0*F(I)*ABS(VR(I,J))*DT/D(I))
     1 +1.0/(RHO*C(I))*PR(I,J)
      XK2(I,J)=1.0/(RHO*C(I))
15    CONTINUE
      DO 16 J=1,N(I)
      VS(I,J)=V(I,J)+(DT/DT0(I))*(V(I,J+1)-V(I,J))
      PS(I,J)=P(I,J)+(DT/DT0(I))*(P(I,J+1)-P(I,J))
      XK3(I,J)=VS(I,J)*(1.0-2.0*F(I)*ABS(VS(I,J))*DT/D(I))
     1    -1.0/(RHO*C(I))*PS(I,J)
      XK4(I,J)=1.0/(RHO*C(I))
16    CONTINUE
14    CONTINUE
C
C     CALCULATION OF CONDITIONS AT END OF NEXT TIME
C     STEP.
C
C     INTERNAL NODES
C
      DO 17 I=1,NPIPE
      DO 17 J=2,N(I)
      PP(I,J)=(XK1(I,J)-XK3(I,J))/(XK2(I,J)+XK4(I,J))
      IF (PP(I,J).LE.PVAP) PP(I,J)=PVAP
      VV(I,J)=0.5*(XK1(I,J)+XK3(I,J))
17    CONTINUE
C
C     CONSTANT PRESSURE RESERVOIR BOUNDARIES
C
      DO 18 I=1,NPIPE
      IF(I.EQ.1.AND.NENT(I).EQ.0) PP(I,1)=UPPRES
      IF(NEXT(I).EQ.0) PP(I,N(I)+1)=PRES(I)
```

```
      IF(I.EQ.1.AND.NENT(I).EQ.0) VV(I,1)=XK3(I,1)+XK4(I,1)*PP(I,1)
      IF(NEXT(I).EQ.0) VV(I,N(I)+1)=XK1(I,N(I)+1)
     1                    -XK2(I,N(I)+1)*PRES(I)
18    CONTINUE
C
      DO 38 I=1,NPIPE
      A(I)=3.142*D(I)**2/4.0
38    CONTINUE
C
      IF(NPIPE.EQ.1) GOTO 62
C
C     JUNCTION BOUNDARY
C
      S1=0.0
      S2=0.0
      DO 19 I=2,NPIPE
      S1=S1+A(I)*XK3(I,1)
      S2=S2+A(I)*XK4(I,1)
19    CONTINUE
      PP(1,N(1)+1)=(A(1)*XK1(1,N(1)+1)-S1)/(A(1)*XK2(1,N(1)+1)+S2)
      IF(PP(1,N(1)+1).LE.PVAP) PP(1,N(1)+1)=PVAP
      VV(1,N(1)+1)=XK1(1,N(1)+1)-XK2(1,N(1)+1)*PP(1,N(1)+1)
      DO 20 I=2,NPIPE
      PP(I,1)=PP(1,N(1)+1)
      VV(I,1)=XK3(I,1)+XK4(I,1)*PP(I,1)
20    CONTINUE
C
62    CONTINUE
C
C
C     EXIT VALVE OR DEAD END BOUNDARY
C
      DO 21 I=1,NPIPE
      IF(NEXT(I).EQ.2) GOTO 21
      IF(NEXT(I).EQ.4) GOTO 55
      IF(NEXT(I).EQ.0) GOTO 21
      IF(NEXT(I).NE.1) GOTO 30
C
C
C
```

```
C     DETERMINE VALVE DISCHARGE COEFF...
C
      DO 22 K=2,NPTS(I)
      IF(TIME.GE.TIMD(I,K-1).AND.TIME.LT.TIMD(I,K)) GOTO 25
22    CONTINUE
25    TAU = TAUD(I,K-1) + (TAUD(I,K) - TAUD(I,K-1)) * (TIME -
     1 TIMD(I,K-1))
     1 /(TIMD(I,K)-TIMD(I,K-1))
      IF(TAU.LT.0.0001) GOTO 30
      XK5=(TAU*QEX(I)/A(I))**2/DPEX(I)
      Z1=XK5/(2.0*XK2(I,N(I)+1))
      Z2=XK5*(PRES(I)-XK1(I,N(I)+1)/XK2(I,N(I)+1))
      ZZ=(Z1**2-Z2)
      IF(ZZ.GE.0.0) VV(I,N(I)+1)=-Z1+SQRT(Z1**2-Z2)
      IF(ZZ.LT.0.0) VV(I,N(I)+1)=Z1-SQRT(Z1**2+Z2)
      PP(I,N(I)+1)=(XK1(I,N(I)+1)-VV(I,N(I)+1))/XK2(I,N(I)+1)
      GOTO 21
30    CONTINUE
C
C     CLOSED VALVE OR DEAD END
C
C
      IF(NEXT(I).EQ.4) GOTO 55
      IF(VOL(I).GT.0.0) GOTO 32
      IF(NEXT(I).EQ.3.OR.TAU.LT.0.0001) GOTO 31
      GOTO 29
31    VV(I,N(I)+1)=0.0
      PP(I,N(I)+1)=XK1(I,N(I)+1)/XK2(I,N(I)+1)
      IF(ORV(I).EQ.1.AND.PP(I,N(I)+1).GT.PORV(I)) GOTO 71
      GOTO 72
71    XK5=QORV(I)**2/(A(I)**2*DPORV(I))
      Z1=XK5/XK2(I,N(I)+1)
      Z2=XK5*(PORV(I)-XK1(I,N(I)+1)/XK2(I,N(I)+1))
      VV(I,N(I)+1)=-Z1/2.0+SQRT((Z1/2.0)**2-Z2)
      PP(I,N(I)+1)=(XK1(I,N(I)+1)-VV(I,N(I)+1))/XK2(I,N(I)+1)
      WRITE (5,*) 'V = ',VV(I,N(I)+1),' P = ',PP(I,N(I)+1)
72    CONTINUE
      IF(PP(I,N(I)+1).LE.PVAP) GOTO 32
      GOTO 21
32    PP(I,N(I)+1)=PVAP
```

```
      VV(I,N(I)+1)=XK1(I,N(I)+1)-XK2(I,N(I)+1)*PVAP
      VOL(I)=VOL(I)-0.5*A(I)*DT*(VV(I,N(I)+1)+V(I,N(I)+1))
      IF(VOL(I).LE.0.0) VOL(I)=0.0
      IF(VOL(I).EQ.0.0) GOTO 31
      GOTO 21
C
C     TRAPPED AIR AT DEAD END BOUNDARY...
C     A FIRST ORDER APPROX IS USED FOR AIR POCKET GROWTH
C     BASED ON THE VELOCITY AT THE END OF THE PREVIOUS
C     TIMESTEP ONLY.
C
55    CONTINUE
29    CONTINUE
      VOL(I)=VOL(I)-A(I)*DT*V(I,N(I)+1)
      PGAS=AIR0(I)*ATM/VOL(I)
      VV(I,N(I)+1)=XK1(I,N(I)+1)-XK2(I,N(I)+1)*PGAS
      PP(I,N(I)+1)=PGAS
      AIRVOL(I)=PGAS*VOL(I)/ATM
21    CONTINUE
C
C     ENTRY VALVE BOUNDARY
C
      I=1
      IF(NENT(1).EQ.0) GOTO 50
      IF(VOL(1).GT.0.0.AND.IGAS.EQ.0) GOTO 44
      IF(VOL(1).GT.0.0.AND.IGAS.EQ.1) GOTO 80
C
C     DETERMINE VALVE DISCHARGE COEFF...
C
      DO 27 K=2,NPTS(1)
      IF(TIME.GE.TIMD(1,K-1).AND.TIME.LT.TIMD(1,K)) GOTO 28
27    CONTINUE
28    TAU = TAUD(1,K-1) + (TAUD(1,K) - TAUD(1,K-1)) * (TIME -
     1 TIMD(1,K-1))
     1 /(TIMD(1,K)-TIMD(1,K-1))
C     IF(TAU.EQ.0.0) GOTO 40
      XK5=(TAU*QEN(1)/A(1))**2/(DPEN(1))
      Z1=XK5/XK4(1,1)
      Z2=-XK5*(UPPRES+XK3(1,1)/XK4(1,1))
      IF(IGAS.EQ.1.AND.P(1,1).LT.ATM) GOTO 80
```

```
      IF(P(1,1).LE.PVAP) GOTO 44
      SI=1.0
      IF(V(1,1).LT.0.0) SI=-1.0
      VV(1,1)=-SI*Z1/2.0+SI*SQRT((Z1/2.0)**2-SI*Z2)
      PP(1,1)=(VV(1,1)-XK3(1,1))/XK4(1,1)
      IF(IGAS.EQ.1.AND.PP(1,1).LE.ATM) GOTO 80
      IF(PP(1,1).LE.PVAP) PP(1,1)=PVAP
      GOTO 50
C
C     CLOSED VALVE AND DOWNSTREAM COLUMN
C     SEPARATION..
C
40    VV(1,1)=0.0
      PP(1,1)=-XK3(1,1)/XK4(1,1)
      IF(PP(1,1).LE.PVAP) GOTO 44
      GOTO 50
C
C     VAPOUR ONLY SEPARATION MODEL
C
44    PP(1,1)=PVAP
      VV(1,1)=XK3(1,1)+XK4(1,1)*PVAP
      DO 271 K=2,NPTS(I)
      IF(TIME.GE.TIMD(1,K-1).AND.TIME.LT.TIMD(1,K)) GOTO 281
271   CONTINUE
281   TAU = TAUD(1,K-1) + (TAUD(1,K) - TAUD(1,K-1)) * (TIME -
     1 TIMD(1,K-1))
     1 /(TIMD(1,K)-TIMD(1,K-1))
      VX1=TAU*(QEN(1)/A(1))*(SQRT((UPPRES-PVAP)/DPEN(1)))
      IF(TAU.LE.0.01) VX1=0.0
      VOL(1)=VOL(1)+0.5*A(1)*DT*(VV(1,1)+V(1,1))-A(1)*DT*VX1
      IF(IRV(1).EQ.1) VOL(1)=VOL(1)-DT*QIRV(1)*SQRT((PIRV(1)
     1          -PVAP)/DPIRV(1))
      IF(VOL(1).LE.0.0) VOL(1)=0.0
      IF(VOL(1).EQ.0.0) GOTO 40
      GOTO 50
C
C     RELEASED GAS COLUMN SEPARATION MODEL..
C
C     A FIRST ORDER APPROXIMATION IS USED, SEE TRAPPED
C     AIR CASE.
```

```
C
80    CONTINUE
      PGAS=PP(1,1)
      IF(AIRVOL(1).EQ.0.0) PGAS=ATM+PVAP
      DO 272 K=2,NPTS(I)
      IF(TIME.GE.TIMD(1,K-1).AND.TIME.LT.TIMD(1,K)) GOTO 282
272   CONTINUE
282   TAU = TAUD(1,K-1) + (TAUD(1,K) - TAUD(1,K-1)) * (TIME -
     1 TIMD(1,K-1))
     1 /(TIMD(1,K)-TIMD(1,K-1))
      IF(TAU.GT.0.0)  VX1 = TAU * (QEN(1)/A(1)) * (SQRT((UPPRES
     1        - PGAS)/DPEN(1)))
      IF(TAU.LE.0.0) VX1=0.0
      IF(VX1.GT.0.0)FLVOL(1)=FLVOL(1)+A(1)*DT*VX1
      IF(IRV(1).EQ.1.AND.PP(1,1).LT.PIRV(1)) FLVOL(1)=
     1 FLVOL(1)+DT*QIRV(1)*SQRT((PIRV(1)-PP(1,1))/DPIRV(1))
      IF(PP(1,1).LE.PMIN) AIRVOL(1) = FLVOL(1) * (ATM - PP(1,1))
     1 * BUNK/ATM
      IF(IRV(1).EQ.1.AND.PP(1,1).LT.PIRV(1)) EXAIR=
     1 (BUNK/ATM)*DT*QIRV(1)*SQRT((PIRV(1)-PP(1,1))/DPIRV(1))
     2 *(PIRV(1)-PP(1,1))
      IF(IRV(1).EQ.1.AND.PP(1,1).GT.PMIN) AIRVOL(1)=AIRVOL(1)
     1 +EXAIR
      EXAIR=0.0
      IF(IRV(1).EQ.1) VOL(1)=VOL(1)-DT*QIRV(1)*SQRT((PIRV(1)
     1 -PVAP)/DPIRV(1))
      BX=XK3(1,1)+V(1,1)-PVAP/XK4(1,1)+VOL(1)/(A(1)*DT/2.0)
     1 -A(1)*DT*VX1
      CX=-PVAP*(XK3(1,1)+V(1,1))-(VOL(1)*PVAP+AIRVOL(1)
     1 *ATM)*2.0/(A(1)*DT)+A(1)*DT*VX1
      BX=BX/XK4(1,1)
      CX=CX/XK4(1,1)
      PP(1,1)=-BX/2.0+SQRT((BX/2.0)**2-CX)
      VV(1,1)=XK3(1,1)+XK4(1,1)*PP(1,1)
      VOL(1)=VOL(1)+A(1)*DT*(VV(1,1)+V(1,1))/2.0
      IF(VX1.GT.0.0) GOTO 50
      IF(PP(1,1).LT.P(1,1).AND.PP(1,1).LT.PMIN) PMIN=PP(1,1)
      GOTO 50
C
C
```

```
C     ASSIGN BASE VALUES FOR NEXT TIME STEP
C
50    CONTINUE
      DO 51 I=1,NPIPE
      DO 51 J=1,N(I)+1
      P(I,J)=PP(I,J)
      PX(I,J)=P(I,J)/1000.0
      V(I,J)=VV(I,J)
51    CONTINUE
C
C     UPDATE TIME AND CHECK ON ELAPSED SIMULATION
C     TIME...
C
      GOTO 100
500   CONTINUE
      CLOSE(1)
      CLOSE(2)
      CLOSE(3)
      END
```

Appendix III:

Bisection technique

Many of the boundary equations met in the application of the method of characteristics to unsteady flow problems, particularly in the gas transient and free surface flow examples found in Chapter 7, are not directly solvable and require some form of iterative solution. In cases where only one root is known to lie within a predetermined range of values, often dictated by the system, for example the maximum available depth to retain a free surface in a partially filled pipeflow case, the bisection technique provides a powerful solution tool.

Consider the case of flow entry, Q, at depth h, to a partially filled pipe flow, maximum depth is the diameter, D. The entry condition is provided by the known flow energy level, E, equation 7.66, and Figure AIII.1

$$f(h) = E - (h_e + (Q_t / A_{h_e})^2 / 2g) \qquad \text{(AIII.1)}$$

In this case the area term, A(h), and the free surface width, T(h), are functions of depth h defined by the channel cross sectional shape, in this case circular, and may be calculated as

$$T(h) = 2\, r\, \theta \qquad \text{(AIII.2)}$$

$$A(h) = r^2\, (\theta - 0.5\, sin2\theta) \qquad \text{(AIII.3)}$$

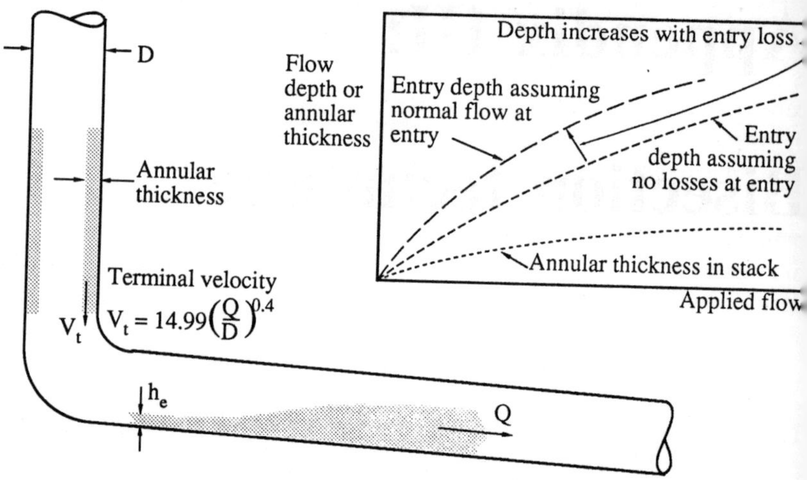

Figure AIII.1 Relationship between entry depth and annular thickness. Note that a normal flow entry condition is equivalent to a total loss of kinetic energy at entry

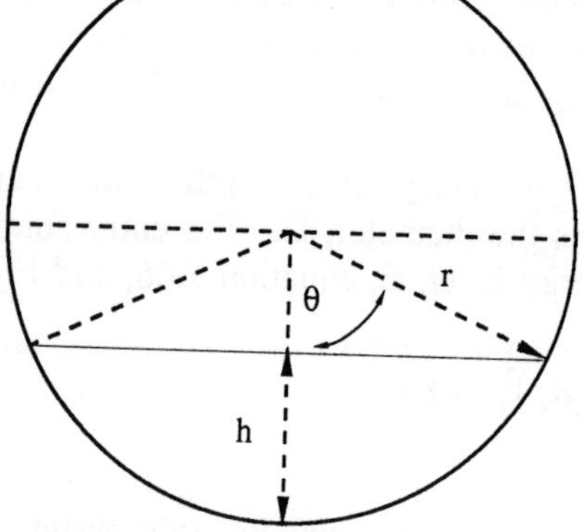

Figure AIII.2 Circular cross section pipe geometry

The root to satisfy the boundary equation must lie between $0 < h < D$, and at the appropriate value of h it follows that $f(h) = 0$. The bisection technique is a process of sequential selection of trial values of h until f(h) becomes zero. The choice of h is determined by utilising trial values which represent the mean of the upper and lower limits possible.

For the example given above, E is known in terms of the flow entry energy, Q is known as the inflow rate at the time instant for which the boundary condition is required and A(h), the flow cross sectional area at pipe entry at depth h, may be calculated from the geometry relationships defined by Figure AIII.1.

The technique involved is as follows :-

1. the upper and lower limits are initially pipe diameter D and zero, thus $h_{max} = D$ and $h_{min} = 0$

2. set the trial $h_t = 0.5 \, (h_{max} + h_{min})$

3. calculate the value of the function, f(h) defined by equation AIII.1

4. if $f(h) < 0$ then reset the lower limit, $h_{min} = h_t$ and repeat the process from step 2

5. if $f(h) > 0$ then reset the upper limit, $h_{max} = h_t$ and repeat the process from step 2

6. when the difference between consecutive values of h_t falls below a preset limit consider the last value of h_t as the required solution.

A satisfactory entry depth solution is normally achieved within 10 iterations. This technique is particularly suitable to partially filled pipe flow conditions, provided that the flow does not become full bore. Its application to the other cases cited in the text is left to the reader.

References:

Abbott, M.B (1966), *An introduction to the method of characteristics*, American Elsevier, New York.

Allievi, L. (1903), 'Notes I-IV', translated as *Theory of Waterhammer* by E. E. Holmes, Ricardo Garoni, Rome, 1925.

Apelt, C.J. (1956), 'Investigations of waterhammer at the University of Queensland', *Journal Institute of Engineering*, Aust., Vol 28.

Angus, R.W. (1935), 'Simple graphical solution for pressure rise in pipes and pump discharge lines', *Journal Eng. Inst. Canada*, Vol 18.

Angus, R.W. (1937), 'Waterhammer in pipes, including those supplied by centrifugal pumps: a graphical treatment', *Procs. I.Mech.E.*, Vol 136.

Angus, R.W. (1939), 'Waterhammer pressures in compound and branched pipes', *Trans. ASCE.*, Vol 104.

American Society of Mechanical Engineers, (1933), 'Symposium on Waterhammer'.

Bergeron, L. (1932), 'Variations in flow in water conduits', *Comptes rendres des travaux de la Soc. Hyd. de France*, Paris

Bergeron, L. (1935), *Etude des variations de regime dans les conduits d'eau: Solution graphique generale*, Revue generale de l'hydraulique, Vol 1, pp12-25, Paris. (Also translated as *Waterhammer in hydraulics and wave surges in electricity*, under ASME sponsorship, John Wiley, 1961, New York.)

Bergeron, L. (1961), *Waterhammer in hydraulics and wave surges in electricity*, J. Wiley, New York.

Boldy, A.P. (1975), 'Analysis of waterhammer in hydroelectric installations' UKSC Conference on Computer Simulation, Bowness-on-Windermere May.

Boldy, A.P. (1976), 'Waterhammer analysis in hydroelectric pumped storage installations', 2nd International Pressure Surge Conference, The City University, London, British Hydromechanics Research Association (BHRA).

Boldy, A.P. and Walmskey, N. (1983), 'Representation of the characteristics of reversible pump turbines for use in waterhammer simulations', Fourth International Conference on Pressure Surges, BHRA, University of Bath, Sept, pp287-296.

Boldy, A.P. (1989), 'Simulation of Hydraulic Transients in Large Hydraulic Systems', International Symposium on Large Hydraulic Machinery and Associated Equipment, Beijing, May 1989, pp674-686.

Binnie, A.M. and Thackrah, M.A. (1951), 'Waterhammer in a pumping main and its prevention', *Procs. I.Mech.E.*, Vol 165.

British Hydromechanics Research Association (1972, 1976, 1980, 1983, 1986, 1989), International Conferences on Pressure Surge, Cranfield, England.

Brown, R.J. (1968), 'Water column separation at two pumping plants', *Trans ASME.*, Series D, Vol 90.

Bunt, E.A. (1953), 'Preliminary study of valve cavitation in a pipeline', *Journal South African I.Mech.E.*, Johannesburg, Vol 2.

Carpenter, R.C. and Barraclough, S.H. (1894), 'Some experiments on the effects of waterhammer', *Trans ASME.*

Chaudhry, H. (1987), *Applied hydraulic transients*, Van Nostrand Reinhold.

Derry, L.D., Evans E.B., Faulkner B.A. and Jelfs E.C.G. (1952), 'Vapour and air release from aviation fuels', *J.Inst. Petroleum,* Vol 38.

Doyle, T.J. and Swaffield, J.A. (1972), 'Evaluation of the method of characteristics applied to a pressure transient analysis of the B.A.C./ S.N.A.I.S. Concorde refuelling system', *Procs. I.Mech.E.,* Vol 186, 40/72, pp509-518.

Driels, M.R. (1973), 'An investigation of pressure transients in a system containing a liquid capable of air absorption', *J. Fluid Eng., Trans. ASME.,* Paper No. 73-FE-28, June.

Driels, M.R. (1974), 'Valve stroking in separated pipe flow', *J. Hyd. Div., ASCE.,* Vol 100, No HY11, Paper 10940, pp1549-1563, Nov.

Driels, M.R. (1975), 'Predicting optimum two stage valve closure', ASME paper 75-WA/FE-2, Houston, Texas, Nov 30-Dec 4.

Duc, J. (1959), 'Negative pressure phenomena in a pump pipeline', Sulzer Technical Review, No 3.

Enever, K.J. (1970), 'The use of the computerised graphical method of surge analysis', The City University Pressure Transient Symposium, London, Nov.

Enever, K.J. (1972), 'Surge pressures in a gas-liquid mixture with a low gas content', Procs. 1st Int. Conf. Pressure Surges, Canterbury, England, BHRA Fluid Eng., Paper C1, Sept.

Evangelisti, G. (1969), 'Waterhammer analysis by the method of characteristics', *L'Energia Elletrica* Vol 10, no. 10-12.

Ezekial, F.D. and Paynter, H.M. (1957), 'Computer representation of engineering systems involving fluid transients', *Trans. ASME.*, Vol 79.

Fox, J.A. (1968), 'The use of the digital computer in the solution of waterhammer problems', *Proc. I.C.E.*, Vol 39.

Fox, J.A. (1977), *Hydraulic analysis of unsteady flow in pipe networks,* Macmillan Press, London

Fox, J.A. (1989), *Transient flow in pipes, open channels and sewers,* Ellis Horwood, Chichester.

Fox, J.A. and McGarry, M.P. (1983), 'Pressure transients in pipelines transporting volatile liquids', Procs. 4th Int. Conf. Pressure Surges, Bath, England, Paper B1, pp39-58, BHRA, Sept.

Galowin, L.S. and Swaffield, J.A. (1989), 'Multistorey building drainage network design - an application of computer based unsteady partially filled pipeflow analysis', *Building and Environment*, Vol 24, No 1, pp99-110, January.

Gayed, Y.K. and Kamel, M.Y.M. (1959), 'Mechanics of secondary waterhammer waves', *Procs. I.Mech.E.*, Vol 173.

Gibson, N.R. (1920), 'Elastic fluid column theory', *Trans. ASCE.*, Vol 83.

Goldberg, D.E. and Wylie, E.B. (1983), 'Characteristics method using time-line interpolations', *J. Hyd. Div., ASCE.*, Vol 109, No 5, May.

Gray, C.A.M. (1953), 'The analysis of the dissipation of energy in waterhammer', *Australian Journal of Applied Science*, Vol 5, No 2, pp125-131.

Gray, C.A.M. (1953), 'The analysis of the dissipation of energy in waterhammer', *Procs. ASCE.*, Vol 119.

Gray, C.A.M. (1954), 'Analysis of waterhammer by characteristics', *Procs. ASCE.*

Graze, H.R. and Horlacher, H.B. (1986), 'Design charts for throttled (by-pass) air chambers', Procs. 5th Int. Conf. Pressure Surges, Hannover. Sponsored by BHRA, Cranfield, pp309-322.

Graze, H.R. and Horlacher, H.B. (1989), 'Design of optimum sized air cushion surge chambers', Procs. 6th Int. Conf. Pressure Surges, Cambridge, UK. Sponsored by BHRA, Cranfield, pp383-397.

Grundy, A.K. and Fox, J.A. (1980), 'The effect of compressible pipe liners on pressure transient generation', 3rd Int.Conf. Pressure Surge, BHRA, Bedford, England, pp241-260.

Halliwell, A.R. (1963), 'Velocity of a waterhammer wave in an elastic pipe', *Procs. ASCE.*, Vol 89.

Harding, D.A. (1966), 'A method of programming graphical surge analysis for medium speed computers', *Proc. I.Mech.E.*, Vol 180, Part 3E, 1965-1966, pp88-103.

Henderson, F.M. (1966), *Open channel flow*, Macmillan Publishing, New York.

Henson, D.A. and Fox, J.A. (1972), 'Transient flows in tunnel complexes of the type proposed for the Channel Tunnel', *Procs. I.Mech.E.*, Vol 188, pp153-167, two papers.

Hope, P. and Papworth, M.U. (1980), 'Fire main failures due to rapid priming of dry lines', 3rd Int. Conf. Pressure Surges, BHRA, Bedford, England, March.

Jenkner, W.R. (1971), 'Uber die Druckstoss-geschwindigkeit in Rohrleitungen mit quadratischen und rechteckigen Querschnitten', *Schweiz Bauzeitung*, 89 Jahrgan, Heft 5, pp99-103, 4 Feb.

Johnson, R.D. (1915), 'Rigid fluid column waterhammer theory', *Trans. ASCE.*, Vol 79.

Joseph, I. and Hammill, F.A. (1972), 'Start up pressures in short pump discharge lines', *J. Hyd. Div., ASCE.*, Vol 98, No HY7, pp1117-1125, July.

Joukowsky, N. (1900), 'Uber den hydraulisher Stoss in Wasser-lietungsrohren', Memoirs de l'Academie Imperiale des Sciences de St Petersburgh, 1900 translated as *Waterhammer* by O Simin, *Procs. AWWA.*, Vol 24, 1904.

Joukowsky, N. (1904), 'Waterhammer', Translated by O. Simin., *Procs. American Water Works Assoc.*, Vol 24, pp341-424.

Kennedy, J.F. (1964), 'A short bibliography of references on waterhammer', *Journal New England Water Works Assn.*, Vol 78.

Kerr, S.L. (1968), 'Surge problems in pipelines - oil and water', *Trans. ASME.*, Vol 98.

Knapp, F. (1937), 'Operation of emergency shut off valves in pipelines', *Trans. ASME.*, Vol 59, pp679-682.

Korteweg, D.J. (1878), 'Uber die Fortphlanzungsgeschindigkeit des Schalles in elastisches Rohren', *Annalen der Physik und Chemie*, 5 Floge, Band 5.

Kranenberg, C. (1974), 'Gas release during transient cavitation in pipes', *Jour. of Hyd. Div., ASCE.*, 100, HY10, pp1383-1398, October.

Lamb, H. (1898), 'On the velocity of sound in a tube as affected by the elasticity of the walls', *Mem. Manchester Lit. Phil. Soc.*, July.

Lamoen, J. (1947), 'Le coup de belier d'Allievi, compte tenu des pertes de charge continues', *Bull. Centre de Etudes, de Recherches et d'Essais Scientifiques des Constructions du Gerrie Civil et Hydraulique Fluviale*, Tome II, Doseor, Liege.

Lawson, J.D., O'Neill I.C., and Graze H.R. (1963), 'Pressure surge in fire services in tall buildings', Procs 1st Australasian Conf. on Hydraulics and Fluid Mechanics, Pergamon Press.

Le Conte, J.N. (1937), 'Experiments and calculations on the resurge phase of waterhammer', *Trans. ASME.*, 59, 691.

Li, Wen-H. and Walsh, J.P. (1964), 'Pressure generated by cavitation in a pipe', *J. Eng. Mech. Div., ASCE.*, Vol 90, No EM6, pp113-133.

Lister, M. (1960), 'The numerical solution of hyperbolic partial differential equations by the method of characteristics', in *Numerical methods for digital computers*, Ralston A. and Wilf H.S. (Eds), John Wiley and Sons, New York.

Loewy, R. (1928), *Druckschwankungen in druckrohrleitungen*, Springer, Wein.

Marchal, M., Flesch, G. and Suter, P. (1965) 'The calculation of waterhammer problems by means of the digital computer', Proceedings Int. Symposium on Waterhammer in Pumped Storage Schemes, ASME, Chicago

Marples, E.I.B. (1966), 'The significance of surge diagrams', *Procs. I.Mech.E.*, Vol 180.

Marsden, N.J. and Fox, J.A. (1976), 'An alternative approach to the problem of column separation in an elevated section of pipeline'. Procs. 2nd Int. Conf. Pressure Surges, BHRA, Bedford, England, Sept.

Martin, C.S. (1976), 'Entrapped air in pipelines', Procs. 2nd Int. Conf. Pressure Surges, Bedford, England, BHRA, Sept.

Martin, C.S. (1982), 'Transformation of pump turbine characteristics for hydraulic transient analysis', IAHR Symposium 'Operating problems of pump stations and power plants', Amsterdam, Sept, 1982, Paper No. 32.

Martin, C.S. (1983), 'Experimental investigation of column separation with rapid closure of downstream valve', Procs. 4th Int. Conf. Pressure Surges, Bath, England, BHRA, Paper B3, pp77-88, Sept.

Martin, C.S. (1988), 'Extrapolation of pump-turbine characteristics towards closed guide vane position', Procs. 14th IAHR Symposium, 'Progress with large and high specific energy units', Trondheim, June, Vol 1, Paper E1.

Massau, J. (1900), 'Memoirs sur l'integration graphique des equations aux derivees partialles', *Ann. Ass. Ingrs. Sortis des Ecoles Speciales de Gand*, 23, pp95-214. Translated as *Unsteady Flow*, H.J.Putman, Rocky Mountain Hydraulic Lab., Colorado, 1948.

Michaud, J. (1878), 'Coup de bellier dans les conduites. Etude des moyens pour en attenuer les effets', *Bull. Soc. Vaudoise Eng. Arch.*, Lausanne.

Parmakian, J. (1955), *Waterhammer analysis*, Longmans Green.

Parmakian, J. (1963), *Waterhammer analysis*, Dover, New York.

Pearsall, I.S. (1966), 'Waterhammer effects due to branched and stepped pipes', *Procs. I.Mech.E.*, Vol 180.

Peek, G.M. (1900), 'Pressure in a pipe due to a stoppage of the flowing fluid', *Trans. ASME.*

Pejovic, S. (1989), 'Similarity of hydraulic transients', 6th International Conference on Pressure Surges, BHRA, Cambridge, Oct.

Perkins, F.E., Tedrow, A.C., Eagleson, P.S. and Ippen, A.T. (1964), 'Hydropower plant transients', Part II and III, Department of Civil Engineering, MIT, Report No. 71, Sept.

Pickford, J.A. (1969), *Analysis of surge*, Macmillan, London.

Price, R.K. (1974), 'Comparison of four numerical methods for flood routing', *Jour. Hyd. Div., ASCE.*, 100, HY7, pp879-899, July.

Rich, G.R. (1945), 'Waterhammer analysis by the Laplace-Mellin transformation', *Trans. ASME.*

Rich, G.R. (1963), *Hydraulic transients*, Dover, New York.

Richards, R.T. (1956), 'Water column separation in pump discharge lines', *Trans. ASME.*, Vol 78.

Safwat, H. and de Kluyver, J.P. (1972), 'Digital computations for waterhammer column separation', Procs. 1st Int. Conf. Pressure Surges, Canterbury, England, BHRA, Paper C5, pp53-68, Sept.

Safwat, H. (1976), 'Water column separation and cavitation in short pipelines', ASME. Fluids Eng. Div. Conf. Minneapolis, May 5-7.

Schnyder, O. (1929), 'Waterhammer in pump discharge lines', *Schweizerische Bauxeitung*, Vol 94, no 22,23.

Sheer, T.J., Baasch, R.J. and Gibbs, M.S. (1973), 'Computer analysis of waterhammer in pumping systems', *The South African Mech. Engineer,* Vol. 23, No 7, pp130-157, July.

Shuy, E.B. and Apelt, C.J. (1983), 'Friction effects in unsteady pipe flows', 4th Int. Conf. on Pressure Surges, Paper D2, pp147-164, BHRA, Cranfield, UK.

Song, C.C.S. (1976), 'Two-phase flow hydraulic transient model for storm sewer systems', Second Int. Conf. on Pressure Surge, BHRA, 1976, Cranfield, England.

St. Venant, A.J.C.B. (1870) 'Elementary demonstration of the propagation formula for a wave in a prismatic channel', *Comptes rendes des seances de l'Academie des Sciences,* 71, pp186-195.

Stoker, J.J. (1957), *Water Waves,* John Wiley and Sons Inc., New York.

Stoner, M.A. (1968), 'Analysis and control of unsteady flows in natural gas piping systems', Ph.D. thesis, Univ of Michigan, Ann Arbor.

Streeter, V.L. and Lai, C. (1962), 'Water hammer analysis including friction', *J. Hyd. Div., ASCE.,* Vol 88, No HY3, pp79-112, May.

Streeter, V.L. (1963), 'Valve stroking to control waterhammer', *J. Hyd. Div., ASCE.,* Vol 89, No HY2, Paper 3452, pp39-66, March.

Streeter, V.L. (1966), 'Computer solutions of surge problems', *Procs. I.Mech.E.,* Vol 180.

Streeter, V.L. (1967), 'Valve stroking for complex piping systems', *J. Hyd. Div., ASCE.,* Vol 93, No HY3, pp81-98, May.

Streeter, V.L. and Wylie, E.B. (1967), *Hydraulic Transients*, McGraw-Hill, New York.

Streeter, V.L. (1969), 'Waterhammer analysis'. *Jour. Hyd. Div., ASCE.,* Vol. 88, HY3, pp79-113 May.

Streeter, V.L. (1972), 'Unsteady flow calculations by numerical methods', *Jour. Basic Eng., ASME.,* 94, pp457-466, June.

Swaffield, J.A. (1970), 'A study of column separation following valve closure in a pipeline carrying aviation kerosene', Thermodynamics and Fluid Mechanics Convention, Steady and Unsteady flows, 57, *Procs. I.Mech.E.,* Vol 184.

Swaffield, J.A. (1972), 'Column separation in an aircraft fuel system', Procs. 1st Int. Conf. Pressure Surges, Canterbury, England, Paper C2, BHRA Fluid Eng., pp13-28, Sept. 6-8.

Swaffield, J.A. (1972[a]), 'A study of the influence of air release on column separation in an aviation kerosene pipeline', *Procs. I.Mech.E.,* Thermodynamics and Fluid Mechanics Group, Vol 186, 56/72, pp693-703.

Swaffield, J.A., Gasiorek, J.M. and Johal, K.S. (1978), 'Mathematical models to predict the celerity of pressure waves in two component gas/liquid flow', Procs. ASME. Symposium on Fluid Transients and Acoustics in the Power Industry, San Francisco, September.

Swaffield, J.A., Gasiorek, J.M. and Johal, K.S. (1978), 'Measurements of the speed of propagation of pressure transients in two component gas/liquid mixtures', Procs. ASME. Symposium on Fluid Transients and Acoustics in the Power Industry, San Francisco, September.

Swaffield, J.A. and Bridge, S. (1983), 'Applicability of the Colebrook White formula to represent frictional losses in partially filled unsteady pipeflow', *Journal of Research*, National Bureau of Standards, Washington DC, Vol 88, No 6, Nov-Dec.

Swaffield, J.A. and Maxwell-Standing, K. (1986), 'Improvements in the application of the numerical method of characteristics to predict attenuation in unsteady partially filled pipe flow', *Journal of Research,* National Bureau of Standards, Washington DC, Vol 91, No 3, May-June.

Swaffield, J.A. and Galowin, L.S. (1992), *The engineered design of building drainage systems,* Ashgate.

Swaffield, J.A. and Campbell, D.P. (1992), 'Air pressure transient propagation in building drainage vent systems, an application of unsteady flow analysis', *Building and Environment,* Volume 27, No 3, pp357-365.

Swaffield, J.A. and Campbell, D.P. (1992), 'Numerical modelling of air pressure transient propagation in building drainage systems, including the influence of mechanical boundary conditions', *Building and Environment*, Volume 27, No 4, pp455-467.

Thorley, A.R.D. and Enever, K.J. (1979) 'Control and suppression of pressure surges in pipelines and tunnels', Construction Industry Research and Information Association, London.

Thorley, A.R.D. (1991), 'Fluid transients in pipeline systems', D and L George Ltd, London.

Tullis, J.P., Streeter, V.L. and Wylie, E.B. (1976), 'Waterhammer analysis with air release', Procs. 2nd. Int. Conf. Pressure Surges, Bedford, England, BHRA, Sept.

Vardy, A.E. (1976), 'On the use of the method of characteristics for the solution of unsteady flows in networks', Proc. 2nd Int. Conf. on Pressure Surge, BHRA, Cranfield, England, pp15-30.

Vensano, H.C. and Warren, M.M. (1915), 'Waterhammer approximations', *Trans. ASCE.*

Weber, E.H. (1865), 'Theorie der durch wasser oder andere incompressible Flussig-Kieten in elastishen Rohren fortgepflanzten Wellen', Liepzig.

Weyler, M.E., Streeter, V.L. and Larsen, P.S. (1971), 'An investigation of the effect of cavitation bubbles on the momentum loss in transient pipeflow', *J. Basic Eng., ASCE*, pp1-10, March.

Wiggert, D.C. and Sundquist, M.J. (1979), 'The effect of gaseous cavitation on fluid transients', *Trans. ASME., Jour. Fluids Eng.*, 101, pp79-86.

Wood, F.M. (1938), 'Application of Heaviside's operational calculus to the solution of waterhammer problems', *Trans. ASME.*, Vol 59, pp703-713, Nov.

Wood, F.M. (1970), 'History of waterhammer', Research Report No 65 Dept. of Civil Eng., Queens University, Kingston, Ontario.

Woodhead, C.A., Fox, J.A. and Vardy, A.E. (1976), 'Analysis of water curtains in transient gas flows in ducts', Proc. 2nd International Conf. on Pressure Surges, London, BHRA.

Wylie, E.B. (1969), 'Control of transient free surface flow', *J. Hyd. Div., ASCE.*, Vol 95, No HY1, Paper 6360, pp347-361, Jan.

Wylie, E.B. and Streeter, V.L. (1978), *Fluid Transients*, McGraw-Hill, New York.

Wylie, E.B. and Streeter, V.L. (1983), *Fluid Transients*, FEB Press, Ann Arbor.

Zielke, W. (1968), 'Frequency dependent friction in transient pipe flow', *Jour. Basic Eng., Trans. ASME.*, Vol. 90, Ser. D, No. 1, pp109-115, March.

Index:

A

B

C

D

E

F

G

N

O

P

R

S

T

U

V

Y

Z